New Media

and Intercultural Communication

Thomas K. Nakayama
General Editor

Vol. 13

The Critical Intercultural Communication Studies series is part of the Peter Lang Media and Communication list.
Every volume is peer reviewed and meets the highest quality standards for content and production.

PETER LANG
New York • Washington, D.C./Baltimore • Bern
Frankfurt • Berlin • Brussels • Vienna • Oxford

New Media

and Intercultural Communication

Identity, Community and Politics

EDITED BY Pauline Hope Cheong,
Judith N. Martin, Leah P. Macfadyen

PETER LANG
New York • Washington, D.C./Baltimore • Bern
Frankfurt • Berlin • Brussels • Vienna • Oxford

Library of Congress Cataloging-in-Publication Data

New media and intercultural communication: identity, community and politics /
edited by Pauline Hope Cheong, Judith N. Martin, Leah P. Macfadyen.
p. cm. — (Critical intercultural communication studies; v. 13)
Includes bibliographical references and index.
1. Intercultural communication. 2. Mass media—Social aspects.
3. Digital media—Social aspects. I. Cheong, Pauline Hope.
II. Martin, Judith N. III. Macfadyen, Leah P. (Leah Pauline).
HM1211.N49 302.23—dc23 2011046129
ISBN 978-1-4331-1365-9 (hardcover)
ISBN 978-1-4331-1364-2 (papaerback)
ISBN 978-1-4539-0532-6 (e-book)
ISSN 1528-6118

Bibliographic information published by **Die Deutsche Nationalbibliothek**.
Die Deutsche Nationalbibliothek lists this publication in the "Deutsche Nationalbibliografie"; detailed bibliographic data is available on the Internet at http://dnb.d-nb.de/.

Cover design by Lisa Barfield

The paper in this book meets the guidelines for permanence and durability of the Committee on Production Guidelines for Book Longevity of the Council of Library Resources.

29 Broadway, 18th floor, New York, NY 10006
www.peterlang.com

Printed in the United States of America

For our families,
at home and abroad

Table of Contents

Foreword xi

1. Introduction: Mediated Intercultural Communication Matters: Understanding New Media, Dialectics, and Social Change 1
PAULINE HOPE CHEONG, JUDITH N. MARTIN, AND LEAH P. MACFADYEN

Section One: Theorizing New Media and Intercultural Communication

2. Designing for Culture: An Ecological Perspective on Indigenous Knowledge and Database Design 21
MAJA VAN DER VELDEN

3. A Structurational Interaction Approach to Investigating Culture, Identity, and Mediated Communication 39
BETH BONNIWELL HASLETT

4. Exploring Cultural Challenges in E-Learning 61
BOLANLE A. OLANIRAN

5. Culture, Context, and Cyberspace: Rethinking Identity and Credibility in International Virtual Teams 75
KIRK ST.AMANT

Section Two: Constructing Identities

6. Producing the Self at the Digital Interface 93
NATALIA RYBAS

7. Who Am I in Virtual Space? A Dialectical Approach to Students' Online Identity Expression 109
PING YANG

8. New Media and Asymmetry in Cultural Identity Negotiation 123
GUO-MING CHEN AND XIAODONG DAI

9. Negotiating a New Identity Online and Off-Line: The HeartNET Experience 139
DEBBIE RODAN, LYNSEY URIDGE, AND LELIA GREEN

10. Inoculating against Invisibility: The Friendly Circle of Cancer Patients' Chinese Blog 155
WEI SUN AND ANDREW JARED CRITCHFIELD

Section Three: Negotiating Community

11. Rite of Death as a Popular Commodity: Neoliberalism, Media, and New Korean Funeral Culture 175
JOONSEONG LEE

12. Far Away from Home . . . With a Mobile Phone! Reconnecting and Regenerating the Extended Family in Africa 193
GADO ALZOUMA

13. When Indian Women Text Message: Culture, Identity, and Emerging Interpersonal Norms of New Media 209
ROBERT SHUTER

14. To Browse or Not to Browse: Perceptions of the Danger of the Internet by Ultra-Orthodox Jewish Women 223
AZI LEV-ON AND RIVKA NERIYA-BEN SHAHAR

15. From the Coffee Table Album to the Mobile Phone: A Portuguese Case Study 237
CARLA GANITO AND CÁTIA FERREIRA

Section Four: Engaging Politics

16. Asian American New Media Communication as Cultural Engagement: E-mail, Vlog/Blogs, Mobile Applications, Social Networks, and YouTube 255
KONRAD NG

Chapter 17: Jamaica and Chile Online: Accessing and Using the Internet in a Developing World Context 275
NICKESIA GORDON AND KRISTIN SORENSEN

Chapter 18: Cultural Peculiarities of Russian Audience Participation in Political Discourse in the Era of New Technologies 291
IRINA PRIVALOVA

Chapter 19: The Vienna Unibrennt Platform: Hidden Pitfalls of the Social Web 307
HERBERT HRACHOVEC

Contributing Authors 321

Acknowledgments 329

Index 331

Foreword

CHARLES ESS, PHD
with
FAY SUDWEEKS, PHD

To be sure, the current volume fulfills its goal of collecting some of the best and most current research and scholarship in intercultural communication vis-à-vis online environments. But much more can justifiably be claimed for this collection—especially from a perspective shaped by our work over the past 14 years or so. During this time, Fay Sudweeks and I have cochaired the biennial conference series on Cultural Attitudes towards Technology and Communication (CATaC), whose inaugural conference took place in 1998 at the Science Museum in London.

To show how, and in what ways, the contributions making up this volume help us move significantly forward in the important work of investigating intercultural communication and new media, I first map out the broad terrains of culture, technology, and communication that we have explored through the CATaC conferences. I then set this volume against that background, so as to bring into sharp relief and focus the way in which the chapters collected here offer significant new contributions. We will see in particular how this volume constitutes a very fruitful reconceptualization and extension of the most recent developments surrounding the questions: What do we mean by "culture"? How can we effectively analyze culture's complex interactions with online communication? And how can we apply the results of such analyses to the design and use of online communication technologies so as to foster more effective cross-cultural communication and learning? In addition, the organization of the volume into what we can think of as culturally variable elements of human identity and society exemplifies and extends a central shift in these domains from attempting to think about culture as such and moving instead to much more fine-grained attention to individual components. As I hope to make clear, this conceptual and theoretical shift is one of the most important developments of more than a decade's worth of work on culture, technology, and communication.

Culture, technology, communication: A brief history

The inspiration for what became the first CATaC conference emerged during my participation in a conference on technology and democracy in Oslo, Norway, in 1997. Although I had enjoyed living and working in Europe previously, this was my first experience outside the United States since the "PC revolution" of the 1980s and the emergence of the Internet and the Web in the late '80s and early '90s. With what now appears as an incredibly naïve ethnocentrism, I arrived in Norway simply assuming that my colleagues there would be making use of their computers and computer-mediated communication (CMC) applications in ways that would be more or less similar to those I was familiar with. To the contrary, of course: Among its many benefits, the conference amounted to a kind of computational and communicative culture shock. As is now obvious—and as this volume helps further document and clarify—culture indeed makes a difference. And so, at the end of the conference, I suggested that possible future conferences should include attention to culture and its interactions with CMC.

It is perhaps not surprising that the suggestion simply evaporated into the air. Indeed, today it may be difficult to appreciate how radical this suggestion seemed to be at the time. In addition to the prevailing presumptions of instrumentalism (i.e., the belief that technologies are neutral tools), it was easy to ignore culture vis-à-vis CMC for a simple demographic reason: As late as 1998, approximately 84% of Internet users were located in North America (Graphic, Visualization, and Usability Center, Georgia Technological University, 1998). And certainly, most scholarship was undertaken in this comparatively homogenous cultural domain. Nonetheless, Teresa Harrison, then the editor of the journal *Electronic Journal of Communication/La Revue Electronique de Communication* (*EJC/REC*), encouraged me to develop a special issue on how culture might interact with CMC. Perhaps Teresa's most important contribution was putting me in touch with Fay Sudweeks, a pioneering scholar in CMC who was also a brilliant hand at organizing conferences, thinking that a conference on these topics would attract presentations that could then be developed into articles for the special issue. Fay happily agreed to organize what we named a conference on Cultural Attitudes towards Technology and Communication (CATaC), thanks to a grant from the Swiss Office of Technology Assessment.

We were simply staggered by the responses to and outcomes of this first conference. Broadly speaking, what we subsequently came to call CATaC '98 was sufficiently successful enough in various ways that we were inspired to continue this work. Hence, CATaC'00 took place in Perth, Australia, with subsequent conferences in Montreal, Canada (2002); Karlstad, Sweden (2004); Tartu, Estonia (2006); Nîmes, France (2008); and Vancouver, Canada (2010). (CATaC'12 will take place in Aarhus, Denmark, [http://www.catacconference.org/]). For its part, CATaC'98 attracted more than 60 participants from 18 countries, representing locales and cultures otherwise separated by the North–South and East–West divides. Moreover, our contributors brought forward groundbreaking empirical research. The research demonstrated, first of all, that *yes*, indeed, culture *does* make a difference; that is, a wide range of culturally variable values, practices, preferences, norms, and beliefs can be shown to shape the design, implementation, use, and responses to communication technologies. In addition to our inaugural issue of *EJC/REC* (Sudweeks & Ess, 1998), much of this work has appeared in journals such as *New Media & Society* and the *Journal of Computer-Mediated Communication*, as well as in our anthology, *Culture, Technology, Communication: Towards an Intercultural Global Village* (Ess, 2001). This body of literature, further expanded by conference participants' individual publications in a range of disciplinary journals and anthologies, thus stands as an extensive and, in some ways, foundational body of work that contemporary scholarship may still profitably build upon.

At the same time, however, awareness of the importance of systematic attention to culture only gradually grew over the first decade or so following CATaC'98. "Culture" is a term that was already notoriously problematic in 1998 and, as we will see, has given way to far more complex understandings of how to explore culturally variable elements in CMC. In this context, one of our primary concerns regarding the rapid spread of CMC was what we thought might be the risk of "computer-mediated colonization." That is, CMC technologies were originally designed and built in ways that were shaped, whether obviously or subtly, by the cultural values and communicative preferences of their builders. As these CMC technologies were distributed beyond the boundaries of their original design and implementation, so they would carry those culturally variable characteristics with them into countries that have perhaps quite different values and communicative preferences. Especially when pushed as a technology of globalization and justified by utopian visions of "the electronic global village," and further shielded from critical scrutiny along these lines precisely by the prevailing assumptions of technological instrumentalism, a worst-case scenario would be that CMC would serve as yet one more subtle but ultimately most powerful weapon in the arsenal of colonization.

As it turned out, CATaC'98 both confirmed and countered these concerns. On the one hand, work presented by Lucienne Rey (2001) on diverse responses to the Web in three cultural regions of Switzerland (German,

French, and Italian) and by Lorna Heaton (2001) on how US-based designs for computer-supported cooperative work systems did *not* fit the cultural values and communicative preferences of Japan, makes clear that, indeed, CMC technologies are *not* culturally neutral, and hence the risk of computer-mediated colonization was real. On the other hand, Soraj Hongladarom (2001) uses examples from Thai chat room practices to argue that we could contrast "thick" local cultures that could effectively resist homogenization with an emerging, globally shared but comparatively "thin" Internet culture. According to this work, local cultures would thereby be able to preserve themselves in something of a hybrid constituted by the addition of a global Internet culture; thus, whatever colonization might be facilitated through CMC technologies could be moderated and contained.

Other research foci emerged at CATaC'98, including conceptual and methodological questions concerning our understandings of what culture might mean and how we could research and reflect on it in conjunction with CMC technologies. In many ways, these initial questions have been supplanted by others in subsequent years. This is in part, obviously, because the applications of CMC continued to be developed and diffused worldwide. Whereas in 1998, the vast majority of Internet users represented a relatively homogenous cultural group who were located in North America, in 2010, the proportion that North Americans constitute of global Internet users had dropped from 84% to 13.5% (Internet World Stats, 2010).

Along with this growth has come more attention to the issues of intercultural communication and CMC. At least within a European context, the publication of the Mohammed cartoons by the Danish newspaper *Jyllands-Posten* in 2005, and the aftermath facilitated by CMC throughout the Islamic world, stands as a watershed moment that demonstrates that we are indeed connected around the world in profoundly new ways. If we have any hope for a peaceful future, among our many tasks and obligations is the necessity to consider far more carefully than we have had to before how to approach intercultural communication in the online contexts that have become increasingly pervasive. Among its many strengths, the current volume contributes significantly to the sorts of knowledge we need in order to foster respectful and fruitful intercultural dialogue and, in this way, the volume directly reflects one of the ethical and political impulses that led to the founding and ongoing work of CATaC.

CATaC and intercultural communication: Contemporary perspectives

In broader terms, as we learned more about culture, technology, and communication, we discovered what should have been obvious from the outset: Culture itself is a highly problematic concept. In particular, although much early and helpful work had been done using the frameworks of Hall (1966, 1976) and Hofstede (1980, 1991) to analyze the cultural dimensions of CMC, by CATaC'04, the limitations of these frameworks and especially their underlying assumptions regarding culture had become abundantly clear (see, for example, Kamppuri & Tukiainen, 2004). And so a considerable portion of our attention at CATaC'06 and CATaC'08 was focused on what was wrong with available notions of culture and their affiliated frameworks. On the one hand, then, it is now more or less a commonplace that culture matters; on the other hand, our ways of studying whatever culture may mean have shifted from the general to the (sometimes, highly) particular—so particular, in fact, that the term "culture" may itself disappear as too general and imprecise to be able to provide useful insight or methodological guidance.

By the conclusion of CATaC'10, the following highlights emerged: These notions of culture both build on and extend previous CATaC work and, at the same time, show just how well the current volume works to articulate, exemplify, and move us forward with regard to core insights in these domains. To begin with, what we can call a critical approach to notions of culture has continued its development. As a primary example, Souleymane Camara and José Abdelnour Nocera (2010) demonstrate that the Hofstede axis of collectivism versus individualism (an axis that otherwise appears to hold up reasonably well in distinguishing between cultures) only goes so far in providing us with a sufficiently fine-grained understanding of culturally variable elements vis-à-vis technologies. Camara and Nocera's investigation of a rural sub-Saharan farming community did demonstrate important collectivist elements, including a primary concern with face-saving. At the same time, they suggest, such collectivism did not capture important differences between farmers with better education, bigger fields, more tools, and so forth, and those with less (2010). As Nocera aptly commented during the closing discussion at CATaC'2010 (cf., Nocera & Camara, 2010), the models developed by Hofstede, Hall, and others are like can openers: They are tools that are appropriate for designers especially, but they barely open the lid on culture. Much more fine-grained analyses are needed.

One way (among many others) to get to such analyses is exemplified in the CATaC contribution on mediated intercultural dialectics (Cheong & Martin, 2010), a theme which is explicated here by Pauline Hope Cheong, Judith Martin, and Leah Macfadyen in the introduction chapter of this volume. Here, a *critical* approach means specifically focusing on the power relationships that emerge in online contexts, especially as these appear to mirror and replicate such relationships as they operate in off-line contexts. In broader terms, Section One of this volume is made up of chapters that explore a range of theoretical approaches that extend the very difficult work of reflecting on theory.

Such criticality and theoretical reflection are all the more crucial, insofar as the threat of computer-mediated colonization remains very real. As one example from CATaC'10, Fiona Brady and Laurel Dyson (2010) document how mobile phones were used by two different communities in rural Australia; namely, the Aboriginal community called Wajul Wajul and the geographically proximate non-Aboriginal community of Bloomfield. On the one hand, as mobile phones are designed first and foremost for talk and mobility, they fit nicely with the primarily oral culture of the Wajul Wajul and the members' characteristic practice of camping away from their primary homes. At the same time, however, Brady and Dyson document how ownership of such phones challenged a foundational cultural value. As Rich Ling and Rhonda McEwen (2010) emphasize, mobile phones are perhaps the most *individualistic* technology yet developed. To keep their expenses down, at least some members of the Wajul Wajul community moved away from the prevailing collectivist ethic and practice of reciprocal sharing toward a more individualist sense of their phones as an exclusive (rather than inclusive) and individual property (Brady & Dyson, 2010). In this way, the introduction of the mobile phone both reinforces and challenges central cultural characteristics and practices in ways that no one may have intended but with outcomes that no one may yet predict. To put it still more broadly, Soraj Hongladarom (2001) has shifted somewhat from his earlier optimism regarding the possibilities of sustaining local cultures as *thick*, and thereby still distinct, from a global but *thin* Internet culture. His more recent work (2008) argues that this ostensible boundary is becoming increasingly blurred—pointing again to the risk of unwanted cultural homogenization, if not colonization, via CMC technologies.

The above point on thin and thick culture helps underscore the importance of both Section Three and Section Four of this volume. Section Two, "Constructing Identities," is foreshadowed by the Wajul Wajul example from Brady and Dyson (2010), in its highlighting of one of the most significant ways in which CMC interacts with the foundational components of a given culture; namely, our sense of *identity* as human beings, beginning with whether that sense is more individualistic and/or more relational or collectivistic. As I have argued elsewhere (Ess, 2010), the Wajul Wajul case and multiple other examples suggest a strong correlation between our sense of self (i.e., as either relational and/or individual) and the primary communication "technologies" of orality, literacy, print and what Walter Ong identifies as the "secondary orality" of electronic media. In this regard, the focus on identities in Section Two is exactly right: Whatever the limitations of broader theories of culture, it seems quite clear that our identity and sense of selfhood as human beings are intimately interwoven with the cultural contexts into which we are born. Hence, examining how our senses of identity may interact and change through our engagement with new media is, in my view, one of the most fundamental and significant foci of contemporary research on culture and new media. Section Three, "Negotiating Community" (with studies examining taboos, disciplining, and boundaries) includes several fine-grained case studies on how new media interact with specific components of human cultures (e.g., death rituals, the role of women in an ultra-Orthodox community) and the ways that mobile phones are adapted in culturally specific ways to family relationships and practices. Identity is a topic also taken up in chapters in Sections One and Two, highlighting the ways in which it cannot be separated easily from a complex of culturally variable factors.

An important aspect of this book is that it brings to the foreground for us one of the primary *ethical* impulses that underlies much of our interest in both the potentially salutary as well as the potentially destructive impacts of CMC; namely, a foundational sense of responsibility for "the Other." This term was originally used by French ethicist and theologian Emmanuel Levinas (1963) to signal the ultimate importance of respecting "the other as other"; that is, as a distinctive and unique human being who remains irreducibly different from us. This failure to respect the Other as Other has been institutionalized, for example, in racism and colonialism—the frameworks that force us to see a human Other as merely a means or instrument for our own ends and goals. For example, at CATaC'10, part of Maja van der Velden's (2010) argument, which is explicated in Section One of this volume, is that approaches to design that fail to take on board the multiple, complex, and culturally variable beliefs, practices, norms, and sense of identity that define human beings—that is to say, considering the user of a technology as thereby somehow suspended from the web of relationships and connections with larger cultural communities—is a form of violence against the Other. For van der Velden, this means that we have to "undesign design."

It is clear that we would not be concerned with the potential of CMC for fostering greater equality, including gender equality and democratic practices—nor for their potential to change both individuals and cultures in destructive ways—if we did not begin from this primary ethical impulse to respect the Other *qua* Other. In broad terms, this concern is becoming increasingly more central as more researchers highlight the ways in which our mediated engagements appear to foster the ultimate elimination of Otherness; namely, our willingness to reduce both ourselves and others on social networking sites to standardized commodities as we create profiles that we hope will attract others by branding ourselves in terms of our specific patterns of consumption (Jin, 2010).

It is clear that such a reduction in particular, and such transformations in our sense of self, in a more general sense, hold enormous consequences in terms of civic engagement, social cohesion, and especially democratic politics—the precise theme of Section Four. Here again, we can discern competing patterns and insights. On the one hand, as recent events in several countries throughout the Middle East have demonstrated, contemporary technologies, including those facilitated via Internet-enabled smart phones, are enormously powerful tools for organization and demonstration in the service of democratic movements. At the same time, however, the tendencies toward commodification and other factors go hand in hand with the contemporary recognition that the democratizing potentials of CMC have, in fact, manifested themselves less powerfully than enthusiasts of the 1990s might have hoped (e.g., Stromer-Galley & Wichowski, 2011).

Culture, technology, communication: Familiar skills and territory—new directions?

I hope to have made clear here that scholarly attention to culture, technology, and communication no longer must begin de novo, as it did (more or less) in 1997 to 1998. Rather, there is now a reasonably well-established body of work that scholars and researchers may draw and build upon—work that is further extended and developed by the contributions to this volume.

What I also hope is clear is how such work must be wildly interdisciplinary and highly collaborative. No single discipline can hope to encompass the multiple perspectives required for us to even begin to get a grip on the complexities of culture, technology, and communication. At the same time, however, there is much in the contemporary university and research systems that weighs against the sorts of interdisciplinary approaches that have characterized CATaC from the outset and that are at work here. In such work, as one of our colleagues once remarked, we have a lot of depth—it's just mostly horizontal. The willingness to enter into the needed dialogue and collaboration across disciplinary boundaries further requires considerable humility and modesty, beginning with the recognition that whatever our own disciplinary expertise may be, we are rank amateurs in those disciplines represented by our colleagues. As with intercultural communication and dialogue, much patience and good humor is required. It is unfortunate that humility, modesty, and patience are not always the first virtues fostered in the academy.

It is promising that a great many of our colleagues do possess these admirable and necessary virtues, and in abundant measure. And one of the particular pleasures in our engagement with this volume is just in seeing how the work here continues these virtues and their most fruitful outcomes. We are confident that our readers will glean much insight and fresh inspiration for the ongoing work catalyzed by the constant development and diffusion of CMC technologies across ever-larger numbers and cultures of human beings scattered around the globe.

References

Brady, F., & Dyson, L. (2010). A comparative study of mobile technology adoption in remote Australia. In F. Sudweeks, H. Hrachovec, & C. Ess (Eds.), *Proceedings cultural attitudes towards communication and technology 2010* (pp. 69–83). Australia: Murdoch University.

Camara, S., & Nocera, J. A. (2010). Validating cultural and contextual traits of a collectivist community. In F. Sudweeks, H. Hrachovec, & C. Ess (Eds.), *Proceedings cultural attitudes towards communication and technology 2010* (pp. 331–340). Australia: Murdoch University.

Cheong, P. H., & Martin, J. (2010). Online outsiders within: A critical cultural approach to digital inclusion. In F. Sudweeks, H. Hrachovec, & C. Ess (Eds.), *Proceedings cultural attitudes towards communication and technology 2010* (pp. 42–44). Australia: Murdoch University.

Ess, C. (Ed.). (2001). *Culture, technology, communication: Towards an intercultural global village.* Albany: State University of New York Press.

Ess, C. (2010). The embodied self in a digital age: Possibilities, risks, and prospects for a pluralistic (democratic/liberal) future? *Nordicom Information 32*(2), 105–118.

Graphic, Visualization, and Usability Center, Georgia Technological University. (1998). *GVU's 10th WWW User Survey.* Retrieved from http://www.cc.gatech.edu/gvu/user_surveys/survey-1998-10/graphs/general/q50.htm

Hall, E. T. (1966). *The hidden dimension.* New York, NY: Doubleday.

Hall, E. T. (1976). *Beyond culture.* New York, NY: Random House.

Heaton, L. (2001). Preserving communication context: Virtual workspace and interpersonal space in Japanese CSCW. In C. Ess (Ed.), *Culture, technology, communication: Towards an intercultural global village* (pp. 213–240). Albany: State University of New York Press.

Hofstede, G. (1980). *Culture's consequences: International differences in work-related values.* Newbury Park, CA: Sage.

Hofstede, G. (1991). *Cultures and organizations: Software of the mind.* New York, NY: McGraw-Hill.

Hongladarom, S. (2001). Global culture, local cultures and the Internet: The Thai example. In C. Ess (Ed.), *Culture, technology, communication: Towards an intercultural global village* (pp. 307–324). Albany: State University of New York Press.

Hongladarom, S. (2008). Global culture, local cultures and the Internet: Ten years after. In F. Sudweeks, H. Hrachovec, & C. Ess (Eds.), *Proceedings cultural attitudes towards communication and technology 2008* (pp. 80–85). Australia: Murdoch University.

Internet World Stats. (2010). *Internet Users in the World: Distribution by World Regions—2011.* Retrieved from http://www.internetworldstats.com/stats.htm

Jin, D. (2010). Transformation of audience commodities towards user commodities: Free labor in web 2.0 technologies. In F. Sudweeks, H. Hrachovec, & C. Ess (Eds.), *Proceedings cultural attitudes towards communication and technology 2010* (pp. 170–186). Australia: Murdoch University.

Kampurri, M., & Tukianen, M. (2004). Culture in human-computer interaction studies: A survey of ideas and definitions. In F. Sudweeks & C. Ess (Eds.), *Proceedings cultural attitudes towards communication and technology 2004* (pp. 43–57). Australia: Murdoch University.

Levinas, E. (1963). La trace de l'autre [The trace of the other] (A. Lingis, Trans). *Tijdschrift voor Philosophie* (Sept.), 605–623.

Ling, R., & McEwen, R. (2010). Mobile communication and ethics: Implications of everyday actions on social order. *Etikk i praksis: Nordic Journal of Applied Ethics, 4*(2), 11–26.

Nocera, J. A., & Camara, S. (2010). Reflecting on the usability of research on culture in designing interaction. In F. Sudweeks, H. Hrachovec, & C. Ess (Eds.), *Proceedings cultural attitudes towards communication and technology 2010* (pp. 150–162). Australia: Murdoch University.

Rey, L. (2001). Cultural attitudes toward technology and communication: A study in the "multi-cultural" environment of Switzerland. In C. Ess (Ed.), *Culture, technology, communication: Towards an intercultural global village* (pp. 151–160). Albany: State University of New York Press.

Stromer-Galley, J., & Wichowski, A. (2011). Political Discussion Online. In M. Consalvo & C. Ess (Eds.), *The Blackwell Handbook of Internet Studies* (pp.168–187). Oxford, United Kingdom: Wiley-Blackwell.

Sudweeks, F., & Ess, C. (1998). Cultural attitudes toward technology and communication/Comportement culturel en vers le progres technique et la communication. *Electronic Journal of Communication/La Journal Electronique de Communication 8*, 3,4. Retrieved from http://www.cios.org/www/ejc/ v8n398.htm

Velden, M. van der. (2010). Undesigning culture: A brief reflection on design as ethical practice. In F. Sudweeks, H. Hrachovec, & C. Ess (Eds.), *Proceedings cultural attitudes towards communication and technology 2010* (pp. 117–123). Australia: Murdoch University.

1. *Introduction: Mediated Intercultural Communication Matters: Understanding New Media, Dialectics, and Social Change*

PAULINE HOPE CHEONG, PHD
JUDITH N. MARTIN, PHD
LEAH P. MACFADYEN, PHD

At the start of the second decade of the new millennium, there is increasing awareness of the development of newer "smart" and more interactive media that is happening in precipitate speed in many parts of the world. The uprisings in the Arab region in 2011, for instance, have focused attention on using digital social media and acknowledged their role in movements for political engagement and change. Terms such as the "Twitter revolution" and the "Facebook revolution" have been used widely, conceptualizing the notions of "dynamic media" or "Web 2.0" as potentially radical, disruptive, and socially transformative. The concept of change—in contrast to continuity—is thus central to the increasing interest in digital media. This focus has not, however, been vigorously matched by substantive theoretical discussions or by extensive empirical examinations of computer-mediated communication and intercultural communication.

What do we mean by "new media"? Our interest here ranges far beyond the simple proliferation of the "new" technologies, gadgetry, or artifacts that are frequently associated with digital media—and beyond the more recent participatory or social media networks and geo-locational mobile applications. Rather, we view new media in its wider significance as a globally distributed web of sociotechnical relationships, imbricated with culture in its design, interface, reception, and appropriation. The field of new media that we examine in this collection is effectively framed by Lievrouw and Livingstone

(2006), who position new media as information and communication technologies *and* their social contexts, in particular the material devices as well as the activities around device use and development and the larger societal arrangements and organizational forms around devices and their practices.

Preoccupations with and debates about new media are largely structured around the premise that novel technologies will drive changes in the way people relate to each other, within and across cultures. The dominant perspective stresses empowerment, standardization, and assimilation into alleged global norms and the World Wide Web culture that particularly exists among "digital natives" and the millennial generation. Yet what *actually* happens in praxis when digital media are implemented within and across cultures is contested and negotiated within complex local and political conditions. Processes of mediation are not friction free. Quantitative studies demonstrating increasing adoption of Internet and social media and simple counts of worldwide users provide only the skeleton of the technology story. Even though current activities and certain "best" current practices exist for the implementation of digital media, ongoing critiques conceive of changes in intercultural communication as reinforcing boundaries, exclusion, and tensions. These relational tensions or dialectics in established rituals and identities generate both diversity and plurality of opportunities and challenges within distinctive ethnic, racial, and religious communities. Therefore, a growing corpus of investigators in this realm have sought to ask which of the values, practices, belief systems, forces, and structures of different social groups influence the ways in which these groups design, perceive, adopt, or utilize information and communication technologies . . . if they have access to them at all. And they ask how these conditions energize and enervate the ways in which societies are able to participate in contemporary systems of knowledge and wealth generation.

This book seeks to contribute to the growing body of work that is exploring the complex, reciprocal, and evolving interactions between dynamic human cultures and information and communication technologies. A testament to the growing vibrancy of this area of research is the enthusiastic response to our initial call for papers. We received many inquiries and considered more than 60 proposals. This collection represents the best of these submissions. The collection as a whole, and the individual contributions to it, interweave theoretical insights, fresh evidence, and rich applications to assess the nature of digital culture(s) in order to address assumptions about the present state of mediated global society/societies and their trajectory into the future.

Attention is given here to showcasing prominent interpretative and critical research from diverse voices in multiple locations and with varying backgrounds. As such, this volume presents a rich and colorful tapestry that offers opportunities for comparative analyses and deepened international understandings of digital media connections, particularly in the areas of identity, community, and politics. For the rest of this discussion, we turn to the title of this chapter to address several compelling descriptive, conceptual, practical, and normative reasons why mediated intercultural communication matters. Here, we offer our reflections on the past, present, and needed future research areas of mediated intercultural communication, leaving the task of summarizing each individual chapter to the introduction of each section (in this volume).

Why mediated intercultural communication matters

The present picture of intensified digital media uptake resoundingly underscores the growing practices and importance of mediated intercultural communication. The widespread and seemingly meteoric growth in the numbers of Internet users is one manifestation of the globalization of digital media and its practices and infrastructures. As of April 2011, aggregated statistical reports show that there are almost two billion Internet users in the world (Internet World Stats, 2011). Contrary to earlier decades in which Internet users were predominately located in North America and parts of Europe, the largest number of Internet users now reside in Asia, with sizable growth rates also seen in Latin America, the Middle East, and Africa. It is important to point out that the growing use of social media by people in all regions of the world has profound communicative implications and consequences. One critical component of this contemporary process of change is the potential offered by new media for the formation of new identities, ties, and activities that express and represent non-Western values, norms, and systems, which is occurring as a greater diversity of social and political actors around the world interact and work within and across cultures using a spectrum of digital media.

A short history of scholarship in mediated intercultural communication

To bring to the foreground further reasons why mediated intercultural communication matters, it is instructive to briefly consider the landscape of scholarship in the field of mediated intercultural communication research that has prevailed in the recent past. A survey of this landscape reveals two parallel *and* intersecting lines of inquiry. *Parallel* because the scholars in each line come from disparate disciplines, often not building on each others' research; *intersecting* because their scholarship often investigates similar topics, resulting in limited cross-pollination between the two fields. Thus, one goal of this collection is to bring together scholarship from different disciplines. As described by Ess and Sudweeks in the Foreword of this volume, one line of inquiry originated in the late 1980s and early 1990s with a diverse group of communication technologies scholars (from subdisciplines such as informatics, human-computer interaction, computer-mediated communication) who realized the importance of culture in the design, implementation, and use of the tools of mediated communication. From such efforts emerged organizations and conferences such as Cultural Attitudes towards Technology and Communication (CATaC) and a range of international, interdisciplinary scholarly exchanges and collaborations.

In the same era, a parallel but less developed line of scholarly inquiry was initiated within the field of intercultural communication. This area of research and study originated in the 1950s and 1960s as scholars and practitioners came together to facilitate postwar rebuilding projects in Europe and Japan. An interdisciplinary effort of anthropologists and linguists eventually came to reside in the field of communication—with various foci in different world regions (for example, linguistics in Europe, social psychology in the United States and Japan (Martin, Nakayama & Carbaugh, 2012). A shared research emphasis in the various regions was on the impact of culture in *face-to-face* (FtF) interpersonal communication between members of different cultures (for example, on how verbal and nonverbal communication patterns vary from culture to culture, the role of communication in the formation and development of intercultural relationships, and identification of dimensions or components of effective communication across cultures). Little if any attention was paid to mediated communication.

Intercultural communication scholars have engaged a rich and varied breadth of questions about culture in various communication contexts, employing a wide range of metatheoretical perspectives—from variable-analytic social psychological research, to interpretive ethnography of communication studies, to critical postmodern and postcolonial studies (Gudykunst, 2005; Martin & Nakayama, 2010). A review of recent research, however, reveals that these scholars have rarely engaged in digital media research. In the broader field of communication studies, *mediated* communication was originally seen as the research domain of the field of media studies, interpersonal communication scholars (e.g., Walther, 1996; Walther & Parks, 2002), and organizational communication scholars (e.g., Poole & Holmes, 1995; Poole, Holmes, & DeSanctis, 1991), few of whom routinely engaged questions of *culture* in their studies (for an exception, see Thurlow, Lengel, & Tomic, 2004).

Recent developments, however, signal a blossoming interest in new media by intercultural communication scholars. For example, in our 2009 communiqué with Professor Ling Chen, then Chair of the Intercultural Communication Division of the International Communication Association (ICA), she encouraged scholars engaging in new media research to submit papers to her division, as the members of the Intercultural Communication Division had recently voted to include digitally mediated aspects of intercultural communication in their conference call for papers. This move reflected an increasing desire to integrate new media contexts into traditional research trajectories of intercultural communication scholarship. A textbook by Wood and Smith (2005) focusing on cultural elements of mediated communication was published in the middle of the current decade. Moreover, an upcoming issue of the *Journal of International and Intercultural Communication* hosted a special forum with commentaries on new media (Cheong & Gray, 2011), edited by Professor Robert Shuter (who has also contributed to a chapter in this book). Professor Shuter, the current chair of the International and Intercultural Communication Division at the (US) National Communication Association (NCA) also launched the Center for Intercultural New Media Research in June 2011 (www.interculturalnewmedia.com).

In the past, intercultural communication scholars and CATaC scholars shared a common interest in exploring the role of culture in human interactions, the traditional conceptualization of which was first borrowed from anthropology (e.g., E. T. Hall, Geertz, Kluckhohn and Strodtbeck), and then from social psychology (e.g., Hofstede). It is not surprising that the first efforts of intercultural communication scholars involved theoretical speculations about how cultural values would influence interpersonal communication in mediated contexts (Olaniran, 2001). More recently, intercultural communication scholars, influenced by European critical theory, have stressed the importance of incorporating an understanding of power relations and historical impacts into their scholarship on neocolonialism (e.g., Ono, 2009). Gajjala (1999) was one of the first to examine the intersections of culture, gender, and power inequities in online encounters, from a postcolonial perspective. Olaniran continues this line of research by illustrating the impact of varying cultural values in e-learning contexts (chapter 4), particularly in relation to unequal access to technology in less economically developed countries (LEDCs), when compared with economically developed countries (EDCs).

The importance of dialectics

We introduce the perspective of mediated intercultural dialectics here to draw attention to the complex, dynamic, and intertwined relationships between new media and culture and as an attempt to move beyond the static, dichotomous, and essentializing conceptualizations of culture and intercultural communication (Chuang, 2003). The dialectical intercultural perspective is a metatheoretical framework that focuses on the simultaneous presence of the two relational forces of interaction and recognizes their opposing, interdependent, and complementary aspects. This notion, well-known in Eastern philosophies, is based on the logic of *soku* (not one, not two), which places emphasis on assumptions that the world is neither monistic or dualistic but on the yin-yang principle or completion of relative polarities (Yoshikawa, 1987). Drawing from Mikhail Bakhtin's work on language and culture to advance research beyond traditional paradigms, Martin and Nakayama (1999, 2010) have explicated dialectics in intercultural communication to refer to the processual, relational, and contradictory logics of intercultural knowledge and practices. These include cultural–individual, personal–contextual, differences–similarities, static–dynamic, history/past–present/future, and privilege–disadvantage dimensions.

In particular, mediated intercultural dialectics refer to the fluid relationality between opportunities and challenges, tensions, and uneven gains within virtual world experiences, given the emergent cyber-cultures and the paradoxical and dynamic culturally variable beliefs, practices, and preferences in the design of and responses to technology, as well as our knowledge about communication with cultural "Others." Even as digital media facilitates more frequent and faster intercultural communication, present and persisting cultural rhetorical differences may amplify zones of contention and contradictions between different cultural audiences (St.Amant, 2002). Intercultural dialectics help highlight differences experienced within groups and between individuals that may account for identity and relational complexities (Collier, 2005).

A particular area of research interest in intercultural communication is *identity* and the role of *cultural* identity in intercultural encounters (Collier, 2005; Ting-Toomey, 2005). Several chapters in this volume extend this research focus and illustrate the dialectical dimensions underpinning instances of mediated intercultural communication, either by offering more nuanced theoretical frameworks for the exploration of culture in mediated communication or by investigating selected instances of culturally influenced mediated encounters. Boniwell Haslett (chapter 3) combines Giddens's structuration theory with Goffman's notions of identity and applies this new synthetic framework to the analysis of a range of studies on online communications, including avatars in Second Life and other virtual worlds. Rybas (chapter 6) explicates how minority online users experience fluid and enduring tensions as they attempt to produce an authentic self on Facebook while negotiating gender and class differences to fit the expectations of their imagined audiences. Extending the work of Martin and Nakayama (1999), Yang (chapter 7) identifies additional intercultural communication dialectics that undergird these authors' identity presentations of college students in online intercultural interactions. Chen and Dai (chapter 8) discuss the ways that online encounters may bring about changes in cultural identities, examining in particular the processes of developing individualistic and collective identities and a sense of belonging. Rodan, Uridge, and Green (chapter 9) meanwhile illustrate how medical conditions shape one's identity and interaction with others online.

The study of mediated intercultural communication also matters because cultural stereotypes persist. The impacts of cultural stereotyping perpetuated by traditional media on FtF intercultural communication have been extensively investigated (e.g., Merskin, 2001; Vargas, 2000; Ono & Pham, 2009). Recent intercultural communication media research extends these analyses by exploring how digital (re)presentations can be constructed in response to/ in resistance to dominant media portrayals. For example, Ng (chapter 16) describes and highlights the Asian American community's proactive use of contemporary communication technologies (e-mail, Web logs, and social network sites) to reshape cultural and political imagery in the US national context.

It follows that intercultural communication matters also in the grounded, geographic sense. We have moved away from the separate virtualization of third spaces and purely virtual communities to increasingly integrated and interoperable mediated worlds (Cheong & Poon, 2009). In other words, *matter matters*, as online communication is rooted in biographical and physical histories and realities. Several readings in this volume reflect this focus. Lee (chapter 11) examines the impact of neoliberal politics on traditional South Korean funeral culture with the introduction of cyber-memorial zones; Gordon and Sorenson (chapter 17) compare the influence of the differing geopolitical contexts of Jamaica and Chile on Internet use and access. Privalova (chapter 18) describes how citizen journalists using new media technologies are changing the nature of political engagement in postcommunist Russia—and the ways in which Russian values and sociopolitical norms undergird Russian usage patterns of YouTube as well as other Russian social media that differ from their Western and American analogues. In terms of transnational ties, Alzouma (chapter 12) discusses how the use of mobile technologies among migratory workers in West Africa presents relational dialectics as the use of mobile phones help overcome family fragmentation but also bring unwanted contacts and communal obligations to contribute remittances back to extended family members. Likewise, Ganito and Ferreira (chapter 15) explore the changing cultural rituals as women—the "tellers" and preservers of family stories—increasingly use mobile phones as family photo albums. Again, a dialectic tension exists, as these mobile family photos are useful in preserving family stories but are also more fragile and fragmented than traditional photo albums.

Last but not least, mediated intercultural communication matters in a normative sense, in that mediated dialectics and digital divides are supposed to matter in contemporary societies concerned about social participation, justice, and cohesion. Because of the uneven opportunities and challenges that shape the design and appropriation of new media in intercultural communication, digitally linked stratification exists and is operant at many levels, in spite of and in light of the grounded corporeality of social and mobile media use. As such, mediated intercultural communication is a crucial issue for scholars, as well as policy makers and practitioners. In this volume, van der Velden (chapter 2) describes the challenges of representing various "knowledges" (e.g., Western or indigenous) in constructing Web-based databases and digital archives and proposes a strategy that locates design as a thoroughly ethical practice, which should be accessible to indigenous and Western users. Hrachovec (chapter 19) cautions us about the uneven gains that come from communication technology adoption for activist groups, being that the mainstream media are relatively more eager to report on the innovative digital dissemination of their political event than to cover the substantive issues at hand. The resulting technical hype thereby allows the entertainment and media conglomerates to assert their influence "by the back door."

Intercultural communication scholars have also recently begun to investigate social and global inequities, influenced by critical theory and postcolonial sensibilities (Halualani & Nakayama, 2010). Several authors here extend this work into mediated contexts. For example, Shuter (chapter 13) examines the impact of existing gender relations on the gender text-messaging divide in India, which poses unique challenges for women who aspire to professional careers. Sun and Critchfield (chapter 10) highlight how some cancer patients (and some of their family members acting as proxy Internet users) communicate grief, depression, encouragement, and health advice to those marginalized and shunned by health providers who repress health diagnoses in China. Lev-On and Neriya-Ben Shahar (chapter 14) describe the dilemma of a group of Jewish ultra-Orthodox women: They use the Internet, although it is forbidden by their community, admitting its negative effects (on others) and struggling to negotiate their own media use in their privileged position at the crossroads between modernity and Orthodoxy.

Pragmatics of mediated intercultural communication

Intercultural communication research has a strong history of emphasizing the pragmatic. This emphasis was established by early studies whose goals were the facilitation of interactions between construction project leaders and policy makers in post-World War II Europe and Asia (Leeds-Hurwitz, 1990). Later, investigators sought to understand (and offer solutions to) the intercultural challenges faced by host communities, immigrants, and refugees streaming into Europe in search of a better life (Kramsch, 2001). Subsequently, intercultural research has broadened to include studies of structural and social inequities among ethnic and racial groups in the US (Gonzalez, Houston, & Chen, 2004) and other countries, of the experiences of international students fanning out all over the world, and of efforts at promoting a resolution to ethnic strife (Broome, Carey, De La Garza, Martin & Morris, 2005). Intercultural communication scholars now also emphasize the pragmatic in their studies of new media (Cheong & Martin, 2009). Without oversimplifying, and within the space constraints of each chapter, several authors in this collection offer prescriptive insights and intelligent actionable proposals on how to ameliorate intercultural relations and offer more fruitful intercultural communication exchanges. For example, St.Amant (chapter 5) extends research on intercultural communication competence with particular reference to the question of "credibility," and proposes context-related strategies to employ when collaborating on multicultural virtual teams. Chen and Dai (chapter 8) identify common asymmetries as the source of tension in mediated communication between the West and non-West and propose strategies for easing these tensions.

Future directions

The dialectical landscape of the Internet's third age

Contributions to this collection add to the growing body of literature exploring the complexity (and, often, the unpredictability) of relationships between culture and mediated communication in this "third age" of the Internet (Wellman, 2011). They reveal the thinness of the utopian and technological determinist hopes of the Internet's "first age," which proclaimed the arrival of a new technological enlightenment and the inevitable development of McLuhan's long-forecasted "global village." At the same time, these studies deny the pessimistic predictions of dystopian naysayers who insisted (as did commentators on earlier technologies) that the Internet and new media would fragment human relationships and cultures and divide us from each other. Instead, the studies give evidence to the multiple, often contradictory and apparently muddled ways in which our online and offline lives interweave and interact with one another, as communication technologies become ever more seamlessly integrated into our lives.

Taking a dialectical perspective assists in illuminating our global "current state" and permits us to acknowledge and embrace the opposing forces and themes emerging from the data. We can therefore recognize, for example, the equally "real" logic of nationalist and internationalist projects promoting new technologies in the service of human development and poverty reduction and also the continuing evidence of a global digital divide—an acknowledgment that highlights the weakness of simple instrumentalist assumptions. We can acknowledge (and celebrate!) the ways in which new media are facilitating new modes of citizen representation and the creation of "new intercultural forms," (Ng, chapter 16)—a shift that Wellman (2011) characterizes as a move from objectivity to subjectivity. But, at the same time, we must also recognize that in many digital contexts, *objective* representation is still all that exists for the poor and marginalized—if they are represented at all. We can acknowledge and explore novel modes of civic and political engagement facilitated by new media and the ways in which these technologies are subverting tightly regulated mainstream media (Gordon & Sorenson, chapter 17; Privalova, chapter 18) and transforming the world of "mass media" to hybrid developments that Castells (2009) calls "mass self-communication." But we must also consider evidence provided by studies that illuminate the mass integration of commercial interests into mediated communication and the degree to which corporate and capitalist interests are shaping mediated communication praxis in different cultural contexts, (Lee, chapter 11; Hrachovec, chapter 19). Others argue that mediated communications are diffusing and de-centering political power and may in some contexts be reducing the effectiveness of social action.

This increasingly complex picture demands that we move beyond narrow visions guided by simplistic theoretical, ideological, or sociopolitical agendas. Following the mediated intercultural dialectics perspective as discussed above, we must embrace (or develop) theoretical and methodological approaches that can accommodate and offer greater insight into the "processual, relational, and contradictory logics" of *mediated* inter- (and intra-) cultural communication and its local and global consequences.

Problems of paradigm: Functionalist models of culture

In the past decade, we can observe a new convergence of research in the fields of intercultural communication and the field of Internet and technology studies on questions of power, participation, justice, and social cohesion. Across these disciplines, scholars are exploring the ways in which new media may facilitate increasingly globalized participation by groups of people around the world in the so-called information society and the possible consequences for human development, conceptions of local identity, ethnic differences, and regional subcultures. Such investigations call for attention to theories and models of culture.

At the start of this second decade of the new millennium, there is increasing awareness of the need to move beyond the theoretical models of "culture" whose main thrust is the classification of people into "cultural groups" (the labels predicting they will interact with information and communication technologies in particular ways). Increasingly evident is the need to instead examine the dynamic processes by which social, institutional, technical, economic, and political forces at any given moment "yield intelligible meanings, enter the circuits of culture—the field of cultural practices—that shape the understandings and conceptions of the world of men and women in their ordinary everyday social calculations" (Hall, 1989).

Models of culture founded on functionalist assumptions about "kinds of people" (Hacking, 2006) may have a certain utilitarian value—and may fruitfully offer an early tool for opening up "what is otherwise a black box of cultural factors" (Williamson, 2002, p. 1391). But at least three difficulties of such models are worthy of elaboration. One is the assumption that culture is a "national" phenomenon, a classification that rests entirely on (relatively recent) history and vagaries of the modern political state and ignores both internal cultural diversity within political states and national cultures that span multiple political states. A second difficulty of such models is their static and essentializing nature. They seek to classify kinds of people, and they permit nonalert investigators to view individuals as "cultural dopes," lacking agency and carrying uniform cultural attributes. Last, functionalist models of culture are inherently reductionist and determinist. Geneticists would say, wisely, "you get what you select for" (Shuman & Silhavy, 2003); that is, if you create conditions to test for an expected difference, that difference is the only difference you will see.

In the context of culture, overreliance on functionalist models of national culture fools us into identifying and studying only a small number of a priori cultural dimensions and blinds us to the myriad cultural and noncultural conditions that influence values and behavior in specific places and times, as well as to emic conditions that are specific only to an individual culture. In an era in which we are increasingly acknowledging that societies can be best understood as a dynamic formation of competing truth regimes rather than a mythical unity (Hall in Grossberg, 1996, p. 136), this is a critical weakness. Such "closed paradigms" (Hall in Grossberg, 1996, p. 19) may be seductive in their ease of application and their apparent existence as objective, value-free tools, but by definition they will make new phenomena—which arise out of continuously evolving "new conditions"—difficult to interpret. Such theories and models simply "let you off the hook, providing answers which are always known in advance" (Hall in Grossberg, 1996, p. 19). The increased interest in and focus on social and cultural "change" in relation to new media and information and communication technologies (ICTs) therefore calls for theoretical approaches that can describe, explain, and accommodate newly emerging sociocultural conditions and practices.

Theory, methodology, method: Moving forward

Williamson (2002) reminds us that it is methodology that underlies our choice and justification of research methods. Methodology depends both on the theoretical models employed and on the values and beliefs of the researcher, including the assumptions and beliefs about epistemology, ontology, and human nature (Burrell & Morgan, 1979). Selection of a methodology is therefore essentially political in nature (Llewellyn, 1992); it cannot and should not be adopted for simple utilitarian reasons. For these reasons, it is important that we continue to examine and make overt the assumptions of our methodologies, beginning with our assumptions about what culture is and how it might best be studied; that we create, seek out, and embrace countermethodologies or counterparadigms; and that we understand theorizing as "an open horizon, moving within the magnetic field of some basic concepts, but constantly being applied afresh to what is genuinely original and novel in new forms of cultural practice" (Hall in Grossberg, 1996, p. 138).

Certainly, by giving up on functionalist models of culture, we lose the speed and neatness they provide. We lose the credibility currently bestowed upon (apparently) "objective" methods and "scientific" approaches. And—at no small cost—we lose the easy communicability and applicability of parsimonious models. But meaningful investigation of the complex landscape of mediated intercultural communication in societies and cultures around the globe requires a shift to alternative learning and discovery paths that will allow us to build on the first-stage understandings revealed within the functionalist research paradigm, and by earlier theories of culture, that we are coming to understand as insufficiently dynamic. Such a shift will allow us to flesh out the skeleton of the global technology story revealed by quantitative work and investigate emerging and as-yet-unpredicted phenomena in this rapidly evolving terrain. In our continuing investigation of culture and communication in the dialectical landscape of new media, we might take for guidance Hall's (in Grossberg, 1996) strenuous rejection of closed theoretical paradigms. "I am not interested in Theory," he insisted. "I am interested in going on theorizing" (p. 150).

Similarly, whereas we tend to think of a "method" as implying the application of "rigid templates or practical techniques to organize research," we propose instead a perspective of "method as practice"—which views method both as "research techniques" and as the activity of "practising or trying out" (Slack, 1996, p. 114). While quantitative studies will continue to provide basic figures about new media penetration rates and access to technology, understanding praxis and impacts requires that we adopt methodologies and research methods that will allow us to examine the dynamic relations of power that structure the many social worlds we study, in pursuit of contextualized understandings of human behavior. This calls for investigation, elaboration, and comparison of specific cases (beyond a search for universals) and an assumption that while individuals are organized and constrained by their intersubjective social realities, they are also actively involved in reproducing these realities, emphasizing the important role of culture within the wider context (Packer, 1999). Moreover, we cannot abstract culture "from its material, technical and economic conditions of existence" (Hall in Grossberg, 1996, p. 139). Indeed, many contributions to this collection lay heavy emphasis on the differing conditions created by variations in political and socioeconomic context—a critical consideration if we seek to meaningfully explore the digital divide.

Future directions?

At the time of this writing, debate was raging over Wikipedia's decision to petition UNESCO for status as a World Heritage Site, based on the argument that the Web site meets UNESCO's key criteria for representation: The site claims that it can be considered an endangered "world cultural treasure" and "a masterpiece of human creative genius" (Keller, 2011). That the proposition has arisen at all is evidence, we believe, that new media are now valued and perceived by some as an important cultural resource. Against such a backdrop, this volume represents a step in the direction of deepening understanding of mediated intercultural dialectics to further knowledge in the areas of digital media and intercultural communication. What insights can it offer about future directions? The varied contributions in this book that come from a range of geographic and scholarly standpoints illustrate both the necessity of comparative research and the value of new media and intercultural communication being international in nature—emphasizing the need not only for the study *of* different cultures around the globe but also for the study *by scholars from* different cultural contexts.

In particular, contributions here have begun to open up our understanding of culture as it may operate in a range of mediated contexts. Expanding our paradigmatic range and exploring cultures in their many dimensions that go beyond their definition as "national," with affiliations that are not only characterized as "ethnic" or "racial" will, one hopes, offer greater insight into social contexts whose relations with new media have, as yet, been dramatically understudied. Examples include social groups whose shared culture may have emerged out of a shared language, shared differences in physical ability, or common religious beliefs and practices. Embracing the impact of socioeconomic and political contexts on cultural phenomena should also allow a more careful investigation of subcultures within national or regional settings and the different mediated communicative beliefs and practices of a variety of people whose identities position them outside mainstream or dominant cultures: the poor and the marginalized, locally and globally. In particular, contributions to this volume are overt in noting the dearth of research on certain populations outside the Anglo-American sphere of influence: women in the Middle East (Lev-On & Neriya-Ben Shahar, chapter 14) and South Asia

(Shuter, chapter 13), diasporic migrants in Africa and elsewhere (Alzouma, chapter 12) and the cultures of less economically developed nation states (Olaniran, chapter 4) to name but three.

Moreover, as both technologies and cultures continue to rapidly evolve and interact, and as revealed by studies in this volume and elsewhere, the diversity and differential impacts of differently mediated digital contexts calls for continuing attention. Mediated communication contexts can vary dramatically, and new tools and contexts are proliferating as we write—for example, geo-locational mobile or fourth-generation cellular wireless (4G) media. Thus, this volume does not constitute the terminal intelligence in this complex arena. Instead, its diversity also reveals gaps in current scholarship and points to myriad paths of further study to enrich this field. It is our hope that this book will inspire new intersecting bricks and clicks to propel scholarship, praxis, and policy going forward.

References

Broome, B. J., Carey, C., De La Garza, S. A., Martin, J., & Morris, R. (2005). "In the thick of things": A dialogue about an activist turn in intercultural communication. In W. J. Starosta & G. M. Chen (Eds.), *Taking stock in intercultural communication: Where to now?* (pp. 145–175).Washington, DC: National Communication Association.

Burrell, G., & Morgan, G. (1979). *Social paradigms and organizational analysis: Elements of the sociology of corporate life.* London, United Kingdom: Heinemann Educational.

Castells, M. (2009). *Communication power.* New York, NY: Oxford University Press.

Cheong, P.H. & Gray, K. (2011) Mediated Intercultural dialectics: Identity Perceptions and Performances in Virtual Worlds. *Journal of International and Intercultural Communication,* 4(4), 265–271.

Cheong, P. H., & Martin, J. N. (2009). Cultural implications of e-learning access (& divides): Teaching an intercultural communication course online. In B. A. Olaniran (Ed.), *Cases on successful e-learning practices in the developed and developing world: Methods for global information economy* (pp. 78–91). Hershey, PA: IGI Global.

Cheong, P. H., & Poon, J. P. H. (2009). Weaving webs of faith: Examining Internet use and religious communication among Chinese Protestant transmigrants. *Journal of International and Intercultural Communication, 2*(3), 189–207.

Chuang, R. (2003). Postmodern critique of cross-cultural and intercultural communication. In W. J. Starosta & G. M. Chen (Eds.), *Ferment in the intercultural field: Axiology/Value/Praxis—International and Intercultural Communication Annual* (Vol. 26). Thousand Oaks, CA: Sage.

Collier, M. J. (2005). Theorizing cultural identification: Critical updates and continuing evolution. In W. B. Gudykunst (Ed.), *Theorizing about intercultural communication* (pp. 235–256). Thousand Oaks, CA: Sage.

Gajjala, R. (1999). Third-world critiques of cyberfeminism. *Development in Practice, 69*(5), 616–619.

Gonzalez, A., Houston, M., & Chen, V. (Eds.). (2011). *Our voices: Essays in culture, ethnicity and communication* (5th ed.). New York, NY: Oxford University Press.

Grossberg, L. (1996). On postmodernism and articulation: An interview with Stuart Hall. In D. Morley & K.-H. Chen (Eds.), *Stuart Hall: Critical dialogues in cultural studies* (pp. 131–150). London, United Kingdom, and New York, NY: Routledge.

Gudykunst, W. B. (Ed.). (2005). *Theorizing about intercultural communication.* Thousand Oaks, CA: Sage.

Hacking, I. (2006). Kinds of people: Moving targets. The Tenth British Academy Lecture, April 11, 2006 [Electronic version]. *The Proceedings of the British Academy.* Retrieved from http://www.proc.britac.ac.uk/tfiles//151p285.pdf

Hall, S. (1989). Ideology and communication theory. In B. Dervin, L. Grossberg, B. J. O'Keefe, & E. Wartella (Eds.), *Rethinking communication: Vol. 1. Paradigm issues* (pp. 40–52). Newbury Park, CA: Sage.

Halualani, R. T., & Nakayama, T. K. (Eds.). (2010). *Handbook of critical intercultural communication.* Malden, MA: Blackwell.

Internet World Stats. (2010). Usage and population statistics. *Internet Users in the World: Distribution by Word Regions 2011.* Retrieved from http://www.internetworldstats.com/stats.htm

Keller, J. (May 23, 2011). Is Wikipedia a world cultural repository? [Electronic version]. *The Atlantic.* Retrieved from http://www.theatlantic.com/technology/archive/2011/05/is-wikipedia-a-world-cultural-repository/239274/

Kramsch, C. J. (2001). Intercultural communication. In R. Carter & D. Nunan, (Eds.), *The Cambridge guide to teaching English to speakers of other languages* (pp. 201–206). New York, NY: Cambridge University Press.

Leeds-Hurwitz, W. (1990). Notes on the history of intercultural communication: The Foreign Service Institute and the mandate for intercultural training. *The Quarterly Journal of Speech, 76,* 262–281.

Llewellyn, S. (1992). The role of case study methods in management accounting research: A comment. British Accounting Review, 24 (1): 17–31.

Lievrouw, L. A., & Livingstone, S. (Eds.). (2006). *Handbook of new media: Social shaping and social consequence.* London, United Kingdom: Sage.

Martin, J. N., & Nakayama, T. K. (1999). Thinking dialectically about culture and communication. *Communication Theory, 9,* 1–25.

Martin, J. N., & Nakayama, T. K. (2010). Intercultural communication and dialectics revisited. In R. T. Halualani & T. K. Nakayama (Eds.), *Handbook of critical intercultural communication* (pp. 51–83). Malden, MA: Blackwell.

Martin, J. N., Nakayama, T. K., & Carbaugh, D. (2012). The history and development of the study of intercultural communication and applied linguistics. In J. Jackson (Ed.), *The Routledge handbook of intercultural communication.* New York, NY: Routledge.

Merskin, D. (2001). Winnebagos, Cherokees, Apaches, and Dakotas: The persistence of stereotyping of American Indians in American advertising brands. *Howard Journal of Communications, 12,* 159–169.

Olaniran, B. A. (2001). The effects of computer-mediated communication on transculturalism. In V. H. Milhouse, M. K. Asante, & P. O. Nwosu (Eds.), *Transcultural realities:*

Interdisciplinary perspectives on cross cultural relations (pp. 83–105). Thousand Oaks, CA: Sage.

Ono, K. A. (2009). *Contemporary media culture and the remnants of a colonial past.* New York, NY: Peter Lang.

Ono, K. A., & Pham, V. (2009). *Asian Americans and the media.* Cambridge, UK: Polity.

Packer, M. (1999). *Interpretive research.* Retrieved from The Duquesne University Web site: http://www.mathcs.duq.edu/~packer/ IR/IRlogic.html

Poole, M. S., & Holmes, M. E. (1995). Decision development in computer-assisted group decision making. *Human Communication Research, 22*, 90–127.

Poole, M. S., Holmes, M., & DeSanctis, G. (1991). Conflict management in a computer-supported meeting environment. *Management Science, 37*, 926–953.

Shuman, H., & Silhavy, T. (2003). The art and design of genetic screens: Escherichia coli. *Nature Reviews Genetics, 4*(6), 419–431.

Slack, J. D. (1996). The theory and method of articulation in cultural studies. In D. Morley & K.-H. Chen (Eds.), *Stuart Hall: Critical dialogues in cultural studies* (pp. 112–127). London, United Kingdom: Routledge.

St.Amant, K. (2002). When cultures and computers collide: Rethinking computer mediated communication according to international and intercultural communication expectations. *Journal of Business and Technical Communication, 16*(2),196–214.

Thurlow, C., Lengel, L., & Tomic, A. (2004). *Computer mediated communication: Social interaction and the Internet.* Thousand Oaks, CA: Sage.

Ting-Toomey, S. (2005). Identity negotiation theory: Crossing cultural boundaries. In W. B. Gudykunst (Ed.), *Theorizing about intercultural communication* (pp. 211–233). Thousand Oaks, CA: Sage.

Vargas, L. (2000). Genderizing Latino news: An analysis of a local newspaper's coverage of Latino current affairs. *Critical Studies in Mass Communication, 17*, 261–293.

Walther, J. B. (1996). Computer-mediated communication: Impersonal, interpersonal, and hyper-personal interaction. *Communication Research, 23*, 3–43.

Walther, J. B., & Parks, M. R. (2002). Cues filtered out, cues filtered in: Computer-mediated communication and relationships. In M. L. Knapp & J. A. Daly (Eds.), *Handbook of interpersonal communication* (pp. 529–563). Thousand Oaks, CA: Sage.

Wellman, B. (2011). Studying the Internet through the ages. In M. Consalvo & C. Ess (Eds.), *The Blackwell handbook of Internet studies* (pp.17–23). Oxford, United Kingdom: Wiley-Blackwell.

Williamson, D. (2002). Forward from a critique of Hofstede's model of national culture. *Human Relations, 55*(11), 1372–1395.

Wood, A. F., & Smith, M. J. (2005). *Online communication: Linking technology, identity and culture* (2nd ed.). Mahwah, NJ: Lawrence Erlbaum Associates.

Yoshikawa, M. (1987). The double-swing model of intercultural communication between the East and the West. In M. Kinkaid (Ed.), *Communication theory: Eastern and Western perspectives* (pp. 319–29). London, United Kingdom: Academic Press/Harcourt Brace Jovanovich College.

Section One: Theorizing New Media and Intercultural Communication

The current theoretical knowledge at the nexus of culture, mediated communication, and context is limited. The chapters in this section develop theoretical notions related to a variety of foci and new media contexts, with each author offering a unique theoretical lens through which to study and understand intercultural communication and new media, as well as offering possible future research trajectories.

The first chapter of the section, by Maja van der Velden, discusses the distinction between Indigenous knowledge and Western science and proposes a de-centered perspective in which the relationship between different local knowledges can be explored. She identifies two approaches to knowledge management: *knower-centered* and *knowledge-centered*. She argues that a knower-centered approach is most appropriate for accommodating the particular structures of Indigenous knowledges and illustrates this argument with examples of Web-based databases and digital archives for Indigenous knowledges. She also introduces the notion of *script* as an approach to inscribing and analyzing use and users in a technology design. She uses findings from her research in India and Kenya to discuss the possible risks of using scripts in design and to introduce the concept of *contact zone*. Van der Velden argues that the meaning and matter of a type of technology is not perceived as the effect of its use only but that these factors emerge unpredictably in each iteration of a design. If meaning and use cannot be predicted, she asks, how can we do justice to the knowledge of Indigenous communifties? Based on the ethical philosophy of Emmanuel Levinas and the work of Brigham and Introna (*Ethics and Information Technology*, 2007), she presents a design strategy that locates design as a thoroughly ethical practice. Design for the contact zone, this chapter concludes, is an intra-active and adaptive process for creating databases that are meaningful for Indigenous knowers.

In the second chapter of the section, Beth Bonniwell Haslett applies her approach—*structurational interaction* (a synthesis of Goffman's and Giddens' theoretical notions)—to examine the intersections of culture, communication, identity, and new media. She argues that structurational interaction offers a nuanced view of culture and the ways in which media are involved in cultural change and development. Building on the sociohistorical context of globalization, she uses structuration interaction as a lens through which to examine the specific challenges brought about by the increased opportunities for human interaction brought about by mediated communication: the multiple and unprecedented opportunities for framing identity offered by avatars, social networking services (SNSs), and other new media; the important role of social presence in impression management in mediated contents; the establishment of trust across time and space in virtual contexts; and the cultural differences in mediated communications, including differences afforded by race, ethnicity, age, or religion. Last, she explores the notion of *third-space interactions*: a new construct that examines intercultural communication as an interactional space that may potentially transcend cultures. Bonniwell Haslett concludes by identifying future topics worthy of further theorizing and research: hybrid relationships (online/offline) and identity; negative social behaviors resulting from Internet use and the related implications for identity; and issues of privacy in relation to trust, risk, and ethics, which becomes a concern as unprecedented information about others becomes available online.

The next two chapters extend traditional theoretical constructs to examine intercultural communication in new media contexts. Bolanle Olaniran explores how culture and power inequities impact communication activities and access in e-learning contexts. As a theoretical lens, Olaniran first employs Hofstede's cultural value framework, in specific the *power distance* and *individualism/collectivism* dimensions, to explore challenges encountered by online learners who have cultural differences from those who develop digital media and their learning contents. He identifies and describes various language and value differences that impact e-learning and then takes a critical look at the implications of the digital divide on e-learning. E-learning technologies are predominantly designed and marketed by companies in economically developed countries, allowing imposition of their cultural ideologies on Internet and communication technology (ICT) consumption and use in less economically developed countries. Noting the ripple effect of limited financial resources—and the implications for new media access, infrastructure, institutional resources, and technical knowledge—Olaniran highlights this particularly problematic situation at the nexus of culture and economics.

Kirk St.Amant's chapter examines changing ideas of context, culture, and communication in online encounters regarding the context of international work teams. Extending E. T. Hall's theoretical framework of *context* (initially developed through an investigation of face-to-face encounters) to online exchanges, he explores how the plastic nature of identity and locality created by cyber-space creates a new and problematic situation related to context-based communication. In specific, he examines how certain conventions associated with establishing authority and credibility in online exchanges can conflict with, or become a point of contention in, various cultural expectations of credible behavior. Building on this analysis, St.Amant proposes a range of context-related strategies that individuals might employ when collaborating on work projects with international virtual teams.

2. *Designing for Culture: An Ecological Perspective on Indigenous Knowledge and Database Design*

MAJA VAN DER VELDEN, PHD

Introduction

Various forms of new media have transformed the way we organize and share information. Media scholar Lev Manovich (2002) calls the database the key *cultural form* of the computer age, as it is now possible to separate content and interface: We can make different interfaces to access the same content. From this perspective, we can understand content management and knowledge management software as interfaces that organize the elements in a database in different ways and that provide access to content.

In this chapter, I consider the management of Indigenous knowledges. Before we start looking at database design, in particular knowledge management software, we need to ask in what ways Indigenous knowledge differs from non-Indigenous knowledge. If we agree that there are different forms of knowledge, we have to contemplate another question: Can we use the same kind of database design for different cultures or do we need to develop different designs? This question brings the relationship between technology and knowledge to the foreground.

Feminist and postcolonial techno-science focus their analysis of this relationship on the mutual shaping of knowledge and technology, on the situatedness of these relations, and on how these relations affect people and bodies. In this chapter, I locate myself in this current field of inquiry by asking: Can we design for culture?

Knowledges?

Let's begin with the question about knowledge. There are scholars who maintain that there is only one form of knowledge that counts and that it is that of modern Western science. Only knowledge that can be separated from its local place of production to become universal, objective, and true can be called science. Some proponents of this perspective present modern Western science as superior knowledge (Gross & Levitt, 1994; Koertge, 1998; Nanda, 2003). Local, traditional, or Indigenous knowledge—knowledge that is bound by its place and its relations, such as culture, religion, and community—is considered mere belief.

There is a growing community of people who maintain that all knowledge, including modern Western science, is local. Turnbull (1997) describes two different perspectives in which this localist position can be expressed:

1. Science is value-laden and should therefore let go of its value-free and universalist stance and adopt a set of quasi-universal values.
2. All knowledges are situated within a particular set of values. Turnbull refers to this perspective as the de-centering of science, the recognition that there are other ways of knowing the world besides our Eurocentric and egocentric way. (Cunningham & Williams, 1993)

De-centering science

Feminist theory plays an important role in the de-centering of science by questioning the frameworks in which science is produced. Sandra Harding (1995) proposes the notion of *strong objectivity* to counter understandings of objectivity based on the subject/object dichotomy, detachment, and value neutrality and argues that knowledge that includes the experiences of those who have been traditionally left out of the production of knowledge, such as women and other subordinate groups, may in effect be more objective because such groups are motivated to understand the views of the people and institutions that are more powerful.

Donna Haraway's (1988, 1991) work plays an important role in understanding how we can talk about *knowledges* in science; she opposes science *and* relativism, which both claim the ability to see everything from "nowhere" and proposes situated knowledges, which present a located, embodied objectivity based on partial perspectives. A partial perspective is not based on identity but on a partial connection with the "other." It is partiality, not universality—a view from "somewhere" rather than from "nowhere"—that offers, according to Haraway, the preferred position for making rational knowledge claims.

The de-centering of science is also supported by postcolonial theory, which queries questions of knowledge and power in wider social and economic terms, by locating it in former colonized societies. For example, Shiva (1993) shows how scientific forestry and agriculture split the plant world in two separate, nonoverlapping domains. From an ecological perspective, the plant world connects forest and agriculture, providing food, fodder, and fertilizer. In the categories of science, following the global commodity markets, only timber is seen as a product of the forest, whereas food is confined to the category of agriculture. In a later publication, Shiva (1997) argues that only through the cultivation of diversity, both in terms of biodiversity and knowledge diversity, and the use of decentralization and local democratic control will we be able to recover the possibility of self-organization.

Feminist and postcolonial theory have inspired the study of science and technology since the 1980s. Their cross-fertilizations have resulted in science studies that include the discussion of Indigenous knowledge traditions (e.g., Hess, 1995; Turnbull, 2000; Verran, 2002; Visvanathan, 2000, 2007; Figueroa & Harding, 2003). Watson-Verran and Turnbull (1995) describe a variety of Indigenous knowledge systems and propose a *symmetric* treatment of all knowledge systems, which enables them to describe these systems, on the one hand, as very different from science, but on the other hand, as knowledges that are systematic and innovative. They discuss how local innovation is the implicit basis of all knowledge systems. In recognition of the localness of modern Western science, they argue that all knowledges can be understood as Indigenous knowledges.

How to deal with difference

When we take the de-centering of science position, we are faced with an important question: How do we deal with the relations between these different, local, situated knowledges? Turnbull (1997) proposes two positions: The first one argues for incommensurability between these knowledges and stresses the uniqueness of a particular local knowledge. The second states that despite the differences between these knowledges, it is important to find ways in which these knowledges can coexist. This perspective is not based on a relativist stand, calling for automatic justification of situated knowledges. Such relativism, based on the equality of positioning, is a denial of responsibility and critical inquiry: It is a way of being nowhere while claiming to be everywhere (Haraway, 1988). The second position is a responsible, mobile, and split position, always partial, never whole. The proponents of this position refer to concepts such as "symmetry" (Watson-Verran & Turnbull, 1995), "cognitive justice" (Santos, 2007; Visvanathan, 2000), or the "postcolonial moment" (Verran, 2002) to explore how these knowledges can coexist.

Distinguishing between Indigenous and non-Indigenous knowledge

The recognition of the localness and situatedness of all knowledges brings up a second question: Can we and should we distinguish between Indigenous and non-Indigenous knowledge? Scholars who have addressed this issue argue that there are no simple or universal criteria that can be deployed to separate Indigenous knowledge from Western scientific knowledge (Agrawal, 2002; Ellen, Parkes, & Bicker, 2000). Should we make separate categories of knowledge if we can't formulate strong distinctions between Indigenous and non-Indigenous knowledge?

The increased use of the term "Indigenous knowledge" since the 1960s has both romantic and practical reasons (Ellen & Harris 2000). The rediscovery of Indigenous knowledge in the "1960s counterculture" was based on the romantic notion of primitive people living in harmony with nature. In a practical sense, the increased use of the term is connected with initiatives to promote socioeconomic development and environmental conservation (Agrawal, 2002). For example, there was no explicit mentioning of the role of Indigenous knowledge in international development projects at the UK branch of the Department for International Development before 1990 (Sillitoe, 1998). On the contrary, traditional knowledge and practices were often seen as obstructing development.

In the 1990s, analyses of development discourse and practices, in particular the work by James Ferguson (1990) and Arturo Escobar (1995), encouraged new debates concerning whose knowledge and what kind of knowledge should inform development practices. These questions opened up the development sector for the insights of feminist theory and postcolonial theory, in which the question of knowledge and power had been critically addressed. Nevertheless, the dominant perspective in development thinking continues to be the inclusion of Indigenous knowledge on the level of artifacts and particular practices. This perspective assumes a dichotomy between Indigenous and Western scientific knowledge. The category of Indigenous knowledge is added to the development discourse as a resource that can be mined to improve development efforts. This approach does not challenge the development practice or engage the ontologies underlying these Indigenous artifacts and practices.

Protecting Indigenous people

The focus on *Indigenous* knowledge is also used as a lobbying strategy by and for Indigenous peoples and has become an important concept in legitimating local practices. Babidge et al. (2007) describe, for example, how Indigenous knowledge provides a management process for engaging with state agencies in Australia. Also, non-Indigenous awareness of Indigenous land issues and bio-piracy has resulted in a variety of initiatives to conserve and protect Indigenous knowledge as a way to protect Indigenous peoples and cultures. This has resulted in a growing number of Indigenous knowledge management initiatives that propose to do this work of conservation and protection.

Structures of Indigenous knowledges

Agrawal (1995) proposes examining the specific practices of investigation and knowledge creation in different countries and of different groups of people. This will make the existence of diversity visible within what is often perceived as the "homogeneous" categories of "Western"and "Indigenous." At the same time, we can find commonalities when we concentrate on the ways in which Indigenous and Western scientists create knowledge.

Christie (2004), Verran (2005), and Watson-Verran and Turnbull (1995) describe a similar approach as they situate their work as both within the social sciences and within the Yolngu Indigenous community in Australia. Their descriptions are strictly symmetric, as neither side is privileged in producing true or good knowledge. Their work shows that instead of focusing on Indigenous knowledge as a resource frozen in time and place, a look at the structures of Indigenous knowledges is preferable. They suggest that we can investigate the frameworks and methods for knowledge production. We can look at where knowledge is located (who can have knowledge about what), where knowledge is stored or archived, how knowledge is shared, how knowledge evolves over time, and how indigenous knowledges are based on *connectedness* with the land and on the *performance* of knowledge (e.g., Onwu & Mosimege, 2004. For example, Richards (1993) argues in his study of agricultural practices among Indigenous peoples in Africa that farmer practices are not based on a *static* body of Indigenous knowledge but rather on "a set of improvisational capacities called forth by the needs of the moment" (p. 62). A question we thus need to address is whether we can manage something that is connected, evolving, heterogeneous, and social?

Managing indigenous knowledges

The descriptions of Indigenous knowledges as dynamic, heterogeneous, social, and distributed, and experimental, collective, and in the process of continuous adaptation and negotiation seem incompatible with the notion of managing Indigenous knowledge. The idea that we can manage knowledge can be traced to the early 1990s and was initially closely connected to the introduction of digital information and communication technologies supporting the archiving, organizing, and sharing of information in an organization.

Since then, the management of knowledge approaches, such as knowledge management (KM) and knowledge management for development (KM4D), have been understood as a range of practices to identify, create, represent, distribute, and enable adoption of insights and experiences in an organization. Such insights and experiences comprise knowledge embodied in individuals and embedded in organizational practices. The two main approaches can be distinguished (van der Velden, 2002) as follows:

1. The knowledge-centered approach focuses on the collection and codification of knowledge and depends heavily on information systems, such as expert systems, portals, digital directories, and best practices databases.
2. The knower-centered approach perceives knowledge as a human resource and focuses more on creating enabling situations in which knowledge can be shared in more informal ways, such as in communities of practice, storytelling, tagging, and so forth. The technologies used in this second approach offer tools for collaboration and knowledge sharing, such as discussion forums, blogs, wikis, and social-networking sites.

Indigenous knowledge management

Indigenous knowledge management can be understood as the combined result of the introduction of knowledge management in the international development sector and the digitalization efforts to turn noncodified, tacit traditional environmental knowledge into codified, explicit knowledge. The rationale for the *ex situ* conservation of Indigenous knowledge with the use of digital technologies is proposed in terms of the protection of Indigenous knowledge and benefit for the Indigenous community (e.g., Department of the Environment, Water, Heritage and the Arts [Australia], 2009; Ngulube, 2002; Hunter, Koopman, & Sledge, 2003; Traditional Knowledge Digital Library, 2009). Expectations of what digital tools such as database software can do for Indigenous knowledge are often high. Whereas conventional knowledge management practices are assumed to support archiving, organizing, and sharing knowledge, it is considered that Indigenous knowledge management activities should also contribute to the protection of Indigenous knowledge against knowledge loss and external exploitation.

The question of how to manage something that is messy, evolving, heterogeneous, and social is thus entangled with issues such as ownership, intellectual property rights legislation, cultural protocols, and technical issues in the form of choice of media and access, as well as more mundane issues such as system maintenance and project financing. How does knowledge management software deal with this entanglement and these high expectations? If we look at early Indigenous knowledge management practices (e.g., World Bank database of indigenous knowledge and practices [World Bank, n.d], the UNESCO register of best practices on Indigenous knowledge [UNESCO, n.d.], the Native American Ethnobotany Database [University of Michigan-Dearborn, n.d.], and the Tanzania Indigenous Knowledge database [Tanzania Development Gateway, n.d.]), we can see that they take a knowledge-centered approach. The technical, social, and legal entanglements were ignored in favor of providing abstract statements about artifacts or practices, what has been referred to as the "museumization of local knowledges" (Visvanathan, 2002).

Agrawal (2002) argues that such abstraction is the instrumental logic of development that informs the design of these databases. He suggests that the scientization of Indigenous knowledge strips away the detailed, contextual, and applied aspects of knowledge that might be crucial in the positive effects claimed for a particular piece of Indigenous knowledge. From a systems development perspective, we can argue that in the process of "abstracting away" (Blackwell, Church, & Green, 2008) the inconvenient complexity of Indigenous knowledge, we end up with systems that are not very useful for Indigenous communities.

I experienced this process of scientization and abstraction during field visits to a local knowledge management project in India. Traditional healers had translated their knowledge of practices into texts, writing a description of the practices down in a paper notebook. The local names of the plants and their medicinal characteristics were then translated into the language of botany and medicine by the people working at the research organization. The organization members responsible for the project mentioned that the knowledge of the healers could only be added to the organization's database if the validity of the healers' knowledge claims was established in a "proper" laboratory.

Another, more recent body of practices in Indigenous knowledge management was not initiated by development agencies or research institutions but by the Indigenous communities themselves. The main motivation for these initiatives is to archive, protect, and preserve Indigenous knowledge. These initiatives actively involve the community and often employ interactive and participatory multimedia tools based on Web 2.0 applications that can be used in a web browser. Examples include Mukurtu (www.mukurtuarchive.org) and the Ara Irititja Project (www.irititja.com), two Aboriginal digital archives in Australia that enable the recording and presentation of Indigenous knowledges in ways that allow some of the relations and performances to be visible and audible.

The role of technology in archiving and protecting indigenous knowledges

The discussion in the previous sections may have given the impression that we can create database software that can archive and protect Indigenous knowledges, even if there are some incompatibilities between knowledge management perspectives, databases, and Indigenous knowledges or the idea that design processes and technology designs are determinist, with linear, causal relations between method, design, and use.

The notion of *script* is often used to describe the phenomenon in which technology is perceived as a kind of film scenario in which the story, the way the technology is supposed to be used, and the actors, who are the users of the technology, seem to be given by the design. According to Akrich (1992):

> Designers [thus] define actors within specific tastes, competences, motives, aspirations, political prejudices, and the rest, and they assume that morality, technology, science, and economy will evolve in particular ways. A large part of the work of innovators is that of "inscribing" the vision of (or prediction about) the world in the technical content of a new object. (p. 208)

Script is a productive term to inscribe and analyze users and uses in design (e.g., Rommes et al., 1999; Shade, 2007; van Oost, 2003). Van der Velden et al. (2008) examined inscriptions of gender and knowledge diversity in design and identified a tension between the desire to design for gender and diversity and the risk of "freezing" particular conceptualizations of gender and knowledge in design. They located the discussion of this tension in the dichotomy between "design-from-nowhere" and "design-from-somewhere." Lucy Suchman (2002) describes design-from-nowhere as closely tied to the goal of designing technical systems as commodities that can be stabilized and untethered from the sites of their production long enough to be exported en masse to the sites of their use. Thus, designers, who see their technologies as objects and themselves as neutral designers, deny the possibility for locating responsibility for the design.

An example of the design-from-somewhere context arose during observations I made on a field trip to Kenya. Here, I looked at the classification work of Jonathan, a Maasai knowledge worker from Enkirgirri in southern Kenya. Jonathan's style of organization was part of a global network in which local knowledge for local development was shared in a global, distributed database system. There was no central system in which to upload items to share globally. The global network consisted of smaller databases, which were able to communicate with each other. The database software also enabled local classification work of each of the networked databases. Because the software was supposed to be used in different cultural settings, each with their own way of knowing the world, the software had a default classification system, which could be adapted to the particular needs of the local communities and organizations. The software was developed by software engineers in New Delhi, India, but was perceived as neutral global technology that would allow for local cultural inscriptions. The global classification did not contain categories such as "Maasai" or "pastoralist," and it was up to Jonathan to add those subcategories to his local version of the classification. Jonathan did not create these subcategories and therefore found it difficult to classify his items. When I asked him about it, he responded that he did not see it as his task or responsibility to change something in the software (van der Velden, 2008). Here, a scenario was scripted in the software, in which it was considered the role of the knowledge worker to localize the classification system—to create a design-from-somewhere out of a design-from-nowhere. In fact, none of the knowledge workers I met in India and Kenya considered it their role to localize the classification system. Because none of them had been part of the software design and its default classification system, they felt it was the responsibility of the designers, and those higher up in the project hierarchy, to adapt the classification system.

Knowledge management projects in general tend to ignore the agential and generative role of technology when it comes to managing and protecting knowledge. Technology is treated as a neutral tool. However, if knowledge is understood as the result of a direct material engagement with the world (Barad, 2007), it can be asked how knowledge management software affects Indigenous knowing. What happens to the evolving, heterogeneous, and entangled Indigenous knowledge when it meets knowledge management software? One way to investigate this question is to look at knowledge management software as a knowledge representation itself. Instead of considering technology as a neutral tool, we can understand it as a particular form of Western techno-scientific knowledge. In the next section, I will look closer at the meeting of these different knowledges.

Contact zones

Mary Louise Pratt's (1998) notion of *contact zone* can help us think about designing spaces in which knowledges can meet on the basis of cognitive justice. Anthropologist James Clifford (1997) discussed museums as functioning as contact zones between Indigenous peoples and non-Indigenous museum people. Clifford described a contact zone as a space where knowledge systems *do not* meet as "sociocultural wholes" but as "systems already constituted relationally, entering new relations through historical processes of displacement" (p. 7).

Clifford's notion of relating knowledge systems can also be described as *assemblages* (Latour, 2005; Watson-Verran & Turnbull, 1995), as *webs of interdependence* (Tsing, 2005) and as *high-risk zones* (Star, 1991). Haraway (2003, 2008) uses the notion of contact zones to discuss overlapping ontologies, the interdependencies of species, and companion species and warns us that such a contact zone is not about methods but about communication across irreducible differences.

The contact zone seems an appropriate metaphor for the meeting between Indigenous knowledges and the techno-scientific knowledges of knowledge management software. Whereas the determinist notion of a script assumes a particular scenario of events in which two autonomous entities come together and interact, the notion of a contact zone implies a more open-ended perspective. "Design," "user," and "Indigenous knowledge" do not pre-exist before their "meeting as 'sociocultural wholes.'" The contact zone is a space in which subjects come in and through their relations.

Physicist Karen Barad's (2003) concept of *intra-action* is useful for understanding how people and things do not pre-exist their relationship: They become. Their characteristics and boundaries *become determinate* in their intra-actions: "Intra-action underscores the sense in which subjects and objects emerge through their encounters with each other" (Suchman, 2007, p. 267). For example, the design process of an Indigenous knowledge database is a series of iterations, which Barad would call "iterative cuts." In each cut, the ontological inseparability of "subject" and "object" becomes disentangled, and their characteristics and boundaries become determinate. (Barad stresses that this *cut* is not a Cartesian cut, based on an inherently distinctive subject and object but a Bohrian cut, effecting a separation between subject and object). In such an iteration or cut, some possibilities are opened up and others are closed off. Such becoming is not an unfolding in time, argues Barad, "Rather the past and future are enfolded participants in matter's iterative becoming" (2007, p. 234). Thus, a database design does not determine use, thereby excluding all other uses, nor does the database acquire meaning through use. Intra-action thus refers to the dynamic reconfiguration of subject and object. Indigenous knowledge and database software are entangled in the design process. In each iteration of the design, new agencies in terms of possibilities and constraints emerge. Possibilities for action are not inscribed in a software program, as we saw in the example of Jonathan, nor restricted to humans but are enacted in the contact zone where Jonathan and the software meet. Intra-actions create new realities in which new and different possibilities open up.

Contact zone design

I have described the relations between humans and technology as a material-discursive practice (Barad, 2007; Haraway, 1991), in which materiality and meaning come into being when humans and technologies do not interact but *intra-act*. In this perspective, the characteristics, properties, and meaning of technology emerge from the intra-actions with other artifacts and with humans. The culture of an artifact is the effect of a particular configuration of humans and things. Technologies have culture but not as an intrinsic characteristic nor as something solely given by its users. In other words, cultural meaning is not given in the design but is the property of an assemblage, which includes the design. This perspective presents us with a new question: How can we do justice to culture in design if we cannot specify culture in a design?

Undesigning the design

Martin Brigham and Lucas Introna (2007) argue for a radical understanding of the division between humans and technology, similar to the one proposed by Barad (2007), Haraway (1991), and Suchman (2002). That is, Brigham and Introna argue that maintaining the ontological division between humans and technology prevents us from having an understanding of the role of politics and ethics in design, and they propose that technologies are relational effects, transforming as "they 'travel' between places and over time and refashion the context into which they are introduced in ways that surpass intentions and that cannot be predicted completely in advance" (p. 5).

Brigham and Introna (2007) examine ethics, especially in technological situations concerned with "Others," such as in the case of users from cultures different than designers or users who are not specified during the design process. They call upon the ethical philosophy of Levinas, whose ethics of the Other addresses our responsibility for the Other and the relationship between our Self and the unique, unknowable Other. In specific, they emphasize two aspects from Levinas's ethical philosophy, which are important for our discussion. The first one is the difference between *need* and *desire*; the second, the difference between *saying* and *said*. Need, according to Levinas, is an instrumentalist assumption, as in the possible suggestion "We need to do justice to culture in technology design." Such a need is a self-centered need, the fulfillment of one's own wants (i.e., as a designer) and is about the love for Self, contrasted with desire, which cannot be satisfied and is about the love for the Other. A desire cannot be fulfilled because one can never fully know the Other. When we design something on the basis of what we think is the Other's desire, we design a representation of this desire, being that we can never fully know the Other.

Related to need and desire are the notions of saying and said. Saying refers to the meaningful communication between the Self and the Other, which is reduced by the Self to the said. The said is what remains of the meaningful communication after it has been ordered and classified by the Self (i.e., the designer). Levinas proposed an *unsaying* of the said, in which the saying is revealed again. In a similar manner, Brigham and Introna propose *undesigning the design*. A design is always a representation of the desire of the Other and the saying of the Other. Undesigning the design is the ongoing questioning of the inscription of the Other in the design in order to reveal who and what is made invisible in the design.

If we go back to our original concern, how to design for culture, we can now see how every attempt to design for culture may result in harm of the Other, as we can only represent the culture of the Other, because we can never fully know the Other.

Metadesign

Undesigning the design does not simplify the design practice, as it involves us in a design practice that is never finished. It complicates our work, as it confronts us with our infinite responsibility toward the Other as well as with the unknowable effects of our design decisions. Maybe that is exactly the strength of *undesigning the design*: It slows us down, makes us think and rethink, and makes us postpone certain design decisions in order to keep certain possibilities open as long as possible. It makes us aware that design is an ethical and political practice and that we, as designers, are fully interconnected with this practice.

An emergent design approach that places important design decisions in the hands of the Other is metadesign, which aims to define and create sociotechnical infrastructures in which new forms of collaborative design can take place (Fischer et al., 2004; Giaccardi, 2005). Metadesign builds on participatory and user-centered design approaches, but it radically shifts the focus from designers to users as designers or codesigners. It postpones certain design decisions by designing systems without preset functions and possibilities in order to allow the users to participate in the evolution of the system. The system becomes like a seed with a mechanism for growth and reseeding (Fischer, 2009).

The metadesign framework proposes environments and applications that enable ongoing emergent processes of communication, collaboration, and creation. Metadesign centers on the design of authoring software that enables Indigenous practitioners to design their own systems for archiving and preserving their knowledge, creating an open-ended and infinite flexible design process.

TAMI

All second-generation Indigenous databases are based on conventional systems development, in which the outcome of a design process is an end product, such as an archive. There are some examples of database designs, still in prototype stage, which exemplify a radical application of the metadesign approach. Text, Audio, Movies, and Images (TAMI), an Aboriginal database developed by Michael Christie and Helen Verran, their colleagues, and an Indigenous community in Northern Australia, is one example; another is Story Weaver. TAMI is a database design based on a perception of reality as not yet described in the database ontology. In contrast with more conventional database designs, such as the database software Jonathan worked with, there is no metadata set to structure a representation of reality in the database. The so-called flat ontology of the database enables practitioners to author community-based and cultural ontologies (Srinivasan, Pepe, & Rodriguez, 2009), also called fluid ontologies (Srinivasan & Huang, 2005), based on the connections they make with the items they upload and organize with the database software.

TAMI can be understood as an attempt to prevent the database from becoming a representation of the saying or a representation of the needs of the designers, instead of those of the Indigenous knowers. A user of the database can upload digital objects, such as a story, video, or photo and organize them in four folders according to their format (text, audio, movie, or image). Each item can be given a file name but not tags. Users can browse through the four folders and pull some of the items into the central screen. Through the process of selecting and organizing items in one collection, a story about a particular event or place is told.

In TAMI, the meaning of an item emerges when it is connected with other items. These connections are made on the central screen, which can be understood as a kind of third space, the contact zone in which Western scientific knowledge and Indigenous knowledge meet and communicate across irreducible differences (Haraway, 2003). When these situated connections are made, database design, knower, and knowledges emerge.

Conclusion

Stewart Brand (1994) uses the term *ecopoiesis* to mean "the process of a system making a home for itself" (p.164), describing how a building and its occupants jointly become the new system. Every time new people move into the building, a new process of ecopoiesis begins, in which the building and the occupants shape and reshape themselves and each other until a tolerable or comfortable fit has become possible. We can think in a similar way about the relations between design and designers/users. When we design for the contact zone, we give form to the processes of ecopoiesis.

In this chapter, I outlined some of the positions in which I ground my design perspective—a de-centered and localist position on knowledge and knowledge production, according to which concerns for dealing with difference inspired me to discuss the design of knowledge management software that matters for Indigenous peoples. I presented a view in which the structures of Indigenous knowledges and the need for cultivating the diversity of knowledge and cognitive justice were contrasted with information and communication technology (ICT)-based knowledge management practices and designs. I discussed the risk of scripting particular scenarios in database software and proposed Barad's notion of intra-action to understand the relations between humans and technology.

A practical expression of Brigham and Introna's undesign perspective can be found in metadesign in which the Other becomes a codesigner and a practitioner and the database becomes a living system that evolves over time. In metadesign, the database is also an authoring tool, giving practitioners the possibility for self-representation and self-organization. Such a database can be understood as a contact zone for different ways of knowing the world, for different becomings of the world.

Metadesign enables design and practitioners to shape and be shaped by each other. Different practitioners will make different design choices, each time creating new processes for ecopoiesis and different fits, users, and uses. Based on Barad, we can understand each decision in the design process as an iterative cut, creating foreseen and unforeseen new inclusions, exclusions, possibilities, and restrictions for known and as-yet-unknown uses and users. In this ecological perspective on design, we can see how the politics and ethics of database design move to the foreground. Designers are not located outside the complex interrelations of design, use, and users but are mutually constituted in the ongoing intra-actions between people and things.

Barad (2007) argues that ethics is about matter and mattering: "We need to meet the universe halfway, to take responsibility for the role we play in the world's differential becoming" (p. 396). In a similar way, we need to meet the realities we cocreate halfway, extending our responsibility from the design process toward the role we play in the foreseen and unforeseen effects of our designs.

Note

1. This chapter is based on the integration of two articles:
"Design for the Contact Zone: Knowledge Management Software and the Structures of Indigenous Knowledges" (pp. 1–18) and "Undesigning Culture: A Brief Reflection on Design as Ethical Practice" (pp. 117–123), which are published in F. Sudweeks, H. Hrachovec, and C. Ess (Eds.). (2010). *Cultural attitudes towards technology and communication 2010, Proceedings 2010*. Available at http://blogs.ubc.ca/catac/proceedings/proceedings-2010/

References

Agrawal, A. (1995). Indigenous and scientific knowledge: Some critical comments. *Indigenous Knowledge and Development Monitor, 3*(3), 333–336.

Agrawal, A. (2002). Indigenous knowledge and the politics of classification. *International Social Science Journal, 173*, 287–297.

Akrich, M. (1992). The de-scription of technical objects. In W. Bijker & J. Law (Eds.), *Shaping technology* (pp. 205–224). Cambridge, MA: MIT Press.

Babidge, S., Greer, S., Henry, R., & Pam, C. (2007). Management speak: Indigenous knowledge and bureaucratic engagement. *Social Analysis, 51*, 148–164.

Barad, K. (2003). Posthumanist performativity: Toward an understanding of how matter comes to matter. *Signs, 28*(3), 801–831.

Barad, K. (2007). *Meeting the universe halfway: Quantum physics and the entanglement of matter and meaning*. Durham, NC: Duke University Press.

Blackwell, A. F., Church, L., & Green, T. R. X. (2008). The abstract is 'an enemy': Alternative perspectives to computational thinking. In *Proceedings PPIG '08, 20th annual workshop of the Psychology of Programming Interest Group* (pp. 34–43). Retrieved from http://www.ppig.org/papers/20th-blackwell.pdf

Brand S. (1994). *How buildings learn: What happens after they're built.* New York: Viking.

Brigham, M., & Introna, L. (2007). Invoking politics and ethics in the design of information technology: Undesigning the design. *Ethics and Information Technology, 9*(1), 1–10.

Christie, M. (2004). *Words, ontologies, and aboriginal databases.* Retrieved from Charles Darwin University Web site: http://www.cdu.edu.au/centres/ik/pdf/ WordsOntologiesAbDB.pdf

Clifford, J. (1997). *Routes: Travel and translation in the late twentieth century.* Cambridge, MA: Harvard University Press.

Cunningham, A., & Williams, P. (1993). De-centring the "big picture": The origins of modern science and the modern origins of science. *British Journal for the History of Science, 26*(4), 407–432.

Department of the Environment, Water, Heritage and the Arts (Australia). (2009). Indigenous Knowledge Management Systems (databases): Guide for Indigenous Communities. Retrieved from http://www.environment.gov.au/indigenous/publications/ knowledge-management-guide.html

Ellen, R., & Harris, H. (2000). Introduction. In A. Bicker, R. Ellen, & P. Parkes (Eds.), *Indigenous environmental knowledge and its transformations: Critical anthropological perspectives* (pp. 1–31). Amsterdam, Netherlands: Harwood Academic.

Ellen, R., Parkes, P., & Bicker, A. (2000). *Indigenous environmental knowledge and its transformations: Critical anthropological perspectives.* Amsterdam, Netherlands: Harwood Academic.

Escobar, A. (1995). *Encountering development: The making and unmaking of the Third World.* Princeton, NJ: Princeton University Press.

Ferguson, J. (1990). *The anti-politics machine: "Development", depoliticization, and bureaucratic power in Lesotho.* Cambridge, United Kingdom: Cambridge University Press.

Figueroa, R., & Harding, S. (Eds). (2003). *Science and other cultures: Issues in philosophies of science and technology.* New York, NY: Routledge.

Fischer, G. (2009). Meta-design: Expanding boundaries and redistributing control in design. *Proceedings of the 11th International Conference on Human-Computer Interaction—INTERACT 2007, Rio de Janeiro, Brazil,* 193–206. Berlin, Germany: Springer.

Fischer, G., Giaccardi, E., Ye, Y., Sutcliffe, A. G., & Mehandjiev, N. (2004). Meta-design: A manifesto for end-user development. *Communications of the ACM, 47*(9), 33–37.

Giaccardi, E. (2005). Metadesign as an emergent design culture. *Leonardo,* 38(4), 342–349.

Gross, P. R., & Levitt, N. (1994). *Higher superstition: The academic left and its quarrels with science.* Baltimore, MD: Johns Hopkins University Press.

Haraway, D. (1988). Situated knowledges: The science question in feminism and the privilege of partial perspective. *Feminist Studies, 14*(3), 575–599.

Haraway, D. (1991). *Simians, cyborgs and women: The reinvention of nature.* New York, NY: Routledge.

Haraway, D. (1997). *Modest_Witness@Second_Millenium.FemaleMan©_Meets_Oncomouse™: Feminism and Technoscience.* New York, NY: Routledge.

Haraway, D. (2003). *The companion species manifesto: Dogs, people, and significant otherness.* Chicago: Prickly Paradigm Press.

Haraway, D. (2008). *When species meet.* Minneapolis: University of Minnesota Press.

Harding, S. (1995). "Strong objectivity": A response to the new objectivity question. *Synthese, 104*(3), 331–349.

Harding, S. (1998). *Is science multicultural? Postcolonialisms, feminisms, and epistemologies.* Bloomington: Indiana University Press.

Hess, J. D. (1995). *Science and technology in a multicultural world: The cultural politics of facts and artifacts.* New York, NY: Columbia University Press.

Hunter, J., Koopman, B., & Sledge, J. (2003). Software tools for indigenous knowledge management. Paper presented at the conference Museums and the Web. Retrieved from The University of Queensland–Australia Web site: http://espace.library.uq.edu.au/ view/UQ:7896

Indigenous Knowledge Commons. (2011). StoryWeaver. Retrieved from http://indigenousknowledge.org/tools-and-resources/storyweaver

Koertge, N. (1998). *A house built on sand: Exposing postmodernist myths about science.* New York, NY: Oxford University Press.

Latour, B. (2005). *Reassembling the social: An introduction to actor-network theory.* Oxford, United Kingdom: Oxford University Press.

Manovich, L. (2002). *The language of new media.* Cambridge, MA: MIT Press.

Nanda, M. (2003). *Prophets facing backward: Postmodern critiques of science and Hindu nationalism in India.* Rutgers, NJ: Rutgers University Press.

Ngulube, P. (2002). Managing and preserving indigenous knowledge in the knowledge management era: Challenges and opportunities for information professionals. *Information Development, 18*(2), 95–102.

Onwu, G., & Mosimege, M. (2004). Indigenous knowledge systems and science and technology education: A dialogue. *African Journal of Research in Mathematics, Science and Technology Education, 8*(1), 1–12.

Oost, E. van. (2003). Materialized gender: How shavers configure the users' femininity and masculinity. In N. Oudshoorn & T. Pinch (Eds.), *How users matter: The co-construction of users and technology* (pp. 193–208). London, United Kingdom: MIT Press.

Pratt, M. L. (1998). Arts of the contact zone. In V. Zamel & R. Spack (Eds.), *Negotiating academic literacies: Teaching and learning across languages and cultures* (pp. 171–186). Mahwah, NJ: Lawrence Erlbaum Associates.

Richards, P. (1993). Cultivation: Knowledge or performance? In M. Hobert (Ed.), *An anthropological critique of development: The growth of ignorance* (pp. 61–78). New York, NY: Routledge.

Rommes, E., Oost, E. van, & Oudshoorn, N. (1999). Gender in the design of the digital city of Amsterdam. *Information, Communication & Society, 2*(4), 476.

Santos, B. de Sousa. (2004). A critique of lazy reason: Against the waste of experience. In I. Wallerstein (Ed.), *Modern world-system in the long-durée* (pp. 157–197). London, United Kingdom: Paradigm.

Santos, B. de Sousa. (2007, June 29). Beyond abyssal thinking: From global lines to ecologies of knowledges. *Eurozine*. Retrieved from http://www.eurozine.com/articles/2007-06-29-santos-en.html

Shade, L. R. (2007). Feminizing the mobile: Gender scripting of mobiles in North America. *Continuum, 21*(2), 179.

Shiva, V. (1993). *Monocultures of the mind: Perspectives on biodiversity and biotechnology.* Dehradun, India: Natraj.

Shiva, V. (1997). *Biopiracy: The plunder of nature and knowledge.* Boston, MA: South End Press.

Sillitoe, P. (1998). The development of indigenous knowledge: A new applied anthropology. *Current Anthropology, 39*(2), 223–252.

Srinivasan, R., & Huang, J. (2005). Fluid ontologies for digital museums. *International Journal on Digital Libraries, 5*(3), 193–204.

Srinivasan, R., Pepe, A., & Rodriguez, M. A. (2009). A clustering-based semi-automated technique to build cultural ontologies. *Journal of the American Society for Information Science and Technology,60*(3), 608–620.

Star, S. L. (1991). Power, technologies and the phenomenology of conventions: On being allergic to onions. In J. Law (Ed.), *A sociology of monsters: Essays on power, technology, and domination* (pp. 26–56). London: Routledge.

Suchman, L. (2002). Located accountabilities in technology production. *Scandinavian Journal of Information Systems, 14*(2), 91–105.

Suchman, L. (2007). *Human-machine reconfigurations: Plans and situated actions* (2nd ed.). Cambridge, United Kingdom: Cambridge University Press.

Sudweeks, F., Hrachovec., H., & Ess, C. (Eds). *(2010). Cultural attitudes towards technology and communication, Proceedings 2010.* Retrieved from http://blogs.ubc.ca/catac/proceedings/proceedings-2010/

Tanzania Development Gateway. (n.d.). Tanzania indigenous knowledge database. Retrieved from http://www.tanzaniagateway.org/ik/

Traditional Knowledge Digital Library. (2009). Retrieved from http://www.tkdl.res.in/tkdl/langdefault/common/home.aspTsing, A. (2005). *Friction: An ethnography of global connection*. Princeton, NJ: Princeton University Press.

Tsing, A. (2005). *Friction: An ethnography of global connection*. Princeton, NJ: Princeton University Press.

Turnbull, D. (1997). Reframing science and other local knowledge traditions. *Futures, 29*(6), 551–562.

Turnbull, D. (2000). *Masons, tricksters, and cartographers: Comparative studies in the sociology of scientific and indigenous knowledge*. Amsterdam, Netherlands: Harwood Academic.

UNESCO (n.d.). Register of best practices on indigenous knowledge. Retrieved from http://www.unesco.org/most/bpikreg.htm

University of Michigan–Dearborn. (n.d.). *Native American Ethnobotany*. Retrieved from http://herb.umd.umich.edu/

Velden, M. van der. (2002). Knowledge facts, knowledge fiction: The role of ICTs in knowledge management for development. *Journal of International Development, 14*, 25–37.

Velden, M. van der. (2008). Situated Classification Work on the Web. *Webology* 5(3). Retrieved from http://www.webology.ir/2008/v5n3/a60.html

Velden, M. van der., Mörtberg, C., & Elovaara, P. (2008). *Tensions in Design*. Retrieved from http://www.informatik.uni-bremen.de/soteg/gict2009/proceedings/GICT2009_vanderVelden.pdf

Verran, H. (2002). A postcolonial moment in science studies: Alternative firing regimes of environmental scientists and Aboriginal landowners. *Social Studies of Science, 32*(5-6), 729–762.

Verran, H. (2005). *Knowledge traditions of Aboriginal Australians: Questions and answers arising in a databasing project*. Retrieved from http://www.cdu.edu.au/centres/ik/pdf/knowledgeanddatabasing.pdf

Visvanathan, S. (2000). *Environmental values, policy, and conflict in India*. Carnegie Council for Ethics in International Affairs. Carnegie Council–sponsored project: Understanding Values: A Comparative Study on Environmental Values in China, India, Japan, and the United States. Retrieved from http://www.carnegiecouncil.org/resources/articles_papers_reports/709.html/_res/id=sa_File1/709_visvanathan.pdf

Visvanathan, S. (2002). The future of science studies. *Futures, 34*, 91–101.

Visvanathan, S. (2007). Knowledge, justice and democracy. In M. Leach, I. Scoones, & B. Wynne (Eds.), *Science and Citizens* (pp. 83–94). London, United Kingdom: Zed.

Watson-Verran, H., & Turnbull, D. (1995). Science and other indigenous knowledge systems. In S. Jasanoff, G. Markle, J. Petersen, & T. Pinch (Eds.), *Handbook of science and technology studies* (pp. 115–139). London, United Kingdom: Sage.

World Bank Group. (n.d.). Database of Indigenous knowledge and practices. Retrieved from http://www.worldbank.org/afr/ik/datab.htm

3. A Structurational Interaction Approach to Investigating Culture, Identity, and Mediated Communication

BETH BONNIWELL HASLETT, PHD

Introduction

As one of the world's foremost social theorists, Anthony Giddens (1984, 1990a, 1991) has explored the evolution of cultures/societies as they have moved from oral to written stages and then through phases of modernity to late modernity. Communications have been a transformational force in each of these sociohistorical stages by altering the ways in which people interact with one another. Because communication technologies allow us to synchronize and dis-embed/re-embed actions and events across time and space, we are able to be both "here and there" and "now and then" (Giddens, 1984).

The current unprecedented, rapidly expanding knowledge base and the increased opportunities for human interaction present many challenges. One major challenge is that of identity. Identity develops through our relationship with others, especially our core group memberships, such as family and culture (Gudykunst & Kim, 2003). Through communication technologies, we now have access to multiple models of identities, lifestyles, and beliefs. Indeed, avatars, virtual worlds such as Second Life, and similar Internet applications allow us to frame our identity in multiple new ways. In addition, these media play a substantial role in shaping how we view others and thus have profound effects on our encounters with them. The concern with identity also leads directly to concerns about trust in personal interactions, as we trust others to respond appropriately in context and to demonstrate basic competency as human beings (Giddens, 1984; Goffman, 1983). Our vulnerability leads us to monitor each other in face-to-face (FtF) interactions. Establishing trust across time and space presents distinct challenges with the current advances in communication technology, as it becomes possible that our interpersonal connections might only be virtual and fabricated.

In this chapter, I introduce the concept of *structurational interaction*, based on the theoretical integration of both Goffman and Giddens, as an illuminating framework for exploring the interconnections between communication technology, culture, and identity.

Structurational interaction theory: Cultures as reflexively constructed sociohistorical contexts

Giddens' structuration theory

First, it is important to establish an overview of the social context in which identity concerns are currently being negotiated. Giddens's structuration theory emphasizes the importance of sociohistorical context for understanding social relationships, social order, and change. It maintains that all human action is performed within, and at least partly predetermined by, the context of a pre-existing social structure that is governed by a unique set of norms. It is important that the structure and rules are not permanent and externally imposed but are instead sustained and modified over time through a process of reflexive feedback.

In modernity, Giddens argues, the pace of change is so rapid and global that it is "discontinuous" with prior changes. His structuration approach proposes three sources of this dynamism: (a) *distanciation*, in which time and space are "emptied" and are no longer connected to place, as exemplified by computer-mediated communication (CMC); (b) *dis-embeddedness*, whereby CMC lifts social interaction and relationships out of local contexts; and (c) the *reflexivity* of modernity, in which humans continuously evaluate social practices in light of new information and may thereby alter their practices. These conditions, of course, heighten uncertainty and risk; ontological security becomes more fragile, with knowledge being questioned and the self being subject to increasing reflexivity.

For Giddens (2005), globalization is directly related to mediated experience and increased dependence on other people, being that "we are constantly in communication with them all" (quoted in Rantanen, p. 73). Because many of our experiences are mediated, we need to appreciate the subtleties and complexities that different modalities of mediation bring to the experiences available to human beings—leading us necessarily to the question of culture.

In light of modern cultural complexity, Giddens presents an identity-based view of societies (social systems). He views societies (cultures) as systems of shared communication (language) and life/values, with control over an identified territory and a common group identity—which seems to imply a construct of culture as nation-state. This view is consonant with the traditional view espoused by many scholars (Hofstede & Hofstede, 2005). Of course, others have raised significant criticism of this model of culture (see Macfadyen, 2011), and other discussions in this collection critique such depictions of culture on the basis of their being historically based, static, and homogenizing.

In addition to pointing out that identification with and commitment to a social group is an essential component of culture, Giddens also notes that modern telecommunication systems have been essential in forming and maintaining those commitments and that cultures have different social practices regarding CMC. In sum, Giddens's insights prompt us to address new issues in defining culture and identity. His work has been instrumental in highlighting the role of mediated communication in the reflexive evolution of culture.

Goffman and "interaction order"

In Giddens's structuration theory, social structure and agency are intertwined through interaction. Giddens relies on Goffman's concept of the *interaction order* (1974, 1983)—which describes "systems of enabling conventions that...provide a basis of social order" (Jacobs, 2007)—as a foundation for the creation, maintenance, and extension of social systems such as institutions and cultures. Goffman's work provides a very cogent analysis of FtF interaction and the presentation of self in everyday encounters. Although his work examines FtF interactions, Goffman was clearly aware of different social cues in differing contexts of social presence (e.g., telephones, radio). By synthesizing Giddens' and Goffman's perspectives, a rich analysis of identity and distanciated relationships is possible, moving from the interpersonal bases of identity to broader mediated influences on identity.

It is within this globalized, distanciated context that cultures (social systems) can be viewed as belonging to nation-states, as well as being features of social groups and virtual communities categorized by age, gender, religion, education, and so forth. The role of communication, both mediated (CMC) and face to face (FtF), is central to the development of the self and social systems. For example, Walther's (1997) theory of social information processing (SIP) argues that through CMC, users are able to selectively edit their online personae and craft desirable self-images.

Constructing identity

The theory of structurational interaction, developed by Haslett (2011), integrates the work of Giddens and Goffman. Giddens extends Goffman's work to apply to distanciated, mediated contexts, and Goffman provides an in-depth analysis of FtF encounters and analyzes the ways in which people manage impressions and develop a mutual understanding that allows them to interpret ongoing interactions. Through Giddens, we enrich our understanding of space, time, and social presence across social systems, and Goffman provides us with the analytic tools for exploring how people adapt to one another in their encounters, develop relationships, and sustain social orders. For example, Giddens's work explores how the global becomes the local—discussing the globalized nature of economy, education, and knowledge and how this globalization impacts local educational practices. Before exploring how structurational interaction can inform our understanding of identity, I briefly discuss the development of the self and the construction of multiple identities.

Initially, the self emerges through FtF interaction with significant others, like caretakers, siblings, and other members of one's primary social group (e.g., Shotter & Gergen, 1989). The self has three dimensions: (a) a basic security system (the ontological trust developed in infancy via caregiving relationships), (b) practical consciousness (an awareness of how to "go on" (act) in social situations), and (c) discursive consciousness (what we are consciously aware of and can discuss). For Giddens (1984, p. 43), interpretive schemes are embedded within each dimension of the self. These interpretive schemes entail a control of the body and a developed knowledge of how to go on in the plurality of contexts of social life. As he concludes, "Routine is integral both to the constitution of the personality of the agent as he or she moves along the paths of daily activities, and to the institutions of society, which are such only through their continued reproduction" (p. 60). For Giddens (1991, p. 53), "Self-identity is not a distinctive trait, or even a collection of traits, possessed by the individual. It is the self as reflexively understood by the person in terms of her or his biography" and "in the capacity to keep a particular narrative going" (p. 54).

Giddens and Pierson (1998) also acknowledge Goffman's insights into "the spatiotemporal features of interaction processes" (p. 93) and the importance of face work in interaction. Both Goffman and Giddens focus on the body and its location in time–space, and both regard the body as the locus of the self and view gestures, facial expressions, and bodily positioning as key in social interactions. We shall see this emphasis on the body and nonverbal gestures in the research exploring how identities are presented, maintained, and altered online.

For Giddens, "the level of time-space distinction introduced by high modernity is so extensive that, for the first time in human history, 'self' and 'society' are interrelated in a global milieu" (McPhee, 1998, p. 32). However, given the dis-embedding of relationships and the reflexivity of modern life, in large part due to communication technologies, ontological insecurity may occur, and trust is balanced against risk as part of modern life (Giddens, 1990a). The increasing range of lifestyle choices, more open and multiple contexts of action, and varied sources of authority provide agents with complex daily choices for constructing identity (Giddens, 1991).

Social positioning is also critical for identity. Social systems exist through the continuity of social practices, whose structure may be characterized as "position–practice" relations. A social position involves the "specification of a definite 'identity' within a network of social relations, that identity, however, being a 'category' to which a particular range of normative sanctions is relevant" (Giddens, 1984, p. 83). Social positions are thus social identities, and people may be positioned differently within different cultures. As such, positioning refers also to movement through positions as individuals move through the life cycle and interact with different institutions, organizations, regions, and intersocietal systems.

Social positioning reflects one's own body-life activities as well as interactions. These activities and relationships are contextualized within time–space and within a specific sociohistorical moment and culture. One's identities are formed and re-formed within and across these social positions, reflecting both local and global influences along the life-span trajectory and are influenced by both proximal FtF relationships and mediated relationships. Shared beliefs are present in groups ranging from nation–states to groups sharing ethnicity, age, gender, educational level, and so forth. Indeed, one's identity reflects movement through complex social, institutional structures in global cultural systems (Giddens, 1984). Through structurational interaction, then, we have a rich theoretical perspective from which to explore identity.

Identity in the context of mediated communication

In discussing identity in mediated global contexts, we should note several qualifiers: First, access to communication technologies is a function of economic wealth, as captured by the idea of the "digital divide." Second, political attitudes shape the use of communication technologies (i.e., the type of access our government allows us). Last, the cultural characteristics of various Web sites also influence the capacity of users to form relationships. For example, Web users perform information-seeking tasks faster when using Web sites designed by a native of their culture. Segev, Ahituv, and Barzilai-Nahon (2007) find that culture influences the local home pages of MSN and Yahoo!, suggested by the finding that MSN reveals more cultural heterogeneity and localization of content and form than does Yahoo! Barnett and Sung (2005) find that nation–state culture was significantly related to Internet network centrality and its overall structure. Würtz (2005) argues that high-context cultures use graphic elements and indirect messages extensively whereas low-context cultures are more static and direct in their messages. Such differences (e.g., in colors, graphical density, etc.) are considered aspects of cultural marking (Callahan, 2005; Cyr, Bonanni, Bowes, & Ilsever, 2005). Thus, culture (viewed as a feature of nation–states) already seems to influence the way in which information is presented in CMC and how it is subsequently used.

Others argue for a more holistic view of culture, however, in order to integrate information technology (IT) and culture (both of the nation–state and organizational). As Gallivan and Srite (2005) note:

> The consensus of scholars who study cross-cultural factors in IT use is that understanding NC [national culture] is critical. . . . Researchers have increasingly recognized the importance of articulating the cultural assumptions that are embedded into IT and explicitly evaluating whether these assumption are congruent with potential adopters in other parts of the world. (p. 296)

For initial uses of IT, national identity and age/generation will be the most significant factors, and after experience, one's occupational group might be the most important factor. Group interaction and collaboration, as well as interpersonal relationships—and their concomitant influence on professional and personal identity—would appear to be substantially influenced by these factors. Although the concern here is with personal identity through individuals' impression management and self-presentation, it is important to realize that professional (Sangwan, 2005), group, organizational, and cultural identity are also developed on the Internet.

Social presence and CMC

Social presence is an important aspect of mediated communication, particularly impression management and identity. Zhao (2004) developed a typology of differing conditions of communication in space and time. *Consociates* are those who interact with one another in real time and space—that is, they are physically copresent and thus share a "world within reach" and can influence one another in a physical "zone of operation" (pp. 91–94). *Contemporaries* are those who share world time (real time) but are spatially distant from one another—for example, interacting with individuals from another country with whom we have no direct, copresent contact. Contemporaries have typified knowledge of one another and establish trust across "objective meaning contexts"—leading us to trust that the postal worker will post our mail, for example, and that it will eventually reach its destination in the same way that we trust the bank to accurately deposit and dispense our money (Giddens, 1990a, 1990b, 1991).

However, various telecommunication devices have created new spatiotemporal conditions of human contact (referred to as face-to-device communication, or FtD, or in more general terms, CMC; Giddens, 1984; Meyerowitz, 1985; Tomlinson, 1997). Through inventions from the telegraph to instant messaging, mediated communication, especially electronic communication, a process of "tearing space away from place" has occurred (Giddens, 1990a). Zhao (2004) points out that mediated communication extends human perception as well as the zone of operation (opening up a secondary zone), although time is somewhat "elastic" because of the varying times taken to respond to messages. When two or more "worlds within mediated reach" coincide, a new connection is forged, which Zhao terms *telecopresence*. He suggests that when people are in a situation of telecopresence, they are simultaneously in two streams—one in real time and real space (*geospace*) and the other in real time but distanciated space (*cyberspace*).

In many forms of CMC, social presence is a fundamental issue, being that most CMC systems try to emulate the social presence of FtF communication, although the cues we use to gauge deception, interest, attractiveness, and involvement are missing or limited in them. Goffman's analysis of face, face work, bodily cues, and use of space have been used as a basis for evaluating and creating a more social presence in CMC. The emotional intensity found in FtF interactions are now characterized by online flaming or emoticons. Blogs and social network sites/services (SNSs) are enhanced by color, music, and videos. Online games and avatars are increasingly humanlike in an effort to simulate the social presence of FtF interactions. Much of the research on SNSs, for example, deals with how various textual cues are used to manage impressions and identity online.

Distributed identity

Turkle (1997, 1999) has discussed the blurring of identity between *real life* (RL) and *virtual* life, and talks about *distributed identity*—acknowledging the reality that identity/identities are negotiated both offline and online. For example, Leppänan, Pitkänen-Huhta, Piirainen-Marsh, Nikula, and Peuronen (2009) conducted a study of how young Finnish people used new media in a translocal context, blending connectivity across multiple locales (including different cultures), and determined culture to be "outward-looking, exogenous, focused on hybridity, translation and identification" (p. 1082). This study is compatible with the tenets of structurational interaction and with Turkle's notion of distributed identity.

Moreover, Aarsand (2008) used Goffman's concepts of *activity frames* and participation *frameworks.* Activity frames define the type of activity one is involved in, its implicit practices, its boundaries, and how the activity is structured. In Aarsand's study, students shifted between online and offline identities in classroom activities—for example by using identity tags (online nicknames) in FtF interactions. He found that there was not one dominant activity frame but rather a "borderwork" in which students shifted between online and offline frames and used identity tags strategically.

Turkle argues, as do Goffman (1967, 1974, 1983) and Giddens (1984, 1990a), that identity/identities are flexible and multiple. Giddens's concept of social positioning reflects the shifting, changing identities of individuals as they move through time and space and occupy different roles. A female, for example, may move through different identities as a child, friend, daughter, mother, and professor. The dynamics of social positioning will also vary from culture to culture. Moreover, Turkle (1999) suggests that we *coconstruct* identities online through interaction with others, using different channels to develop different aspects of identity. She proposes that some media, like avatars and multiuser domains (MUDs), may be used for self-exploration, whereas others, like blogs (and an increasing range of Web 2.0 technologies) may integrate more directly into RL (see also Fischer, 2010). Thus, CMC is increasingly permitting individuals to lead parallel lives (Turkle, 1999).

Face and identity

As identities have become more fluid and multiple, issues of identity, trust, and risk have become more complex. In addition to Giddens's concept of *social positioning* and multiple identities, Goffman's concept of *face* also helps us address the complexity of identity. Goffman (1967) defines face as the "positive social value a person effectively claims for himself by the line others assume he has taken" (p. 5) in a given interaction. As such, face is interactional and relational and reflects different facets of identity, including the self as social, expressive, and communicative (Goffman, 1967, 1974, 1983; Haslett, 2011; Branaman,1997). Hecht, Jackson, and Lindsley (2006) similarly conclude that identity reflects an individual, relational, and collective construal of the self. And culture—reflecting core values and aspects of identity—is incorporated into all these levels.

Spencer-Oatey (2007) suggests that both face and identity are related to "'self-image' (including individual, relational and collective construal of self) and both comprise multiple-self aspects" (p. 644). It is also important to note that face involves *emotion*, because violation of one's face results in feelings of shame and/or embarrassment for participants and threatens the smooth flow of interaction (Goffman, 1967). Thus studies of face, identity, and politeness that examine cultural differences add to our understanding of identity and self-presentation (see for example, Graham [2003, 2007] on conflict in CMC; Haugh [2010] on contentious e-mail exchanges; and Locher [2010] and Androutsopoulos [2006] on key issues of politeness online).

Structurational interaction captures the breadth and complexity of identity, trust, and risk. With distanciation, one extends relationships and thus trust across a wide network of individuals and organizations. Individuals, through their self-presentations and negotiation of identity, are vulnerable to others' evaluations (Goffman 1967, 1983). Structurational interaction is an integrated, coherent theory of social interaction that captures the complexity of identity/identities in modernity. In particular, structurational interaction incorporates mediated communication and thus provides a theoretical base for exploration of identity in mediated communication and its relationship to culture and identity. It is to these issues we now turn.

Impression management and identity in cyberspace

We begin by exploring the role of online communication in developing and maintaining family ties and friendships. A number of studies support the proposition that these are both important aspects of identity. Walther (1997) notes, for example, that CMC enables people to overcome barriers and limitations experienced in FtF interaction. Quan-Haase (2007) finds that university students used online and mobile communication to maintain their local and distant social ties and that students used instant messaging (IM) most for social communication. Within a *massively multiplayer online game* (MMOG), trust among players was highest within close social circles and self-disclosure was positively associated with trust of one's teammates and other gamers (Ratan, Chung, Williams & Poole, 2010). Shao (2009) finds that user-generated media (UGM), like YouTube, are used for information seeking, entertainment, and social connections, as well as for self-expression (identity) and self-actualization. Similar reasons are given for participation in virtual communities (Ridings & Gefen, 2004). In addition, social support networks like HeartNet, provide information and anonymous sources of support online (Rodan, Uridge, & Green, 2010). Thus, CMC appears to provide an important forum for the development of friendships and relationships that are instrumental in identity construction.

Scholars have also linked social-psychological characteristics to participation and risk in online communication (Livingstone & Helsper, 2007). An interesting finding is that the online presentation of self reflects a "pretend self" as well as a true self. While much of the research has shown that users solidify their offline relationships online, other research is exploring relationships initiated online that move to offline FtF interaction (Gennaro &

Dutton, 2007) or the effects of modality switching on interpersonal relationships (Ramirez & Zhang, 2007). These findings suggest that CMC creates opportunities for development of identity and social skills and that e-relationships are blurring and reconfiguring online and offline social boundaries and relationships.

Social network sites

Social network sites (SNSs) support existing human/social networks and also offer an opportunity to participate in new social networks (boyd & Ellison, 2007). In an exploration of SNSs, Tufekci (2008) uses Goffman's presentation of the self to explore SNS usage. Tufekci distinguishes between the expressive Internet (i.e., technologically mediated sociality, such as self-presentation, social monitoring, and the maintaining of social ties) and the instrumental Internet (i.e., uses of the Internet for information seeking and nonsocial transactions such as banking). Use of SNSs is influenced by two factors: attitudes toward social grooming (i.e., small talk, general social information about others) and concerns about privacy. Users of SNSs were heavier users of the expressive Internet, but there were no differences among students in terms of their usage of the instrumental Internet. Nonusers of SNSs were open to the possibility of genuine social interaction via online connections but had concerns about privacy. Similar patterns of motives for using the Internet—information seeking, relaxation/entertainment, and social utility—were found among East Asian students studying in the United States (Ye, 2005).

Pearson (2009) finds that participants on SNSs create performances that play with identity and alternate between the front stage/back stage (Goffman's conceptualization) and public/private aspects of social connections. Users also negotiate the complex tensions between positive impression management and authenticity in their identity claims online (Ellison, Steinfeld, & Lampe, 2007; Utz, 2010). Identity may also be directly expressed online on SNSs. Liu (2007) argues that users' statements of interests (about music, movies, books, etc.) operate as an expression of identity and outlines four types of expressive identity statements: those conveying prestige, differentiation, authenticity, and theatrical persona. Taste statements also connect users with groups expressing the same interests and thus express identification with that group.

Avatars

Bailenson and Blascovich (2004) suggest that "avatar interactants possess the ability to systematically filter their physical appearance and behavioral actions in the eyes of their conversational partners, amplifying or suppressing features and nonverbal signals in real-time for strategic purpose" (p. 9). Avatars also allow users to create and experience a shared virtual environment in which they share space, tools, and interactions with others (Whittaker, 2002). Bente, Ruggenberg, Kramer, and Eschenburg (2008) also note that social presence for avatars may reflect a number of different dimensions, such as nonverbal cues, copresence, relational information, emotional and physical closeness, and behavioral contingencies—all within a shared virtual environment. Social presence is viewed as important in building interpersonal trust. Cyr, Hassanein, Head, and Ivanov (2007) suggest that through the re-embedding of social cues and content, avatars can help establish trust through enhanced social presence. However, Bente et al. caution that the artificial nature of avatars may create a loss of trust and relatedness as well as activate stereotypes.

Studies about CMC contexts for avatars range from creating avatars and interacting in virtual environments (Soukup, 2004) to their use in net-based collaboration (Bente et al., 2008). The latter study investigated the effects of incorporation of a real-time window, including a special avatar interface, into Internet collaboration, and the influence of the avatar on social presence and interpersonal trust. Audio, audiovisual, and avatar conditions ranked significantly higher on perceived intimateness, copresence, and emotionally based trust than in the text chat mode. Bente et al. (2008) note that "virtual worlds and avatars could thus be a means to contextualize social interaction and foster the salience of nonverbal communication" (p. 310). To exploit their social-communicative potential, however, they argue that avatar communication needs to be embedded into relevant tasks and social settings.

A fascinating study by Soukup (2004) analyzes a comprehensive range of avatar features, choices, social presence, and the creation of an interactive environment. Using Goffman's dramaturgical perspective, he analyzes multimedia performance in a virtual community. Participants were able to creatively define and establish social settings and collaboratively built an online community.

Another aspect of avatars that reflects identity issues is that avatar characteristics—their physical features and degree of perceived "humanness" (perceived social presence)—are qualities that influence the choice of an avatar, an avatar's influence, and its credibility (Nowak & Rauh, 2005). Gong and Nass (2007) find that humanoid figures and faces are evaluated using different criteria from those used to evaluate actual human faces and voices, with inconsistent pairings (a humanoid face but an actual human voice) resulting in more comprehension time and less credibility. Nowak & Rauh (2005) also finds that users choose human avatars that match their gender and that anthropomorphic avatars were viewed as more attractive and credible (Nowak & Rauh, 2005). In a study contrasting avatar use in blogging, dating, and gaming, Vasalou and Joinson (2009) find that "avatar attributes drew on particular self-images and thus avatars were perceived by their owners as highly similar to themselves" in terms of physical appearance, lifestyle, and preferences (p. 510). Avatars also present anonymity as well as an opportunity to create unique self-representations (Cheng, Farnham, & Stone, 2002) and enhance self-awareness (Vasalou, Joinson, & Pitt, 2007). The progressive embodiment of avatar representation is a rich area to continue to explore as animation, avatars, and other representations of online identity become increasingly sophisticated and popular (see, for example, Lee, Park, & Song, 2005).

In sum, structurational interaction theory presents a strong theoretical base for exploring online interaction generally and for avatars in particular. Giddens's social positioning helps anchor identity/identities contextually in terms of time–space, sociohistorical context, and an individual's use of CMC modalities. Goffman's dramaturgical metaphor sharpens our analyses of online self-presentation by incorporating avatar characteristics and the use of space and props (i.e., in creating different online environments and persona). Soukup's study is an example of how Goffman's concepts can be applied. Last, structurational interaction provides an enriched understanding of the complexity of identity and the multiple contexts in which identity/identities are negotiated.

Cultural influences on CMC and identity

As CMC extends our social and informational world, intercultural interactions become more frequent in professional and personal contexts. Thus far, cultural differences in CMC have primarily been measured using nation-state cultural identity, despite the substantive criticisms leveled against such models. For example, Walther (1997) explores the effects of group identity versus individual identity on the social and intellectual responses of geographically dispersed partners. He reports that for long-term relationships, a "heightened salience of similarity and group belonging is preferred, even among teams whose members comprise different nationalities and backgrounds" (p. 360). Careful use of the conditions in which CMC is used can shift interactional outcomes, so attention needs to be given to the setting in which CMC is utilized, and the cultural attitudes and norms associated with technology and communication will undoubtedly play a role in such arrangements.

Hargittai (2007) finds that gender, race, ethnicity, and parental educational background influenced users' choice of SNSs. Existing offline relationships also influenced users choice of SNSs in that they connected with their existing networks, rather than seeking new acquaintances. Hispanic students were significantly more likely to use Myspace while Asian and Asian American students were more likely to use Xanga and Friendster (perhaps due to the service popularity of these sites in the Philippines, Malaysia, and Indonesia; boyd & Ellison, 2007). Users often have preferred ways of interacting through their choice of CMC media (Kline & Liu, 2005). Gender also influenced the expression of social presence through empathy and immersion in a computer-gaming simulation (Nicovich, Boller, & Cornwell, 2005). In addition, gender influenced the blogs of male and female teenagers, with males using more emoticons and active, direct language (Huffaker & Calvert, 2005). Race, culture, and gender also influenced the use of Internet media and predicted different social uses of the media (Jones, Johnson-Yale, Millermaier & Perèz, 2009; Kim, 2010; Leppänen, 2007).

Kim and Yun (2008) explore the influence of CMC, specifically in the South Korean SNS Cyworld, on relationships among Koreans and find a tension between offline and online interactions. They conclude, "Cyworld users routinely negotiate multiple dialectical tensions that are created in the online world, transferred from face-to-face contexts, or imposed by interpersonal principles that relate to Korea's collectivist culture" (p. 298). *Minihompies*—private sections on a site whose access controlled by the site's creator—function as a forum for private thoughts and self-relations that are free from distressing social contexts. Interviewees referred to them as "another me."A minihompy thus provides a place for exchanging and expressing elaborate emotional communication—lacking in FtF interpersonal interactions. What is especially noteworthy in this study is the online expression of emotions that are not usually expressed in FtF and the use of minihompies for identity construction and maintenance of close offline relationships in a fashion consistent with Korean communicative norms.

Another study by Yum and Hara (2006) explores the influence of self-disclosure on relationship development across Korea, Japan, and the US. In online relationships, self-disclosure and trust were positively related for US participants, inversely related for the Koreans, and a nonfactor for the Japanese. Across all cultures, participants reported that greater self-disclosure was associated with greater love, liking, commitment, and a willingness to adjust their communication style to match that of their partners. Overall, results indicate that people have confidence in cyberspace relationships and use cyberspace to enhance their FtF relationships. Differences in online interaction have also been found across collectivist and individualistic cultures (Seo, Miller, Schmidt, & Sowa, 2008). Yum and Hara conclude that when participants from high-context and low-context cultures interact, more attention to face (identity) concerns may be needed to prevent misunderstanding.

A number of studies answer the call for more complex and sophisticated treatments of culture. For example, Lee and Choi (2005) find that cultural orientation (horizontal and vertical individualism and collectivism) and ethnicity are correlated with Web skill and attitudes toward Web advertising.

Brock (2005) explores multiple, intracultural differences in CMC. Africana.com (now Blackvoices.com) was found to reflect the more fluid, dynamic characteristics of African American communication and match the interests, cultural values, and views of self-identity proposed by Du Bois. Religion has also been explored from cross-cultural perspectives and explored in CMC (see Ess, Kawabata, & Kurosaki, 2007). A very subtle cultural effect has been found by contrasting features of Web design: Users prefer Web sites designed by members of their own culture (Gevorgyan & Manucharova, 2009), and they identify collectivist/individualist aspects of Web design (Callahan, 2005; Kims, Coyle, & Gould, 2009). Age has also been explored as an issue in texting (Ling, 2010). Giddens's concept of social positioning provides an insightful way of exploring these intracultural intersections as well as the time–space edges of different times and social systems (e.g., the clash of traditional versus secular cultures).

In sum, culture influences users' preferences in CMC modalities, their differing social purposes, the design of Web sites, and the styles of self-presentation. Structurational interaction provides a strong theoretical base for exploring these cultural differences but also has the additional value of providing a more complex, nuanced view of culture. Cultures are viewed as social systems that are unified by core values, identities, and social practices and located in specific sociohistorical contexts. As such, cultures (social systems) may vary in size and continually evolve over time. As Giddens notes, CMC is heavily implicated in the transformation of societies over time. Goffman's concepts of face, presentation of self, impression management, and use of bodily images, space, and props provide a range of strategies to use in examining how identities are negotiated both offline and online and their potential interactions.

Third space interactions

A fourth way to explore the interrelationship between culture and CMC is to look at attempts, primarily in educational contexts, to create culture-free or "third space" interactions online—interactional spaces that transcend culture. A study by Bretag (2006) looks at e-mail exchanges between an Australian lecturer and 10 international ESL students. An analysis of their exchanges demonstrates a movement toward "claiming common ground" and sharing intimate information that will lead toward an interactional third space that will negate cultural and status differences. Richter, Pawlowski, and Lutze (2008) advocate an internationally compatible solution to e-Learning that adapts learning to the learning contexts and the available technology. Hewling (2005) argues for "intercultural communication where the focus is on interaction among participants identifying simultaneously with multiple cultural frames of reference." Such interactions appear to be transformative for users in overcoming cultural barriers to understanding and facilitating open exchanges.

Conclusion

As can be readily seen, structurational interaction offers a rich perspective on culture, communication, and technology. Structuration theory itself offers an in-depth analysis of the multiple changes communication technologies have made in social and cultural systems. However, each social system has a common set of core values, established over time that individuals identify with.

The theory of structurational interaction, a synthesis of the work of Goffman and Giddens, provides a rich perspective with which to explore identity, both off- and online and the increasingly "hybrid relationships" (in which off- and online interactions mutually influence one another and subsequently influence our identity). It offers a nuanced view of culture and of how media are involved in cultural change and development, and it develops a fine-grained understanding of identity/identities and articulates how relationships (and identities) may be dis-embedded from one context and reembedded in another. Furthermore, it suggests that the issue of social presence—clearly a central factor in online communication and experienced differently across different media—is based on embodied cues, dis-embedded from FtF contexts and reembedded in online interactions. Goffman's work on the presentation of self, the use of space, bodily movement, and gestures is instrumental in understanding various aspects of social presence.

Significant issues concerning culture and CMC need further research. For example, the issue of social presence is becoming so complex that we may reach the point at which, as one Internet user indicated, "RL [real life] is just another window" (Turkle, 1999). As individuals interact across different cultures and within different intracultural groups, issues of social presence, identity, and interaction become significantly more complex. Structurational interaction, with its nuanced view of culture and sociohistorical view of context, is well positioned to provide a strong base for exploring the interrelationships among CMC and cultures.

In addition, hybrid relationships—the blending of offline and online relationships—will be an important issue to explore for the construction of identity. Research demonstrating problematic use of the Internet leading to and reflecting negative social behaviors (Caplan & High, 2010) and its implications for identity will also be important. Last, the issue of privacy needs to be looked at—in relation to both trust and risk—in terms of what information about others is available online (Debatin, Lovejoy, Horn & Hughes, 2009; Livingstone & Helsper, 2007). Not only do we have unprecedented surveillance by the state, but hackers and others may have fairly open access to personal information (Giddens, 1990a, 1991).

With its extensive analysis of the role of communications in the development of social systems (cultures), time–space distanciation, the conceptualization of social presence and dis-embedded and re-embedded relationships (as reflected in off- and online relationships), the structurational interaction approach presents a robust perspective for analyzing interconnections between communication technology, culture, and identity.

References

Aarsand, P. (2008). Frame switches and identity performances: Alternating between online and offline. *Text and Talk, 28*(2), 147–165.

Androutsopoulos, J. (2006). Introduction: Sociolinguistics and computer-mediated communication. *Journal of Sociolinguistics, 10*(4), 419–438.

Bailenson, J., & Blascovich, J. (2004). Avatars. In W. S. Bainbridge (Ed.), *Berkshire encyclopedia of human-computer interaction* (pp. 64–68). Great Barrington, MA: Berkshire Publishing Group.

Barnett, G., & Sung, E. (2005). Culture and the structure of the international hyperlink network. *Journal of Computer-Mediated Communication, 11*(1), article 11. Retrieved from http://jcmc.indiana.edu/vol11/issue1/barnett.html

Bente, G., Ruggenberg, S., Kramer, N., & Eschenburg, F. (2008). Avatar-mediated networking: Increasing social presence and interpersonal trust in net-based collaborations. *Human Communication Research, 34*, 287–318.

boyd, d. m., & Ellison, N. B. (2007). Social network sites: Definition, history, and scholarship. *Journal of Computer-Mediated Communication, 13*(1), article 11. Retrieved from http://jcmc.indiana.edu/vol13/issue1/boyd.ellison.html

Branaman, A. (1997). Goffman's social theory. In C. Lemert & A. Branaman (Eds.), *The Goffman Reader* (pp. xiv–lxxiii). New York, NY: Rowman & Littlefield.

Bretag, T. (2006). Developing 'third space' interculturality using computer-mediated communication. *Journal of Computer-Mediated Communication, 11*(4), article 5. Retrieved from http://jcmc.indiana.edu/vol11/issue4/bretag.html

Brock, A. (2005). "A belief in humanity is a belief in colored men": Using culture to span the digital divide. *Journal of Computer-Mediated Communication, 11*(1), article 17. Retrieved from http://jcmc.indiana.edu/vol11/issue1/brock.html

Callahan, E. (2005). Cultural similarities and differences in the design of university websites. *Journal of Computer-Mediated Communication, 11*(1), 239–273. Retrieved from http://onlinelibrary.wiley.com/doi/10.1111/j.1083-6101.2006.tb00312.x/pdf

Caplan, S., & High, A. (2010). Online social interaction, psychosocial well-being, and problematic Internet use. In K. Young & N. de Abrue (Eds.), *Internet addiction: A handbook and guide to evaluation and treatment* (pp. 35–53). Hoboken, NJ: John Wiley & Sons.

Cheng, L., Farnham, S., & Stone, L. (2002). Lessons learned: Building and deploying shared virtual environments. In R. Schroeder (Ed.), *Social life of avatars: Presence and interaction in shared virtual environments* (pp. 90–111). New York, NY: Springer.

Cyr, D., Bonnani, C., Bowes, J., & Ilsever, J. (2005). Beyond trust: Website design preferences across cultures. *Journal of Global Information Management, 13*(4), 24–52.

Cyr, D., Hassanein, K., Head, M., & Ivanov, A. (2007). The role of social presence in establishing loyalty in e-service environments. *Interacting with Computers, 19*, 43–56.

Debatin, B., Lovejoy, J., Horn, A., & Hughes, B. (2009). Facebook and online privacy. Attitudes, behaviors, and unintended consequences. *Journal of Computer-Mediated Communication, 15*(1), 83–108. Retrieved from http://onlinelibrary.wiley.com/doi/10.1111/j.1083-6101.2009.01494.x/pdf

Ellison, N. B., Steinfeld, C., & Lampe, C. (2007). The benefits of Facebook "friends": Social capital and college students' use of online social network sites. *Journal of Computer-Mediated Communication, 12*(4), article 1. Retrieved from http://jcmc.indiana.edu/vol12/issue4/ellison.html

Ess, C., Kawabata, A., & Kurosaki, H. (2007). Cross-cultural perspectives on religion and computer-mediated communication. *Journal of Computer-Mediated Communication, 12*(3), article 9. Retrieved from http://jcmc.indiana.edu/vol12/issue3/ess.html

Fischer, K. (2010). Why it is interesting to investigate how people talk to computers and robots: Introduction to the special issue. *Journal of Pragmatics, 42*, 2349–2354.

Gallivan, M., & Srite, M. (2005). Information technology and culture: Identifying fragmentary and holistic perspectives of culture. *Information and Organization, 15*, 295–338.

Gevorgyan, G., & Manucharova, N. (2009). Does culturally adapted online communication work? A study of American and Chinese Internet users' attitudes and preferences toward culturally customized web design elements. *Journal of Computer-Mediated Communication, 14*(2), 393–413. Retrieved from http://onlinelibrary.wiley.com/doi/10.1111/j.1083-6101.2009.01446.x/pdf

Gennaro, C. di, & Dutton, W. H. (2007). Reconfiguring friendships: Social relationships and the Internet. *Information, Communication & Society, 10*, 591–618.

Giddens, A. (1984). *The constitution of society*. Berkeley: University of California Press.

Giddens, A. (1990a). *The consequences of modernity*. Cambridge, United Kingdom: Polity Press.

Giddens, A. (1990b). Structuration analysis and social theory. In J. Clark, C. Modgil, & J. Modgil (Eds.), *Anthony Giddens: Consensus and controversy* (pp. 297–315). Bristol, PA: The Falmer Press.

Giddens, A. (1991). *Modernity and self-identity*. Cambridge, MA: Polity Press.

Giddens, A., & Pierson, C. (1998). *Conversations with Anthony Giddens*. Cambridge, MA: Polity Press.

Goffman, E. (1967). *Interaction ritual: Essays on face-to-face behavior*. New York, NY: Pantheon Press.

Goffman, E. (1974). *Frame analysis*. New York, NY: Pantheon Press.

Goffman, E. (1983). The interaction order. *American Sociological Review, 48*, 1–17.

Gong, L., & Nass, C. (2007). When a talking-face computer agent is half-human and half-humanoid: Human identity and consistency preference. *Human Communication Research, 33*, 163–193.

Graham, S. (2003). *Cooperation, conflict and community in computer-mediated communication.* Unpublished doctoral dissertation. Georgetown University, Washington, DC.

Graham, S. (2007). Disagreeing to agree: Conflict, (im)politeness and identity in a computer-mediated community. *Journal of Pragmatics, 39*, 742–759.

Gudykunst, W., & Kim, Y. (2003). *Communicating with strangers.* New York, NY: McGraw-Hill Higher Education.

Hargittai, E. (2007). Whose space? Differences among users and non-users of social network sites. *Journal of Computer-Mediated Communication, 13*(1), article 14. Retrieved from http://jcmc.indiana.edu/vol13/issue1/hargittai.html

Haslett, B. (2011). *Communicating and organizing in context: The theory of structurational interaction.* New York, NY: Taylor & Francis.

Haugh, M. (2010). When is an email really offensive?: Argumentativity and variability in evaluations of impoliteness. *Journal of Politeness Research, 6*, 7–31.

Hecht, M., Jackson, R., & Lindsley, S. (Eds.). (2006). *Redefining culture: Perspectives across the disciplines.* Mahwah, NJ: Lawrence Erlbaum Associates.

Hewling, A. (2005). Culture in the online class: Using message analysis to look beyond nationality-based frames of reference. *Journal of Computer-Mediated Communication, 11*(1), article 16. Retrieved from http://jcmc.indiana.edu/vol11/issue1/hewling.html

Hofstede, G., & Hofstede, G. J. (2005). *Cultures and organizations: Software of the Mind.* New York, NY: McGraw-Hill.

Huffaker, D., & Calvert, S. (2005). Gender, identity, and language use in teenage blogs. *Journal of Computer-Mediated Communication, 10*(2), article 1. Retrieved from http://jcmc.indiana.edu/vol10/issue2/huffaker.html

Jacobs, M. D. (2007). Interaction order. In G. Ritzer (Ed.), *Blackwell Encyclopedia of Sociology.* doi: 10.1111/b.9781405124331.2007.x

Jones, S., Johnson-Yale, C., Millermaier, S., & Pérez, F. (2009). U.S. college students' internet use: Race, gender and digital divides. *Journal of Computer-Mediated Communication, 14*(2), 244–264. Retrieved from http://onlinelibrary.wiley.com/doi/10.1111/j.1083-6101.2009.01439.x/pdf

Kim, J. (2010). *Acculturation motives and use of the Internet among Chinese and Korean International Students in the US* (Unpublished master's thesis). The Florida State University, Tallahassee.

Kim, H., Coyle, J., & Gould, S. (2009). Collectivist and individualist influences on website design in South Korea and the US: A cross-cultural content analysis. *Journal of Computer-Mediated Communication, 14*(3), 581–601. Retrieved from http://onlinelibrary.wiley.com/doi/10.1111/j.1083-6101.2009.01454.x/pdf

Kim, K., & Yun, H. (2008). Cying for me, cying for us: Relational dialectics in a Korean social network site. *Journal of Computer-Mediated Communication, 13*(1), 298–318. Retrieved from http://jcmc.indiana.edu/vol13/issue1/kim.yun.html

Kline, S., & Liu, F. (2005). The influence of comparative media use on acculturation, acculturative stress, and family relationships of Chinese international students. *International Journal of Intercultural relations, 29*, 367–390.

Lee, K., Park, N., & Song, H. (2005). Can a robot be perceived as a developing creature? Effects of a robot's long-term cognitive developments on its social presence and people's social responses to it. *Human Communication Research, 31*(4), 538–563.

Lee, W., & Choi, S. (2005). The role of horizontal and vertical individualism and collectivism in online consumers' response toward persuasive communication on the web. *Journal of Computer-Mediated Communication, 11*(1), article 15. Retrieved from http://jcmc.indiana.edu/vol11/issue1/wnlee.html

Leppänen, S. (2007). Cybergirls in trouble: Fan fiction as a discursive space for interrogating gender and sexuality. In D. Caldas-Coulthard & R. Iedema (Eds.), *Identity trouble: Discursive constructions* (pp. 156–179). London, United Kingdom: Palgrave-Macmillan.

Leppänen, S., Pitkänen-Huhta, A., Piirainen-Marsh, A., Nikula, T., & Peuronen, S. (2009). Young people's translocal new media uses: A multiperspective analysis of language choice and heteroglossia. *Journal of Computer-Mediated Communication, 14*(4), 1080–1107. Retrieved from http://onlinelibrary.wiley.com/doi/10.1111/j.1083-6101-.2009.01482.x/pdf

Ling, R. (2010). Texting as life phase medium. *Journal of Computer-Mediated Communication, 15*(2), 277–292. Retrieved from http://onlinelibrary.wiley.com/doi/10.1111/j.1083-6101.2010.01520.x/pdf

Liu, H. (2007). Social network profiles as taste performances. *Journal of Computer-Mediated Communication, 13*(1), article 13. Retrieved from http://jcmc.indiana.edu/ vol13/ issue1/ liu.html

Livingstone, S., & Helsper, E. (2007). Taking risks when communicating on the Internet. *Information, Communication & Society, 10*, 619–644.

Locher, M. (2010). Introduction: Politeness and impoliteness in computer-mediated communication. *Journal of Politeness Research, 6*, 1–5.

Macfadyen, L. P. (2011). Perils of Parsimony: The Problematic Paradigm of 'National Culture.' *Information, Communication and Society, 14*(2), 280–293. Retrieved from http://tandfprod.literatumonline.com/doi/abs/10.1080/1369118X.2010.486839

McPhee, R. (1998). Giddens' conception of personal relationships and its relevance to communication theory. In R. Conville & L. Rogers (Eds.), *Personal relationships and communication* (pp. 83–106). Thousand Oaks, CA: Sage.

Meyerowitz, J. (1985). *No sense of place: The impact of electronic media on social behavior.* New York, NY: Oxford University Press.

Nicovich, S., Boller, G., & Cornwell, T. (2005). Experienced presence within computer-mediated communications: Initial explorations on the effects of gender with respect to empathy and immersion. *Journal of Computer-Mediated Communication, 10*(2), article 6. Retrieved from http://jcmc.indiana.edu/vol10/issue2/nicovich.html

Nowak, K., & Rauh, C. (2005). The influence of the avatar on online perceptions of anthropomorphism, androgyny, credibility, homophily, and attraction. *Journal of*

Computer-Mediated Communication, 11(1), article 8. Retrieved from http://jcmc.indiana.edu/vol11/issue1/nowak.html

Pearson, E. (2009). All the World Wide Web's a stage: The performance of identity in online social network sites. *First Monday, 14*(3). Retrieved from http://firstmonday.org/htbin/cgiwrap/bin/ojs/index.php/fm/article/view/2162/2127

Quan-Haase, A. (2007). University students' local and distant social ties. *Information, Communication & Society, 10*(5), 671–693.

Ramirez, A., & Zhang, S. (2007). When online meets offline: The effect of modality switching on relational communication. *Communication Monographs, 74*(3), 287–310.

Rantanen, T. (2005). Giddens and the 'G-word': An interview with Anthony Giddens. *Global Media and Communications, 1*, 63–77.

Ratan, R., Chung, J., Williams, C., & Poole, M. (2010). Schmoozing and smiting: Trust, social institutions, and communication patterns in a MMOG. *Journal of Computer-Mediated Communication, 16*(1), 93–114. Retrieved from http://onlinelibrary.wiley.com/doi/10.1111/j.1083-6101.2010.01534.x/pdf

Richter, T., Pawlowski, J., & Lutze, M. (2008). Adapting e-learning situations for international reuse. In Sudweeks, Hrachovec, & Ess (Eds.), *Proceedings Cultural Attitudes towards Communication and Technology 2008.* Australia: Murdoch University.

Ridings, C., & Gefen, D. (2004). Virtual community attraction: Why people hang out online. *Journal of Computer-Mediated Communication, 10*(1), article 4. Retrieved from http://jcmc.indiana.edu/vol10/issue1/ridings_gefen.html

Rodan, D., Uridge, L., & Green, L. (2010). Using nicknames, pseudonyms and avatars on HeartNET: A snapshot of an online health support community. Paper presented at the Australian and New Zealand Communication Association (ANZCA) 2010 Conference, Canberra, Australia. Retrieved from http://www.canberra.edu.au/anzca2010/ attachments/pdf/Rodan,-Uridge-and-Green_ANZCA2010.pdf

Sangwan, S. (2005). Virtual community success: A uses and gratifications perspective. *Proceedings of the 38th Hawaii International Conference on System Sciences* (pp. 1–10).

Segev, E., Ahituv, N., & Barzilai-Nahon, K. (2007). Mapping diversities and tracing trends of cultural homogeneity/heterogeneity in cyberspace. *Journal of Computer-Mediated Communication, 12*(4), article 7. Retrieved from http://jcmc.indiana.edu/vol12/issue4/segev.html

Seo, K., Miller, P., Schmidt, C., & Sowa, P. (2008). Creating synergy between collectivism and individualism in cyberspace: A comparison of online communication patterns between Hong Kong and US students. *Journal of Intercultural Communication, 18*, 4–14.

Shao, G. (2009). Understanding the appeal of user-generated media: A uses and gratification perspective. *Internet Research, 19*(1), 7–25.

Shotter, J. & Gergen, K. (1989) *Texts of identity.* London, United Kingdom: Sage.

Soukup, C. (2004). Multimedia performance in a computer-mediated community: Communication as a virtual drama. *Journal of Computer-Mediated Communication, 9*(4), article 3. Retrieved from http://jcmc.indiana.edu/vol9/issue4/soukup.html

Spencer-Oatey, H. (2007). Theories of identity and the analysis of face. *Journal of Pragmatics, 39*, 639–656.

Tomlinson, J. (1997). A phenomenology of globalization? Giddens on global modernity. In C. Bryant & D. Jary (Eds.), *Anthony Giddens: Critical assessments: Vol. 4* (pp. 116–136). New York, NY: Routledge.

Tufecki, Z. (2008). Grooming, gossip, Facebook, and Myspace: What can we learn about these sites from those who don't assimilate? *Information, Communication and Society, 11*(4), 544–564.

Turkle, S. (1997). Constructions and reconstructions of self in virtual reality. In S. Kiesler (Ed.), *Culture of the Internet* (pp. 143–156). Mahwah, NJ: Lawrence Erlbaum Associates.

Turkle, S. (1999). Cyberspace and identity. *Contemporary Sociology, 28*(6), 643–648.

Utz, S. (2010). Show me your friends and I will tell you what type of person you are: How one's profile, number of friends, and type of friends influence impression formation on social network sites. *Journal of Computer-Mediated Communication, 15*(2), 314–335. Retrieved from http://onlinelibrary.wiley.com/doi/10.1111/j.1083-6101.2010.01522.x/pdf

Vasalou, A., Joinson, A., & Pitt, J. (2007). *Constructing my online self: Avatars that increase self-focused attention.* Paper presented at the Conference on Human Factors in Computing Systems 2007, San Jose, CA. Retrieved from http://portal.acm.org/citation.cfm?doid=1240624.1240696

Vasalou, A., & Joinson, A. (2009). Me, myself and I: The role of interactional context on self-presentation through avatars. *Computers in Human Behavior, 25*, 510–520.

Walther, J. (1997). Group and interpersonal effects in international computer-mediated collaboration. *Human Communication Research, 23*(3), 342–369.

Whittaker, S. (2002). Theories and methods in mediated communication. In A. Graesser, M. Gernsbacher, & S. Goldman (Eds.), *The handbook of discourse processes* (pp. 243–286). Mahwah, NJ: Lawrence Erlbaum Associates.

Würtz, E. (2005). A cross-cultural analysis of websites from high-context cultures and low-context cultures. *Journal of Computer-Mediated Communication*, 2005, *11*(1), article 13. Retrieved from http://jcmc.indiana.edu/vol11/issue1/wuertz.html

Ye, J. (2005). Acculturative stress and use of the Internet among East Asian international students in the United States. *CyberPsychology & Behavior, 8*, 154–161.

Yum, Y. & Hara, K. (2006). Computer-mediated relationship development: A cross-cultural comparison. *Journal of Computer-Mediated Communication, 11*(1), 133–152. Retrieved from http://jcmc.indiana.edu/vol11/issue1/yum.html

Zhao, S. (2004). Consociated contemporaries as an emergent realm of the lifeworld: Extending Schutz's phenomenological analysis to cyberspace. *Human Studies, 27*(1), 91–105.

4. *Exploring Cultural Challenges in E-Learning*

Bolanle A. Olaniran, PhD

Introduction

The race to prepare learners and workers for the technological requirements of the 21st century has increased the demand for a new type of education that differs from the traditional and institutional basis and emphasizes technological literacy (Olaniran & Agnello, 2008; Olaniran, 2007). The current demand for education has transitioned, so that training is no longer confined to a particular geographic location—instead, education and training can be delivered on a global scale, and convenience is paramount. Information communication technologies (ICTs), such as computer-mediated communication (CMC) systems, have made e-learning—a process in which knowledge is acquired through ICTs—a reality. Furthermore, with the globalization of economies and knowledge, e-learning offers opportunities to take greater control of the learning needs of potential workers (Olaniran & Agnello, 2008; Stewart & Kagan, 2005).

The increased emphasis on and usage of ICTs has led to the development of online, or virtual, communication and the formation of virtual teams—most of which occur in contexts in which participants are noncollocated or geographically dispersed. As learners and workers transcend their geographic boundaries with the aid of ICTs, greater importance needs to be placed on how culture impacts communication activities within CMC encounters (Olaniran, 2007, 2009). There is a large body of research exploring the ways in which ICTs and their features impact usage and provide a benefit for task and goal accomplishments; yet, there has been less emphasis on understanding the role of culture in the implementation of ICTs in e-learning contexts.

The need to recognize and understand the influence of culture is particularly important, as globalization tends to drive universal ideals. As communication transcends national boundaries, ICTs promote the values of the communication realities in which they are entrenched (Mejias, 1999; Olaniran, 2007). Most research emphasizes the technological effects or characteristics of the media employed in constructing online environments and fails to explore the ways in which culture influences the ensuing online interactions and social impacts among communication participants (Anawati & Craig, 2006; Olaniran, 2004). Culture, however, has been identified as an important variable that creates significant challenges for cross-cultural virtual teams (Eggington, 1993). The goal of this chapter, then, is to explore cultural and social impacts and identify some challenges that culture poses for communication that uses ICTs, especially in e-learning. First, though, it is necessary to define culture and how culture is viewed in this paper.

Conceptualizing culture

Culture is germane to how individuals view themselves as part of a unique social collective and its ensuing communicative interactions (Keesing, 1974; Gudykunst & Kim, 2003); thus, culture influences learners in both face-to-face (FtF) and e-learning contexts. Hofstede (1996) contends that culture is the "collective programming of the mind" (p. 51) that differentiates members of one human group from another. In this chapter, culture is taken to represent groups of people, constituting national identity, socioeconomic development, and beliefs. This is important especially when addressing e-learning, in which ICTs are often designed in one country for use in others; thus, the potential for cross-cultural encounters increases.

Emphasis on the potential benefits of e-learning often outweighs consideration of the cultural impacts of ICTs. Perhaps one of the key reasons for this overlooking of culture is that alongside the globalization process, standardization is often the norm (Olaniran, 2001). It has been argued that with increased immigration and globalization, new or "third" identities have developed that represent shifting hybridization of preexisting national cultural patterns (Ess & Sudweeks, 2005; McSweeney, 2002). Others contend that a shift in cultural patterns is restricted to economic changes at best, whereas cultural norms regarding relational patterns have been practically unchanged (Smith, 2002). Some scholars (Olaniran, 2001, 2004; Smith) have argued that the increased immigration and mixing of people from different cultures resulting from globalization does not imply that all cultures are steering toward universal ideals. De Mooij (2003) has similarly suggested that globalization is not converging cultures, but rather that cultures are diverging (Economides, 2008).

Theoretical relevance for e-learning

To understand the importance of culture in e-learning, Hofstede's ***dimensions of cultural variability*** (1980, 1996) offer a useful platform. These dimensions initially consisted of ***power distance***, ***uncertainty avoidance***, ***individualism-collectivism***, and ***masculinity-femininity***; a fifth dimension, ***time orientation***, was added later (Hofstede, 1980, 2001; see also Dunn & Marinetti, 2007; Olaniran, 2007 for overviews of cultural value orientations and cultural dimensions). This chapter focuses on two of the five dimensions: power distance (PD) and individualism-collectivism (IDCV). PD is defined as "the extent to which the less powerful members of institutions and organizations accept that power is distributed unequally" (Hofstede & Bond, 1984, p. 418). IDCV is based on the fact that in individualistic cultures, "people are supposed to look after themselves and their family only," whereas in collectivistic cultures, "people belong to in-groups or collectivities which are supposed to look after them in exchange for loyalty" (Hofstede & Bond, p. 419). There has been extensive critique of Hofstede's dimensions as being too simplistic for meaningful illumination of complex intercultural interaction (see Ess & Sudweeks, 2005; Fougere & Moulettes, 2007; Martin & Cheong, in press; and elsewhere in this volume). Macfadyen (2008, 2011) also points to the challenge of viewing culture as national only, and Fougere and Moulettes focus on the colonial and hegemonic structure surrounding the approach.

Notwithstanding these criticisms, Hofstede's dimensions offer a general and useful framework, especially when one takes into account other variables, such as socioeconomic development and social relationships among other things. The dimensions of IDCV and PD are particularly useful when exploring communication interactions and cultural impacts in e-learning in several ways: First, the dimensions of IDCVand PD directly influence how one communicates with others using ICTs (Olaniran, 2007). Second, both dimensions focus on relationships and relational development within a culture, which are believed to be relatively stable and less subject to profound or sudden change (Smith, 2002). For instance, regardless of the economic changes that might have moved individuals from embracing collectivistic cultures toward individualistic tendencies (for example, accumulation of personal wealth), they may still maintain their sociocultural identities and retain traditional norms that guide different relationships (for example, with co-workers or family members) and communication rules (Smith).

Furthermore, the focus on participant interactions in e-learning allows us to understand how social contexts (that is, culture) influence peoples' use of technology—and perhaps the adoption/selection of technology, which is responsible for the resulting social consequences. One example would be the way that the idea of democracy (an individualistic tendency), which is valued by Scandinavians, is reflected in their information and communication technology (ICT) usage patterns. In CMC participation, Scandinavian users have been found to actively support weak team members and work together as equals (Heaton, 2001). On the other hand, Japanese culture stresses contextual awareness and nonverbal communication for guiding communication, making FtF communication critical for creating the culturally important notion of atmosphere, or *kuuki* (Heaton). As a result, the biggest challenge facing Japanese online communicators is how to nurture kuuki in online environments.

Both individualistic (e.g., Scandinavian and American) and collectivistic (e.g., Japanese) cultures have *preferred* communication styles, an important element of which is the low–high context distinction. In specific, E. T. Hall (1976) indicates that low-context communication is one in which the majority of information is in the explicit code. In contrast, a high-context communication style carries most of the information in the physical context or it is internalized within the person. Thus, a high-context communication preference puts emphasis on the need to understand the meaning embedded in a message, however implicit. By implication, ICTs tend to be more conducive to low-context communication patterns in which information must be explained clearly to avoid confusion, especially in textual form. Illustrating these differences, cultural effects on communication are described in a study examining differences between Taiwanese and US students in online discussions (Teng, 2007). This study found that US students were comfortable in, and enjoyed interacting in, the low-context medium and found it easier than their Taiwanese counterparts to initiate conversations with others and express themselves openly. On the other hand, Taiwanese students were more reserved and tended to post fewer messages. The challenge for global e-learning platforms lies in individuals' ability to discern which communication style is being used to effectively attribute meaning to messages.

Cultural perspectives and challenges in eLearning

National cultures impact—and are impacted by—e-learning in different ways. Social and cultural aspects of education demand that curricula retain specific aspects of a nation's cultural heritage to maintain some of its traditional functions, rather than reflect the universal theme of globalization. With this in mind, scholars have called for more culturally aware e-learning systems (Economides, 2008; Olaniran, 2007). Continuing conversations about e-learning require sensitivity to culture and an exploration of the dimensions of cultural variability along with their implications for e-learning.

The question is not whether to use ICTs, but rather how learners who have cultural differences from the developers of the ICT media and the learning materials utilize ICTs while preserving their own cultural values? In essence, ICTs pose challenges for learners who come from different cultures. In order for e-learning to produce concrete results, some consideration of the effectiveness of the learning process is needed. The effectiveness of e-learning, however, cannot be assessed outside of its cultural underpinnings. Attention to learners' cultures requires a look at the dimensions of cultural differences. The following paragraphs address the ways in which Hofstede's dimensions of culture help identify key cultural differences. Again, particular attention is given to the value dimensions of PD and ICDV.

The PD dimension suggests that high-equality cultures (that is, cultures rated as individualistic and of low PD) tend to be more focused on the self and on individual identities and rights. Within such cultures, an expectation tends to exist that knowledge should be shared or distributed equally. For example, low PD cultures tend to function on the premise that ICTs can empower learners. Web-based instruction and Web 2.0 technologies are promoted as a way to put learning in the control of learners rather than the instructor (Dron, 2007). Low PD democratic ideals are more cherished in the United States and European countries when compared to other cultures and create a foundation for promoting the benefits of e-learning systems. In a high PD culture, on the other hand, greater emphasis is given to relationships within groups or a given context. It is important that the expectation of learners in these cultures calls for a *telling* strategy in which "authorities" (teachers) are the primary sources of knowledge, whose job is to impart knowledge to students. Lanham and Zhou (2003) point to studies that acknowledge distinct differences in PD cultures and students' e-learning experiences and preferences. For example, learning for many Asian students implies the ability to remember, repeat, and reproduce information presented by instructors. This means learning is instructor centered, and students rely on and hold great respect for teachers; thus, ignoring information from the instructor is considered unwise and disrespectful. As in the FtF classroom, this cultural programming influences student beliefs and practices in e-learning contexts (Conlan, 1996; Lanham & Zhou, 2003).

Like low PD cultures, individualistic cultures are similarly focused on the self—promoting a sense of controlling one's destiny, as far as career and work choices go. In collectivistic cultures, however, the success of the group is more important than individual accomplishments. These distinctions are also reflected in the behaviors of online learners. For example, studies have found that both Australian and US students exhibited individualistic cultural tendencies—according to which students are able to challenge and question information from instructors (Economides, 2008; Lanham & Zhou, 2003). Another study identifies differences between US and Finnish students' participation levels in course-based online discussions. American students were more "talkative" than the Finnish, and they tended to talk about social issues, stating their ideas and opinions freely. On the other hand, Finnish students only responded when they felt they had something worthwhile to discuss (Kim & Bonk, 2002). E-learning environments designed using individualistic and low PD principles may therefore well be perceived as "countercultural," creating difficulties for students.

In spite of the opportunities for increased learner participation in e-learning, most high-context and high PD cultures retain these values and communication style preference. In a high-context culture, information is internalized in the person or situation, whereas high PD cultures recognize or accept the fact that power is not evenly distributed (see Hall, 1976; Hofstede, 1980). It seems reasonable that in these high-context/high PD cultures, powerful groups might hold the power to not only design and produce technologies but also to decide how these technologies are used; such groups therefore control the values that become dominant in communication environments (Mejias, 1999). Mejias argues that this process is not necessarily violent or malicious but rather simply exists as a structure within which people are, by way of the hegemony, discouraged to critique it.

The cultural "categories" discussed above have implications for implicit and explicit communication tendencies and the general propensity to use ICTs in e-learning and other global contexts. Cultural factors influence how people use or view communication technologies and the interpretations they draw from messages through technology media. As Devereaux and Johansen (1994) have argued, it might be difficult to get people to use certain technologies, including e-learning technologies, in high PD cultures in which status dictates every aspect of interpersonal communication.

Ess (2002) argues for a *soft deterministic* effect of technology, implying that every culture tends to find ways to adapt technologies to their cultural communication patterns. No one has disputed, however, the fact that cultural differences affect technology adoption and use. For instance, in African cultures, in which significant emphasis is placed on relationships, it was found that when e-mail was used to communicate to start off with, users were in the habit of retreating back to traditional modes of communication (FtF, telephone) as a backup to ensure that the message was received (McConnell, 1998) and interpreted correctly (Olaniran, 2001, 2007). Kawachi (2000) notes the initial widespread reluctance to embrace e-learning by the Japanese; he attributed this resistance to cultural factors—such as the nature of the Japanese language, which is considered to be conducive to the right brain learning mode (visual and memorization skills) and not that of the left brain (analytic and argumentation skills), along with their lack of proficiency in English. As a result, the Japanese used ICTs—particularly the Internet—primarily for searching and printing out information to read or translate and secondarily for entertainment and games (Kawachi, 2000). In addition to the influence of value differences, there are additional challenges affecting e-learning, including language, color symbolism, and status issues, which are addressed below.

Central to cultural challenges in e-learning in the global arena are language differences. In specific, the majority of e-learning materials are in the English language, which leads some non-English speakers to assume that e-learning is not for them (Barron, 2000). Furthermore, language barriers influence an individual's willingness to use or adopt ICTs. For instance, 40% of online users indicated that they would prefer if the technology were in a language other than English (Van Dam & Rogers, 2002). In a study looking at the global e-learning program offered by Cisco in the English language, English as a second language students indicated that they prefer their instructors to first overview the contents of the chapters in local languages before they assign them (Selinger, 2004). For e-learning curriculum designers, it may also help to note that even when a curriculum is made available in languages other than English, there are different variations in languages. For example, Cisco provided a Spanish version of its e-learning course. It is unfortunate that the version solely used South American Spanish, which created problems for European students (Selinger, 2004).

For the most part, the challenge with ICTs is that sometimes there appears to be a technological and cultural "misfit" in the diffusion of these technologies. From the global e-learning standpoint, the learning materials rarely match users' needs. Thus, the key to resolving cultural problems with technology use, especially in the e-learning environment, is to recognize and address cultural differences associated with ICTs in a way that aligns the existing cultural values, structures, and activities with the various technologies. Differences in cultures are not easily reconciled, and sometimes imagined cultural differences create not only physical but also psychological barriers that are very real (Popkewitz, 2001).

Other important influences on e-learning—less directly related to culture—include learners' motivation, instructor feedback, and e-learning curricula design. For instance, ICTs are often presumed to afford people a special status by using them, and people who own or use them are among the elite of that society (Olaniran, 2007). Powerful groups often maintain their status by controlling access to and distribution of ICTs. Again, the ensuing usage can be countercultural and can have profound ramifications for cultures and communication realities. Lanham and Zhou (2003), for example, conducted a study evaluating the effects of culture in collaborative learning environments between Australian and Asian international students and find that most international Asian students expressed a preference to work in teams with friends who have similar cultural backgrounds.

Dunn and Marinetti (2007) point to other "cultural problems" with e-learning technologies, indicating that when giving corrective feedback to learners in some Eastern Asian cultures, the use of the color red is considered inappropriate. They argue that red is a good luck color, unlike in Western culture where it suggests warnings or problems. In a similar sense, they caution against the use of the color white when designing a relaxing entry page for Japanese online modules, because the Japanese culture views white as a symbol for mourning. Furthermore, upon evaluation of the Cisco Networking Academy e-learning platform that spans 149 countries, Selinger (2004) finds that students in Denmark and Sweden feel a need to take greater responsibility for their learning than those in France. This finding is in line with the low PD dimension identified in Hofstede's study, according to which Scandinavians are ranked lower than the French. A similar finding by Edmundson (2004) is that Indian learners (high power distance) have different motivations that affect levels of support for collaborative learning, learner control, and teacher roles as compared to their Western cultural counterparts.

E-learning: Using ICTs and the global digital divide

The motivation for using ICTs starts with access to technology and the willingness to use it—and e-learning is no different. A study on attitudes toward and perceptions of e-learning found easy accessibility to it as the top motivator to participate in it for students at 47%, whereas course relevance to future career and user friendliness came next at a distant 29% (Vaughan & MacVicar, 2004). ICT advances occur rapidly in economically developed countries (EDCs), resulting in a disparity (that is, a digital divide) in access to this technology in less economically developed countries (LEDCs). Furthermore, what "counts" as technology is for the most part biased toward EDCs and digitalization such that technology initiatives from LEDCs are often devalued (Yau, 2004).

World Internet usage statistics, in spite of the tremendous growth in Internet penetration, paint a gloomy picture of the global digital divide. The regions of Africa and Asia, with 71% of the world population, have an Internet penetration rate of 11% and 23.5%, respectively, according to the March 2011 figures (Internet World Stats, 2011). The reality is that ICTs have greater presence in affluent regions. Drori and Jang (2003) explore factors contributing to the level of ICTs usage and find that political and economic factors play a lesser role than cultural features in the global digital divide. Others (Guillen & Suarez, 2005) have attributed cross-national differences in Internet use to economic, regulatory, and sociopolitical characteristics of the nations. These authors argue that the global digital divide is likely to persist because of economic, political, and social gaps that separate EDCs from LEDCs. They conclude that the global digital divide is due to inequitable power relations and dependency in the world system status, such that the rich will continue to get richer while the poor get poorer. Hence, wealthy, high-status countries and IT users stand to benefit the most from ICTs (Guillen & Suarez, 2005). The global digital divide and the resulting inequities in access to and distribution of knowledge further weaken interest in e-learning and its adoption (Ojo, 2009).

This socioeconomic limitation contributes to e-learning problems in the sense that poverty hinders learners from being able to acquire ICTs, limiting access to technology and learning materials. Governments in LEDCs for the most part are unable to provide the needed resources for e-learning. Ojo (2009) points to locations in Africa where teachers are not adequately compensated and are denied opportunities to acquire up-to-date technologies to facilitate e-learning. Furthermore, government incentives to provide resources for the necessary ICT equipment are absent, as well as infrastructure such as reliable electricity, further hindering technology use. At the same time, in some instances when access is provided, the institution's individual goal becomes the only purpose of granting access, such that the students and teachers in that school are the only beneficiaries of the access rather than the greater community (Ojo, 2009).

The implications of the digital divide extend beyond lack of access due to limited financial resources. Other problems include lack of the technological literacy (basic knowledge of commands and protocols) necessary to facilitate adoption and sustainability (Olaniran, 2007; Tiene, 2002). Also, the digital divide and access problems often require users to travel several miles to access required e-learning content, which does not bode well in encouraging would-be users to adopt ICTs.

Along with general access, culture plays a significant role in how people perceive technology. For instance, adoption of new technological innovations (such as e-learning) can be slowed by traditional local cultural beliefs. In high PD cultures, people tend to see technology systems as threatening to their traditional power structures (whether acquired or attributed) and learning modalities. The perceived threat creates anxiety about ICTs and, consequently, negative reactions to using them. For instance, Henning (2003) provides a synopsis of the effects of culture on students in an online course in South Africa. In addition to the physical distance to the cybercafe access point and lack of computer facilities at their schools and homes, the struggle with the new medium is vividly apparent in the statement of this student provided by Henning:

> When I wrote my first discussion posting I was so afraid. Would this get to others? Will they laugh, what will Prof say? . . . I feel I have not the same control as

before. I type and I read and I am scared to click because when I do that I feel I am falling down. (p. 308)

From this statement, one can see the terrifying emotions produced by the loss of control that comes from attempting to adapt to the new mode of learning. These feelings are real to students because they have certain ideals and expectations of how learning ought to occur. Henning (2003), therefore, concludes that these individuals face an inner struggle of trying to decipher who is in charge of the learning environment. Online learning also points to another problem that needs to be addressed by e-learning content providers (usually from industrialized EDCs and low PD cultures), who often tout the convenience of online learning and, more important, the freedom it offers to put learners in control of learning. These factors are certainly not the case in the scenario described above— collectivistic, high PD learners affirm the need for instructors to teach them and reflect particular notions of "collective" and "self," which revolve around the need to be part of a group.

The global digital divide points to a painful reality in which EDCs dominate the design, production, and transfer of ICTs, revealing their motive for ICTs to become predominantly profit based. The interest of EDCs in the ICTs of the LEDCs is typically based more on the profitability of the enterprises than on the development of ICTs and e-learning platforms in these countries. As Ahmed and Nwagwu (2006) put it, LEDCs face "the challenge of either becoming an integral part of the knowledge-based global culture or face the very real danger of finding themselves on the wrong side of the *digital divide*" (p. 87).

The challenges facing LEDCs as a result of the global digital divide are massive, given that they are less developed and have inadequate resources and yet are compelled to integrate ICTs that they may not be able to afford, while facing other infrastructure and social challenges. Learners are therefore even more vulnerable as a result of their inability to access and compete with similar ICTs. The irony of modern ICTs is that whereas they offer the hope of bringing the world closer, they may also contribute to some "despair" globally. According to Drori and Jang (2003), information technology "brings the world together . . . yet simultaneously, it adds another layer to the process of global differentiation and exacerbates global inequalities" (p. 144).

Conclusion

When technologies are designed and implemented, the impacts they will have on communication realities globally are generally not taken into consideration from cultural and ethical standpoints. Instead, what appears to matter is the potential of ICTs to increase productivity and efficiency. However, reliance on ICTs has had some detrimental repercussions for communication realities, as ICTs typically favor text over oral traditions. In contexts where audiovisual technologies do not rely on text, a textual way of thinking is still engaged, such that interactivity with an object—rather than humans—or a separation of the knower and the known is the norm. Therefore, the application of ICTs in itself creates hegemonic discrimination against oral tradition and challenges the way people interact and communicate, as well as the way they view the world (Olaniran, 2009; Mejias, 1999).

By implication, societies are conditioned by the media to believe that when something cannot be digitized, then it is not real, or experienced. However, ancient modes of learning began with oral traditions in which certain cultures excelled. For instance, Mejias (1999) argues that by confusing the consumption of information with *being in the world,* ICTs give users a false sense of an inclusive reality, when they are in fact alienated from it.

From a global e-learning standpoint, it stands to reason that the textual and linguistic bias in many ICTs raises important questions about what counts as learning and how learning should occur. It challenges both instructors and users globally about how to connect with learners in a *hyper-connected* world. This brings us to explore the notion of who is in control of learning: Is it the teachers, the students, or the social networks in which the learners find themselves (Olaniran, 2009)? Furthermore, scholars need to address how best to implement e-learning to bridge the global digital and cultural divides. Dissemination of education or knowledge over ICTs could result in a scenario in which learners from EDCs have a huge advantage over those from LEDCs. Resultant social problems include a continuation of the brain drain from LEDCs or situations in which populations are "educated" but remain technologically illiterate.

As we move toward the future with ICTs, there is a need to consider how to bridge the global divide between EDCs and LEDCs with novel approaches. For example, partnerships between the two as an approach to transfer of ICTs holds potential but has not yielded the projected benefits. Hence, there is a need to figure out ways of making ICTs accessible and affordable to citizens of the LEDCs, to use e-learning in culturally appropriate ways, and to adapt technologies to solve some of their pressing basic social needs (agriculture, health, government sustainability, and cultural preservation). ICTs have been accused of helping to commodify Western cultures and sell their products in LEDCs. There is very little reciprocal exchange and cultural learning when ICTs are designed and accepted as the norm across the globe, creating the impression that resistance to these technologies is futile.

All of this suggests that ICTs have more far-reaching and complex implications globally than initially thought. There are also pressing ethical implications of ICTs and e-learning, as the line between borrowing and plagiarism becomes increasingly blurred, and different cultures have a different take on this issue. It is therefore our task, individually and collectively, to figure out the best way forward and to design the best means of disseminating knowledge and conducting learning in a manner that addresses the different sociocultural needs of learners in their various and unique learning environments.

References

Ahmed, A., & Nwagwu, W. E. (2006). Challenges and opportunities of e-learning networks in Africa. *Development,* 49, 86–92.

Anawati, D., & Craig, A. (2006). Behavioral adaptation within cross-cultural virtual teams. *IEEE Transactions on Professional Communication, 49*(1), 44–56.

Barron, T. (2000). E-learning's global migration. Retrieved from http://www.astd.org/LC/2000/0900_barron.htm

Conlan, F. (1996).Can the different learning expectations of Australian and Asian students be reconciled in one teaching strategy? *Proceedings of the 5th Annual Teaching and Learning Forum,* Murdoch University, Australia. Retrieved from:http://lsn.curtin.edu.au/tlf/tlf1996/conlan.html

Devereaux, M. O., & Johansen, R. (1994). *Global work: Bridging distance, culture, & time.* San Francisco, CA: Jossey-Bass.

Dron, J. (2007). Designing the undesignable: Social software and control. *Educational Technology & Society, 10*(3), 60–71.

Drori, G. S., & Jang, Y. S. (2003). The global digital divide: A sociological assessment of trends and causes. *Social Science Computer Review, 21*(2), 144–161.

Dunn, P., & Marinetti, A. (2007). Beyond localization: Effective learning strategies for cross-cultural e-learning. In A. Edmundson (Ed.), *Globalized e-learning cultural challenges* (pp. 255–266). Hershey, PA, & London, United Kingdom: Idea Group.

Economides, A. A. (2008). Culture-aware collaborative learning. *Multicultural Education & Technology Journal, 2*(4), 243–267.

Edmundson, A. (2004). *The cross-cultural dimensions of e-learning.* (Unpublished doctoral dissertation). Walden University, Minneapolis, MN.

Egginton, B. (1993). Cultural complexity and the improved performance of a large international project. In *Proceedings of Engineering Management conference* (pp. 53–58). Retrieved from:http://ieeexplore.ieee.org/xpls/abs_all.jsp? arnumber=316504&tag=1

Ess, C. (2002). Cultures in collision: Philosophical lessons from computer-mediated communication. *Metaphilosophy, 33*(1, 2), 229–253.

Ess, C., & Sudweeks, F. (2005). Culture and computer-mediated communication: Toward new understandings. *Journal of Computer-Mediated Communication,11*(1), article 9. Retrieved from http://jcmc.indiana.edu/vol11/issue1/ess.html

Fougere, M., & Moullettes, A. (2007). The construction of the modern West and the backward rest: Studying the discourse of Hofstede's *Culture's Consequences. Journal of Multicultural Discourses, 2*(1), 1–19.

Gudykunst, W. B., & Kim, Y. Y. (1997). *Communicating with strangers: An approach to intercultural communication* (3rd ed.). New York, NY: McGraw-Hill.

Guillen, M. F., & Suarez, S. L. (2005). Explaining the global digital divide: Economic, political and sociological drivers of cross-national Internet use. *Social Forces, 84*(2), 681–708.

Hall, E. T. (1976). *Beyond culture.* New York, NY: Doubleday.

Heaton, L. (2001). Preserving communication context: Virtual workspace and interpersonal space in Japanese CSCW. In C. Ess (Ed.), *Culture, technology, communication: Towards an intercultural global village* (pp. 213–240). Albany, New York: SUNY Press.

Henning, E. (2003). 'I click therefore I am (not)': Is cognition 'distributed' or is it 'contained' in borderless e-learning programmes? *International Journal of Training and Development, 7*(4), 303–317.

Hofstede, G. (1980). *Culture's consequences.* Beverly Hills, CA: Sage.

Hofstede, G. (1996). *Cultures and organizations: Software of the mind.* New York, NY: McGraw-Hill.

Hofstede, G. (2001). *Culture's consequences: Comparing values, behaviors, institutions, and organizations across nations.* Thousand Oaks, CA: Sage.

Hofstede, G., & Bond, M. (1984). Hofstede's culture dimensions: An independent validation using Rokeach's value survey. *Journal of Cross-Cultural Psychology, 15,* 417–433.

Internet World Stats. (2011). *Internet Users in the World: Distribution by World Region 2011.* Retrieved from http://www.internetworldstats.com/stats.htm

Kawachi, P. (2000). Why the sun doesn't rise: The impact of language on the participation of Japanese students in global online education. (unpublished masters thesis). Open University, Milton Keynes, Great Britain.

Keesing, R. (1974). Theories of culture. *Annual Review of Anthropology, 3,* 73–97.

Kim, K. J., & Bonk, C. J. (2002). Cross-cultural comparisons of online collaboration. *Journal of Computer-Mediated Communication, 8*(1), article 5. Retrieved from http://jcmc.indiana.edu/vol8/issue1/kimandbonk.html

Lanham, E., & Zhou, W. (2003). Cultural issues in online learning—is blended learning a possible solution? *International Journal of Computer Processing of Oriental Languages, 16*(4), 275–292.

Macfadyen, L. P. (2008). The perils of parsimony. "National culture" as red herring? In F. Sudweeks, H. Hrachovec, & C. Ess (Eds.), *Proceedings, Cultural attitudes towards technology and communication* (pp. 569–580). Nimes, France, and Perth, Australia: Murdoch University.

Macfadyen, L. P. (2011). Perils of parsimony: The problematic paradigm of 'national culture.' *Information, Communication and Society, 14*(2), 280–293.

Martin, J., & Cheong, P. (2011). Cultural considerations of online pedagogy. In K. St. Amant & S. Kelsey (Eds.), *Computer-mediated communication across cultures: International interaction in online environments* (pp. 283–311). Hershey, PA: IGI Global.

McConnell, S. (1998). NGOs and Internet use in Uganda: Who benefits? In C. Ess & F. Sudweeks (Eds.), *Cultural attitudes towards communication and technology* (pp. 104–124), Australia: University of Sydney.

McSweeney, B. (2002). Hofstede's model of national cultural difference and their consequences: A triumph of faith—a failure of analysis. *Human Relations, 55*(1), 89–117.

Mejias, U. (1999). Sustainable communicational realities in the age of virtuality. *Critical Studies in Media Communication, 18*(2), 211–228.

Mooij, M. de. (2003). Convergence and divergence in consumer behavior: Implications for global advertising. *International Journal of Advertising, 22*(2), 1–20.

Ojo, B. A. (2009). E-Learning and the global digital divide: The challenges facing distance education in Africa. *Turkish Online Journal of Distance Education, 10*(3), 68–79.

Olaniran, B. A. (2001). The effects of computer-mediated communication on transculturalism. In V. Milhouse, M. Asante, & P. Nwosu (Eds.), *Transcultural realities* (pp. 83–105). Thousand Oaks, CA: Sage.

Olaniran, B. A. (2004). Computer-mediated communication in cross-cultural virtual groups. In G. M. Chen & W. J. Starosta (Eds.), *Dialogue among diversities* (pp. 142–166). Washington, DC: NCA.

Olaniran, B. A. (2007). Challenges to implementing e-learning in lesser developed countries. In A. Edmundson (Ed.), *Globalized e-learning cultural challenges* (pp. 18–34). New York, NY: Idea Group.

Olaniran, B. A. (2009). Discerning culture in e-learning and knowledge management in the global workplace. *Knowledge Management& E-Learning: An International Journal, 1*(3), 180–195.

Olaniran, B. A., & Agnello, M. F. (2008). Globalization, educational hegemony, and higher education. *Multicultural Education & Technology Journal, 2*(2), 68–86.

Popkewitz, T. (2001). Rethinking the political: Reconstituting national imaginaries and producing difference. *International Journal of Inclusive Education, 5*(2/3), 179–207.

Selinger, M. (2004). Cultural and pedagogical implications of a global e-learning programme. *Cambridge Journal of Education, 34*(2), 213–229.

Smith, P. B. (2002). Culture's consequences: Something old and something new. *Human Relations, 55*(1), 119–135.

Stewart, V., & Kagan, L. (2005). Conclusion a new world view: Education in a global era. *Phi Delta Kappan, 87*(3), 241–245.

Teng, L. Y. W. (2007). Collaborating and communicating online: A cross-border intercultural project between Taiwan and the US. *Journal of Intercultural Communication, 13.* Retrieved from http://www.immi.se/intercultural/nr13/teng-2.htm

Tiene, D. (2002). Addressing the global digital divide and its impact on educational opportunity. *Educational Media International, 39*(3, 4), 212–222.

Van Dam, N., & Rogers, F. (2002). E-learning cultures around the world: Make your globalized strategy transparent. *Elearning.* Retrieved from http://www.elearningmag.com

Vaughan, K., & MacVicar, A. (2004). Employees' pre-implementation attitudes and perceptions to e-learning: A banking case study analysis. *Journal of European Industrial Training, 28*(5), 400–413.

Yau, Y. (2004). Globalisation, ICTs and the new imperialism: Perspective on Africa in the global electronic village. *Review of African political economy, 99,* 11–29.

5. *Culture, Context, and Cyberspace: Rethinking Identity and Credibility in International Virtual Teams*

KIRK ST.AMANT, PHD

Introduction

The growth of global online access now allows individuals to communicate more directly and more regularly with persons living in other cultures. This unprecedented access is eroding traditional barriers of time and distance to create an increasingly international online environment. This environment, however, brings with it new challenges related to culture and communication.

Perhaps the most striking of these challenges involves differing cultural expectations of what constitutes an effective or a credible way to present information. In fact, certain aspects of online media can even conflict with more deep-seated factors that underlie the communication practices in certain cultures. For these reasons, individuals need to understand such differences if they are to interact effectively in international cyberspace. Perhaps one of the more interesting differences to consider in developing such an understanding is that of context—of the setting in which an interaction occurs.

This chapter examines the concept of context as it relates to cross-cultural communication via online media, and in so doing reviews

- how notions of context affect cultural expectations of what constitutes credible communication behavior in a particular setting
- how the notion of identity governs the context and thus the credibility of expectations that guide interactions
- how online media create identity-related challenges that can affect online interactions—particularly for international teams that collaborate via online media

The chapter concludes with a summary of strategies individuals can use to interact more credibly—and thus more effectively—in international online contexts.

The importance of "context"

Context and communication

Human communication is not random. Rather, almost every exchange is governed by rules that dictate what is considered an appropriate—or a credible—way to convey ideas in different situations. The idea is that humans interact in a variety of settings—or contexts (Berry, Poortinga, Segall, & Dasen, 2002). Each of these contexts is associated with a particular purpose or reason for interacting (Walzer, 2000; St.Amant, 2006). The purpose that governs interactions in the context of a classroom, for example, is teaching others how to consider and apply information related to a specific subject. Likewise, the purpose of the context of a courtroom is to assess and determine the guilt or innocence of an accused party.

The purpose that individuals associate with a given context often determines what they consider credible communication behaviors in it (Hall, 1981, 1989; St.Amant, 2006). Discussing class policy, for example, is a credible communication topic in the context of a classroom but not in a courtroom. On the other hand, closing legal arguments are an acceptable and a credible means of conveying ideas in a courtroom but not in a classroom.

What makes these contextual expectations of credibility interesting is the notion of *assumption*. That is, when individuals meet to interact in a given context, they do so based on assumptions that all participants

- share a common understanding of what the purpose of interacting in that context is (i.e., why they are interacting)
- know what roles individuals are expected to play when interacting in that context (i.e., how to interact appropriately/credibly within that context; Hall, 1981, 1989; Hoft, 1995; St.Amant, 2006).

For example, individuals in a classroom context share the assumption that the purpose for interacting in that setting is to learn about the subject of the specific class. To accomplish this purpose, two roles need to be established: that of the teacher/instructor who imparts information and that of the students who learn/internalize that information.

Each role brings with it certain assumptions about what constitutes credible communication behavior for individuals interacting in that context. In the context of a classroom, it is assumed that students who wish to participate credibly in a class discussion will raise their hands and wait for the instructor to call on them prior to speaking. Students who violate this expected behavior in that context will be viewed as being inappropriate or rude by other individuals in that setting (Hall, 1981, 1989; Ng & Van Dyne, 2001; Ulijn & Strother, 1995). What makes such aspects of context and credible communication behavior interesting is how they can vary from culture to culture.

Context and culture

The culture in which individuals are raised exposes them to what the members of their culture consider appropriate and credible behavior in various contexts (Berry, Poortinga, Segall, & Dasen, 2002; Connor, 2001). Cultures, however, evolve in relation to different world views and change in response to different events (Neuliep, 2000; Varner & Beamer, 2010). Thus, over time, different expectations of context (purpose) and how to communicate credibly in it (roles) can evolve. As a result, the assumptions that different cultures hold about similar context-related behaviors can create problems.

In examining these ideas, Hall (1981, 1989) discovered that cultural communication differences often involved *amounts of information*. In specific, Hall noted that cultures differ in terms of how much information individuals were expected to explicitly state about the purpose for communicating in a given context. This difference arose from the fact that cultures had varying expectations of the amount of purpose-related information individuals were expected to intuit from the context itself. Through examining these expectations, Hall developed a context-based system for comparing cultures. Hall (1981, 1989) finds that, in certain cultures, individuals are able to make relatively few assumptions concerning

- how much others knew about the purpose for interacting in a particular context.
- how they were expected to behave (roles) in relation to achieving that purpose.

In such cultures, participants could not rely on the context itself to implicitly provide much of the information needed to guide interactions. Because there were so many context-related unknowns, the members of those cultures expected much of the communication in a given situation to explicitly address the purpose for interacting in that context (e.g., "We are meeting here today in order to . . ."). Individuals from such cultures thus placed a *low* degree of reliance on the context itself to provide much of the information needed to guide interactions.

By contrast, other cultures had evolved relatively standard and widespread (within those cultures) expectations of what the purpose of a given context was and what role each individual was expected to play in that context.

Individuals from such cultures, therefore, placed a relatively *high* degree of reliance on the context to implicitly provide most or all of the information needed about the purpose for interacting in it (Hall, 1981, 1989). As a result, individuals from such cultures rarely, if ever, conveyed context-related information directly or explicitly. (To do so would be to belabor a point all individuals were assumed to already know.) Thus, communication in such settings often involved more implicit references to context-related concepts. (For example, in the context of receiving a gift, instead of saying you disliked a gift you were just given, you talk about everything but the gift—the idea being the omission indirectly indicates your dislike of the gift.) The underlying idea was that individuals would use their shared knowledge of the context to correctly interpret what message others implied in that setting.

Based on these distinctions, Hall developed a system of comparatively classifying cultures. Cultures that relied highly on the context of an interaction to provide information about the purpose of that interaction he classified as *high-context cultures.* Such cultures included those of Asia, the Middle East, and Latin America. Cultures in which individuals could not rely on the context to convey much purpose-related information Hall classified as *low-context cultures* (cultures in the United States, Germany, and Scandinavia).

It should be noted that a culture's status as high or low context is relative. Hall, for example, considers French culture to be more high context than US culture (Hall & Hall, 1990) because the French tend to rely on the setting of an interaction to provide more context-related information than do Anglo Americans. When compared to the Japanese, however, the French appear to be much more of a low-context culture. That is, in comparison to the Japanese, the French appear to explicitly note many more factors related to interacting in a given setting (Hall, 1981, 1989).

Hall believes differences in context-related expectations of what one should state explicitly versus convey implicitly causes miscommunication in many cross-cultural interactions. This miscommunication is related to cultures' expected balance between how much context-related information one should state explicitly and how much direct discussion constitutes belaboring a point. For individuals from low-context cultures, interaction with high-context counterparts often proved frustrating for a number of reasons: First, they often see high-context counterparts as avoiding issues and never getting to the point of an interaction (Ulijn & St.Amant, 2000). Second, they often perceive high-context counterparts as aloof and perhaps dishonest or incapable, in that they see them as generally avoiding providing specific information associated with the purpose for interacting in a given situation even when directly asked for that information (Ulijn & Strother, 1995; Ulijn & St.Amant, 2000).

To individuals from high-context cultures, the behavior of low-context counterparts is often seen as insulting or patronizing (Ulijn & Strother, 1995;

Ulijn & St.Amant, 2000). The low-context desire to explicitly state the purpose of an interaction—and explicitly discuss details related to that interaction—is generally viewed by high-context individuals as repeatedly stating the obvious. In some cases, these self-obvious assertions are considered annoying, lacking in sophistication, rude, and even patronizing. That is, by repeatedly reminding others of the reason for interacting, low-context individuals are seen as treating high-context counterparts like children who constantly need to be reminded of what they "should know" or "should do."

Context and identity

The question then becomes how individuals know what context they are in at any given point in time. One answer is identity—in specific, the identity of the individuals interacting in a particular space. In most cases, the identity of the person of authority in a setting distinguishes the context and lets participants know what behavior is expected (Hall, 1981, 1989; Ng & Van Dyne, 2001; St.Amant, 2006). For example, a classroom can be defined as a "room" (a physical place), but it does not become a "classroom" (a context where learning takes place) simply because individuals are congregated in that space. Rather, the identity of two particular sets of individuals—the students and the teacher—must be established for individuals in that setting to recognize the context as classroom and adjust their communication behavior to the expectations of that setting. Moreover, it is not until the individuals identify the teacher that the setting becomes that of a classroom. Until that individual is identified, the setting is but a collection of individuals who are free to do as they like (e.g., engage in personal conversations, make cell phone calls, etc.). Thus, to know what context one is in—and to know how to communicate in a context-appropriate way—individuals need to know the identity of the participants, particularly the figure of authority (Blumer, 1969; Hall, 1981, 1989; Ng & VanDyne, 2001).

To help with this process of context-related identification, many societies have developed cues or markers that help define individuals—particularly authority figures—quickly and in unspoken ways. In some cultures, these cues involve dress (Neuliep, 2000; Varner & Beamer, 2010). The identity of a judge in the context of a US courtroom, for example, is indicated by that person being the only one wearing flowing black robes. In other cases, these cues involve placement or positioning. In the context of a classroom, for example, one establishes his or her identity as the "teacher" by being the person who stands at the front of the room (Neuliep, 2000; Varner & Beamer, 2010). In a similar way, in the context of a business negotiation, where an individual sits at the negotiating table often provides cues as to who is the lead negotiator/authority figure in that context. The presence of that authority figure, in turn, indicates when the negotiation can begin (Bosrock, 1994; Ulijn & Strother, 1995).

In other cases, such cues can be more subtle. For example, how an individual stands or holds him or herself in a given setting can indicate who is "in charge"/"in authority" in that meeting (Neuliep, 2000; Ulijn & St.Amant, 2000). Key markers of status and authority that help establish context, often nonverbal in nature, can be unknown/unseen in some online environments, creating communication challenges related to identity and thus to context.

Cyberspace and identity

In *some* online interactions, nonverbal cues associated with identity are missing (Derks, Bos, & von Grumbkow, 2007; Morio & Buchholz, 2009). That is, the primary mechanism people have for interacting with one another online is often a keyboard or a keypad (and often involves typed text). As nonverbal cues are an important part of effective face-to-face interactions in spoken exchanges, what then happens when human communication suddenly becomes restricted to just typed text? How will interactions between individuals change?

Many researchers in the field of computers and communication note how concepts of identity can change as a result of interaction through computers. For example, the lack of nonverbal features in text-based online exchanges (e.g., e-mail, Facebook posts) involves many physical cues associated with an individual's identity in terms of race, sex, or physical appearance often seeming to be nonexistent (Morio & Buchholz, 2009; Mesch & Beker, 2010). In a similar way, features such as accent and intonation—cues that often indicate someone is from another culture—are also missing (St.Amant, 2004). In other cases, individuals can create avatars—or online characters—to visually represent themselves in whatever way they wish and hide or misrepresent their actual identity (Ruane, 2007; Yee & Bailenson, 2007). These aspects of online media allow individuals to mask their true identity or even create new identities when online.

Other researchers have noted that problems of online identity can be further complicated by the ease with which online communiqués can be edited or shared with others. In online settings, the author of a given communiqué can quickly become blurred by the "re-posting" of his or her messages to different online groups. That is, when someone receives an online message, he or she has the ability to forward or post that message to an almost unlimited number of individuals or lists. Any of these recipients can then re-post that message to a similarly diverse range of persons (see St.Amant, 2005, for an overview of such problems related to attribution and online messages).

What makes this online forwarding/re-posting interesting in terms of identity is the ease with which one can

- usurp the identity of another by copying someone else's words/ text-based identity and present that text as one's own, in other words steal the online identity of another
- modify the online/text-based identity of another by typing new text into a message from someone else and then forward that edited message to others without noting the changes that have been made, in other words altering or misrepresenting the online identity of another.

It is interesting to note that such practices of altering or forwarding information in this way seem to have been commonplace since the earliest days of cyberspace (Baym, 1997). Yet the notion of a stable, known identity is often essential in correctly identifying the context of an interaction and how to behave credibly in that context. And the more a group depends on context to provide cues for communication, the more problematic the plasticity of online identity becomes. Such factors can be particularly challenging as more organizations begin using online media to coordinate internationally distributed projects.

Context and globalized virtual workplaces

Traditional workplaces were organized around a central physical location. In such cases, one's co-workers were usually located in the same office, as physical proximity was essential to the effective exchanging of information needed to perform tasks and achieve objectives. Today, online media make interaction with overseas colleagues almost as quick and easy as interacting with counterparts in the same building (Brewer, 2007). Moreover, by freeing workplace interaction from physical proximity, online media allow organizations to draw from a much wider pool of expertise to perform different functions. Now, organizations can use virtual workplaces—online environments where co-workers meet to exchange ideas, information, and materials on a project—to complete a range of activities (Brown, Huettner, & James-Tanny, 2007).

The use of such online workplaces allows organizations to tap the abilities of the best skilled employees around the world to collaborate on a task rather than relying on a limited pool of local talent. At the same time, many of these skilled overseas workers can perform specialized tasks for a fraction of what it might cost to perform the same task domestically (St.Amant, 2009; St.Amant & Cunningham, 2009). These benefits have led to a variety of virtual workplace models in which individuals located in different nations use online media to collaborate on a range of projects (St.Amant, 2009).

The success of such globalized virtual workplaces, however, is based upon one key factor—effective access to information. Without the easy and quick exchange of needed data, the benefits of relying on such workplaces cannot be realized, and they can become more cost, time, and labor intensive (Ang & Inkpen, 2008; St.Amant & Cunningham, 2009). Yet the plastic nature of online identity—and the problems it can cause in terms of context—can have an adverse effect on online information sharing in global teams.

Identity and comfort

Individuals are often hesitant to communicate until they know the context-related expectation associated with credible communication behavior (Ng & Van Dyne, 2001). Such hesitation seems particularly acute in high-context cultures where knowing how to communicate effectively is connected to knowing what that specific context/setting is. Without the cues commonly used to identify persons in—and thus the context of—a given situation, individuals from such cultures tend to remain silent. Such silence, in turn, can limit the sharing of essential information out of concern that doing so might violate expectations of credible communication in that context (Ng & Van Dyne, 2001). As a result, the easy access to sources of information allowed by online media also creates a resistance to sharing that information through that media.

The hesitation with which certain cultural groups might share information online seems to be the case in relation to Web 2.0 technologies. One study, for example, found that certain cultural groups displayed different levels of comfort related to particular types of information they shared via social networking sites (SNSs). US subjects, for example, tended to share more personal information and opinions via such sites, while French subjects provided only general information (e.g., general interests or hobbies). And South Korean subjects used SNSs primarily for sharing photos with friends and family (Chapman & Lahav, 2008). Such behavior seems linked to concerns over who might have access to and be participating in online exchanges, such as not knowing the identity of the participants and the nature of the online exchange (Chapman & Lahav, 2008; Marcus & Krishnamurthi, 2009).

Identity and accountability

Hesitancy to share information online might also be compounded by the degree of accountability created by Web-based media. That is, online spaces are often relatively open for others to observe. Whereas some systems limit user access (e.g., through Facebook's "Friends" option), supposedly restricted information can still be shared indirectly with a much larger audience (e.g., one's Facebook friend posting another person's "Friends only" information to a very public wall). Even in the case of closed or proprietary networks, the owners of such systems (e.g., the company sponsoring them or using them) still have the ability to monitor online exchanges on these systems. Thus, individuals interacting online can find themselves highly accountable for any communication "mistakes" they make in cyber-exchanges.

This degree of accountability, however, can create problems in terms of context and international exchanges. In certain cultures, particularly high-context cultures, the social penalties for violating the communication expectations of a given context can be quite severe (Ng & Van Dyne, 2001). These penalties can range from reprimands to being excluded from future group activities. As a result, individuals from such cultures might prefer to remain silent in online exchanges where the lack of a fixed identity means the specific context of the interaction is unknown.

Cultural concerns related to accessibility, accountability, and online media can be seen in certain behaviors and designs affecting SNSs in different nations. Recent research shows that many Chinese SNS users will reveal very different kinds of and degrees of information based on accountability. If, for example, a Chinese SNS user already knows—and is known by—the person with whom he or she is interacting, that user tends to avoid discussing personal feelings or opinions. However, if the identities of the parties are unknown, then users tend to be more open in sharing personal information, ideas, and perspectives (Chapman & Lahav, 2008). Such behavior is in stark contrast to many US-based SNS users who seem comfortable posting a variety of personal information for almost anyone to view (Grasz, 2009).

The reason for this distinction seems connected to accountability. That is, any information shared with a known individual could later be used against the person who provided the information. If, for example, you know who I am and I say something negative about my employer, you could share that information with my employer, and I could lose my job. Such sentiments can be particularly acute in cultures where Internet users are accustomed to having their online communications monitored and even censored by the government (Yang, 2009).

Identity and networks

One final problematic aspect of identity and context involves the nature of the relationships among individuals. In certain cultural systems, communication involves interacting though complex social networks. These networks are often based on the trust formed by the long-term relations that individuals establish over time or by familial ties (Hofstede, 1997; Condon, 1997). Within these cultures, such systems act as essential mechanisms for collecting and disseminating information. And within such systems, context becomes highly important. The idea is that effective information sharing in such systems involves distinguishing between the context of *interacting with someone who is known and trusted* and *interacting with someone who is new and unknown* (Hofstede, 1997). In cultures where social networks are highly important, establishing the context of communicating with someone who is known and trusted is essential to gaining access to needed information.

When individuals in such cultures use SNSs, they often base their decisions to ignore information or respond to requests for information on the identity of the speaker and establishing the context of interacting with someone who is known and trusted. If the information comes from someone whose identity is clearly known, then the desired context can be created and information can be effectively exchanged. If, however, the information comes from an unknown source or a source whose identity is dubious (e.g., if it is unclear whether an online message is from a known source or an imposter of that source), an alternative context is established, and the exchange of information is highly limited or nonexistent.

Such cultural preferences for social networks seem to affect how individuals interact when online. Many US users of SNSs, such as Facebook, seem comfortable using these technologies to quickly generate large-scale online social networks by using their e-mail address book to send mass invitations to "join one's network" (Fogg & Iizawa, 2008). In such cases, other individuals can receive and accept such invitations from persons they might know either superficially or only from online interactions. On the other hand, in cultures where social networks consisting of known individuals are prominent, SNS use is quite different. In China, for example, SNS users tend to rely on oral recommendations from known parties when it comes to deciding which new contacts to add to one's online networks (Guo & Yu, 2009). In Japan, preferred SNS providers such as Mixi do not allow users to employ bulk e-mail to add individuals to one's online network. Rather, each prospective network member must be invited individually—thus reducing the chances that an unknown party might join such networks (Fogg & Iizawa, 2008).

Strategies for coping with issues of identity and context online

Because notions of context and identity can be important for successful cross-cultural communication, individuals engaging in international online interactions need to make sure the correct context for interactions—particularly work-related interactions—can be established. Individuals might use some of the following certain basic strategies to establish both identity and context in international online exchanges.

Strategy 1

Individuals can conduct an introduction session before international teams begin work on joint projects. Such initial meetings and introductions should be coordinated by an individual who is known as a "credible source" to all involved (e.g., someone who is well known by all team members). By having this credible source introduce new individuals, an initial sense of credibility for persons who were previously unknown can be created. The first thing to do at this meeting is to identify the team manager (i.e., the authority figure in the context of the project) and introduce him or her to the group. Next, the newly identified manager should note the objectives of the project and introduce each individual and note what role that person will play in relation to the overall team. In so doing, the manager establishes a common understanding of context for the project (e.g., goals, deliverables, and timelines) and establishes the identity and the roles of each team member.

Strategy 2

Individuals can use voice- and video-based synchronous media regularly. The time spent apart from colleagues combined with the "faceless" nature of online interactions can result in team members forgetting they are interacting with international individuals who might have different communication expectations. Media that allow team members to hear and to see each other when interacting can provide the social cues needed to remind individuals of such differences. For this reason, initial team introductions should be done via online video and audio conferencing programs such as Skype. In a similar way, project managers should conduct regular online synchronous video or audio meetings involving all team members. Such meetings would help remind individuals of the diversity of the team. These meetings would also confirm the contexts of the project (e.g., by having the manager note the progress of the project in relation to its stated goals) and the roles of the participants (e.g., by having the manager ask each team member to report in on his or her progress in relation to the project).

Strategy 3

Individuals can require that all project-related interactions take place via a secure project Web site that employs Web 2.0 media. The use of such a secure site helps confirm that the overall network of participants collaborating on the project is limited to known individuals. Such sites thus maintain the credibility of individuals interacting via that network/on that project. It would be ideal for such a site to allow team members to engage in all project-related communication through the site. These interactions might include sending e-mail to one or more people, providing an online bulletin board for engaging in asynchronous communication, hosting blogging technologies that allow team members to keep individual project logs, and offering chat functions that facilitate quick, impromptu discussions when needed. Such diverse functions are offered by content management systems that only allow group members access to a common project site. Open-source examples of such systems include Drupal, Moodle, and Plone. Having a project manager establish and control access to such systems can enhance the trust and the credibility of individuals associated with them.

Strategy 4

Individuals can create communication protocols that note how team members should address certain items or issues. Such protocols should note how to present questions and provide related answers, how to request information and clarification, how to provide constructive comments on each others' work, and how to address disputes should they arise. By providing these protocols, a manager creates the expectations of how much or how little information needs to be provided within the context of the project. In so doing, these protocols address prospective communication problems that could result from individuals applying their own cultural communication expectations to project interactions. To avoid the concern that team members might see such protocols as restrictive, managers should explain these requirements, facilitate interactions, and provide a mechanism for locating key pieces of project-related information later if needed.

Strategy 5

Individuals should establish requirements for how to attribute sources of information. To avoid the concern that information that might be copied from other individuals, managers should create requirements for using information from other sources. Such requirements should include noting where the information from another source begins and ends and the original source of that information (e.g., team members might write, "<<BEGIN DAN'S IDEA>>" [Present idea.] "<<END DAN'S IDEA>>"). Such requirements can address concerns over the sources of information provided by team members. They can also mitigate concerns about individuals forwarding parts of another team member's message without providing him or her necessary attribution.

Although establishing online identity and online context can be difficult, doing so is essential to effective international exchanges of information. The strategies presented here seem rather basic, but they can greatly affect the context individuals associate with an online interaction. These strategies, moreover, should not be seen as the only approach related to effectively addressing such issues. Rather, readers should view them as a foundation that can be adapted or built upon to communicate more effectively in such contexts.

Conclusion

The rapid evolution of new media and communication technologies is constantly altering how people think about space. Now, the relatively widespread use of online media has changed the concept of *context* from referring to a physical location to referring to a state of mind. As virtual environments expand to include peers and co-workers from different cultures, the future of online communication will almost certainly be shaped by intercultural issues. This chapter has examined how one particular concept—that of context—can affect interactions in international online exchanges. Although this chapter is by no means a comprehensive review of the topic, it does provide foundational information individuals can use to explore and participate in cross-cultural interactions. It is now up to the reader to take the next step and expand upon these ideas by exploring them further. In so doing, the reader can add to our understanding of how cyberspace affects cross-cultural exchanges.

References

Ang, S., & Inkpen, A. C. (2008). Cultural intelligence and offshore outsourcing success: A framework of firm-level intercultural capability. *Decision Science, 39,* 337–358.

Baym, N. K. (1997). Interpreting soap operas and creating community: Inside an electronic fan culture. In S. Kiesler (Ed.), *Culture of the Internet* (pp. 103–120). Mahwah, NJ: Lawrence Erlbaum Associates.

Berry, J. W., Poortinga, Y. H., Segall, M. H., & Dasen, P. R. (2002). *Cross-cultural psychology: Research and applications* (2nd ed.). New York, NY: Cambridge University Press.

Blumer, H. (1969). *Symbolic interactionism: Perspective and method.* Englewood Cliffs, NJ: Prentice–Hall.

Bosrock, M. M. (1994). *Put your best foot forward: Asia.* St. Paul, MN: International Education Systems.

Brewer, P. E. (2007). Researching online intercultural dialogues in business: Using established methods to create new tools. In K. St.Amant (Ed.), *Linguistic and cultural online communication issues in the global age* (pp. 112–122). Hershey, PA: Idea Group.

Brown, M. K., Huettner, B., & James-Tanny, C. (2007). *Managing virtual teams: Getting the most from wikis, blogs, and other collaborative tools.* Sudbury, MD: Jones and Bartlett Learning.

Chapman, C. N., & Lahav, M. (2008). International ethnographic observations of social networking sites. *Proceedings of CHI 2008* (pp. 3123–3128). New York, NY: Association for Computing Machinery.

Condon, J. C. (1997). *Good neighbors: Communicating with the Mexicans.* Yarmouth, ME: Intercultural Press.

Connor, U. (2001). Contrastive rhetoric redefined. In C. G. Panetta (Ed.), *Contrastive rhetoric revisited and redefined* (pp. 75–78). Mahwah, NJ: Lawrence Erlbaum Associates.

Derks, D., Bos, A. E. R., & Grumbkow, J. von (2007). Emoticons and social interaction on the Internet: The importance of social context. *Computers in Human Behavior, 23,* 842–849.

Fogg, B. J. & Iizawa, D. (2008). Online persuasion in Facebook and Mixi: A cross-cultural comparison. In H. Oinas-Kukkonen, P. Hasle, M. Harjumaa, K. Segerståhl, & P. Øhrstrøm (Eds.), *Lecture notes in Computer Science 5033* (pp. 35–46). Berlin, Germany: Springer-Verlag.

Goldsmith, J., & Wu, T. (2006). *Who controls the Internet? Illusions of a borderless world.* New York, NY: Oxford University Press.

Grasz, J. (2009, August 24). 45% employers use Facebook-Twitter to screen job candidates. *Oregon Business Report.* Retrieved from http://oregonbusinessreport.com/2009/08/45-employers-use-facebook-twitter-to-screen-job-candidates/

Guo, C., & Yu, J. (2009). Socializing online in various cultural contexts: A cross nation study of social network service development. *AMCIS 2009 Proceedings,* paper 514. Retrieved from http://aisel.aisnet.org/amcis2009/514

Hall, E. T. (1981). *Beyond culture.* New York, NY: Anchor Books.

Hall, E. T. (1989). *The dance of life.* New York, NY: Anchor Books.

Hall, E. T., & Hall, M. R. (1990). *Understanding cultural differences.* Yarmouth, ME: Intercultural Press.

Hofstede, G. (1997). *Cultures and organizations: Software of the mind.* New York, NY: McGraw-Hill.

Hoft, N. L. (1995). *International technical communication: How to export information about high technology.* New York, NY: Wiley.

Marcus, A., & Krishnamurthi, N. (2009). Cross-cultural analysis of social network services in Japan, Korea, and the USA. In *DID '09 Proceedings of the 3rd International Conference on Internationalization, design and global development* (pp. 59–68).

Mesch, G., & Beker, G. (2010). Are norms of disclosure of online and offline personal information associated with disclosure of personal information online? *Human Communication Research, 36,* 570–592.

Morio, H., & Buchholz, C. (2009). How anonymous are you online? Examining online social behaviors from a cross-cultural perspective. *AI & Society, 23,* 297–307.

Neuliep, J. W. (2000). *Intercultural communication: A contextual approach.* New York, NY: Houghton Mifflin.

Ng, K. Y., & Van Dyne, L. (2001). Culture and minority influence: Effects on persuasion and originality. In C. K. W. De Dreu & N. K. De Vries (Eds.), *Group consensus and minority influence: Implications for innovation* (pp. 284–306). Malden, MA: Blackwell.

Ruane, D. (2007). Weavers & warriors: Gender and online identity in 1997 and 2007. *Transforming Cultures eJournal, 2,* 25–51.

St.Amant, K. (2004). International online workplaces: A perspective for management education. In C. Wankel & R. DeFillippi (Eds.), *The cutting-edge in international management education* (pp. 143–165). Briarcliff Manor, NY: Information Age.

St.Amant, K. (2005). Online ethos and intercultural technical communication: How to create credible messages for international Internet audiences. In M. Day & C. Lipson (Eds.), *Technical communication and the World Wide Web* (pp. 133–165). Mahwah, NJ: Lawrence Erlbaum Associates.

St.Amant, K. (2006). Globalizing rhetoric: Using rhetorical concepts to identify and analyze cultural expectations related to genres. *Hermes: Journal of Language and Communication Studies, 37,* 47–66.

St.Amant, K. (2009). Understanding IT outsourcing: A perspective for managers and decision makers. In K. St.Amant (Ed.), *IT outsourcing: Concepts, methodologies, tools, and applications: Vol. I* (pp. xxxii–lix). Hershey, PA: IGI Global.

St.Amant, K. & Cunningham, R. (2009). Examining open source software in offshoring contexts: A perspective on adding value in an age of globalization. *Technical Communication, 56,* 361–369.

Ulijn, J. M., & Strother, J. B. (1995). *Communicating in business and technology: From psycholinguistic theory to international practice.* Frankfurt, Germany: Peter Lang.

Ulijn, J. M., & St.Amant, K. (2000). Mutual intercultural perception: How does it affect technical communication? Some data from China, the Netherlands, Germany, France, and Italy. *Technical Communication, 47*(2), 220–237.

Varner, I. & Beamer, L. (2010). *Intercultural communication in the global workplace* (5th ed.). Boston, MA: Irwin.

Walzer, A. E. (2000). Aristotle on speaking 'outside the subject': The special topics and rhetorical forums. In A. G. Gross & A. E. Walzer (Eds.), *Reading Aristotle's rhetoric* (pp. 38–54). Carbondale: Southern Illinois University Press.

Yang, G. (2009). *The power of the Internet in China.* New York, NY: Columbia University Press.

Yee, N., & Bailenson, J. (2007). The Proteus effect: The effect of transformed self-representation on behavior. *Human Communication Research, 33,* 271–290.

Zittrain, J., & Palfrey, J. (2008). Reluctant gatekeepers: Corporate ethics on a filtered Internet. In R. Deibert, J. Palfrey, R. Rohozinski, & J. Zittrain (Eds.), *Access denied: The practice and policy of global Internet filtering* (pp. 103–122). Cambridge, MA, & London, United Kingdom: The MIT Press.

Section Two: Constructing Identities

Identity has a profound influence on intercultural communication, but its implications for the formation and expression of the self within and through digital media are less well understood. Virtual environments are commonly described as safe havens for the anonymous expression and manipulation of selves, yet alternative perspectives question the extent to which individuals are able to reconfigure their identities when they are on- and off-line. The five chapters in this section explore the opportunities and tensions in identity constructions as they exist in multiple digital media today.

As Natalia Rybas describes in the first chapter of this section, participants in social media networks perform aspects of their everyday life at the intersections of on- and offline life *because, through,* and *with* Facebook. Using cyber-ethnography, she examines how the individual self is performed on Facebook as it encounters multiple social worlds and communities in which it is invested. Based on her findings, she discusses two contrasting themes that reflect the processes of living at the intersection of on- and off-line contexts and reveal the logic of constructing and managing the self at the digital interface. The first theme, which she calls "producing authentic selves," focuses, on the one hand, on the ways in which Facebook respondents expect and strive for authenticity in the process of profile production. The second theme, "creating/erasing differences at the interface" highlights, on the other hand, the ways in which Facebook users deliberately create, heighten, or erase differences based on various markers of identity like gender and class to fit the expectations of imagined audiences. Performance of virtual identities is fluid, argues Rybas, and specific articulations of the self depend on the design affordances of new media and the settings, as well as the relationship with other participants.

In the second chapter of the section, Ping Yang further explores the complexity of online identity performance. Relying on phenomenology, with a focus on the lifeworld and the lived experiences of American college students from a variety of backgrounds, she identifies several dialectics that undergird the students' identity presentations. The presence of time lags and varying chronemics underlying e-mail responses, for example, demonstrate that the synchronous and asynchronous natures of different online communication contexts and media may be more or less appropriate for some interactants. In the next chapter, Guo-Ming Chen and Xiaodong Dai broaden the discussion of identity by considering the increasingly globalized nature of online communication. These authors examine the potential of new digital media for facilitating cultural appropriation and hybridization through global interconnectivity and the implications for cultural identity development, including the ways in which people develop individualistic and collective identities and a sense of belonging. They discuss ways in which asymmetries in mediated intercultural communication can exacerbate differences in identity negotiation between the West and non-West and propose several approaches for managing these tensions.

The next chapters in this section closely examine the notion of identity in relation to health issues and developments in Australia and China. Authors Debbie Rodan, Lynsey Uridge, and Lelia Green draw upon findings of a five-year project and its associated community-based Web site, HeartNET, which supports people living with the effects of heart disease. They use this case study to illustrate how conditions that challenge an individual's notion of self—such as health conditions—can trigger the emergence of new cultural groups and how online communities can serve as conducive crucibles for the expression of feelings of frustration and despair, which are often censored off-line in interactions with family and health professionals. It is interesting that in their examination of identity negotiation in HeartNET, they uncover how some members actively identify with the online identity of fellow participants, whereas others report feeling pressure to conform to the dominant and overwhelmingly positive conception of the heart patient identity. Their study therefore stands as a fascinating microcosm of cultural evolution and the ways in which online platforms can support the production of new identities, as well as exemplifying the push–pull nature of individual engagements with cultural community.

On the other side of the globe, Wei Sun and Jared Critchfield offer an analysis of communication among cancer patients on a prominent Chinese blog and echo findings from the HeartNET study in the last two chapters of the section. China's complex health care system, as well as practices by medical professionals who repress cancer diagnoses and cures as being "too sensitive for public discussion," render Chinese cancer patients culturally different from and invisible within mainstream culture. Sun and Critchfield report on ways in which blogging appears to empower Chinese cancer patients, by offering a channel for expressing health fears and emotional distress and for discussing the attendant strains on their off-line family and social relationships. Moreover, blogging offers cancer patients and their family members access to alternative health information and treatment suggestions. This fascinating study therefore speaks to Bonniwell Haslett's contention, in Section One, that new media are permitting development of new hybrid forms of on- and off-line relationships and identity.

6. *Producing the Self at the Digital Interface*

NATALIA RYBAS, PHD

Introduction

Online social networks are a fairly recent addition to the spectrum of hot topics investigated in the field of communication studies and other disciplines. As the popularity of Facebook and other social Web applications has been increasing, so has the scholarly attention to the phenomenon (Beer, 2008; boyd, 2007). This chapter contributes to the hyper-interest in the *new* technologies of the Internet and also strives to critique both the technologies as well as approaches to understanding them.

Recent studies exploring online social networks make clear the necessity of paying special attention to culture when examining online socializing. Beer (2008) theorizes about the nature and specifics of online social networks, assuming that these portals have become a part of mainstream culture. Without a doubt, the popularity of Facebook, as well as other Web 2.0 applications, marks a significant cultural presence. A careful investigation into the mode or form of such presence is necessary in order to assess how various constituents of online social networks build and make sense of their life in the context of newer technologies. Much research has focused on younger generations and on the use of online social networks by teenagers and college students. Various practices of use (see, for example, boyd, 2007; Hargittai & Hsieh, 2010) have been the focus of such research initiatives. Among these studies, danah boyd (2007) argues that teenagers join social-networking sites (SNSs) to sustain a cultural requirement of the contemporary generation of youth. A team of researchers lead by Mizuko Ito (2010) concludes that Web-based sharing and creative sites and tools define youth culture as a user-driven collaborative learning environment in which users shape and support their general social skills and specific interests. These researchers argue that by engaging each other when using online social networks, users experience and learn friendship and companionship, work, family, and other aspects of life. Despite the conceptual and methodological biases towards focusing on a particular layer of users—younger people—these studies suggest a need to explore the cultural underpinnings of online social network practices, as well as the social and cultural roles afforded and constrained by contemporary social media.

This chapter strives to provide insights into relationship-based computing that happens at the intersections of online and offline environments in the context of social network systems. I employ theories of technology and culture to examine how identities become constructed within the creative and limiting possibilities of an online social network. In later sections of the chapter, I articulate the theory of techno-culture, briefly describe my method of exploration, and then focus on the themes emerging from the data. I conclude with implications of this research.

Culture/technology/identity

In order to speculate about the cultural meanings that emerge in the process of participating in online social networks, it is necessary to carve out complex notions as culture and technology. In this section of the chapter, I articulate the theories that help me speculate about communication in such networks. This chapter holds culture to be "a whole way of life" and "ordinary" (Slack & Wise, 2005, p. 4). This argument underscores the dynamic nature of culture and suggests that culture is made of the heterogeneous meanings, values, and artifacts that we live by. Culture consists of the processes of arranging, challenging, and enacting these artifacts and ideas in everyday life. In this sense, people's existence within culture is made up of a variety of practices that constitute mundane life, and online social networks form a part of living in culture. Similar to other media, commodities, and technologies, online social networks merge with everyday reality and become invisible and taken for granted. As of August 2011, the social network of Facebook claimed a membership of more than 750 million users, who visit the site at least once a month (Facebook, 2011). Indeed, the platform is a part of culture, yet what communication patterns does participation in the online community of a social network suggest?

Slack and Wise (2005), as well as Bijker (2006), help conceptualize the relationship between culture and technology. These authors use the terms *technological culture* and *techno-culture* to mark the pervasiveness of science and technology in the shaping of modern societies and everyday life. This approach emphasizes that social interactions are mediated by technologies and that technologies function when embedded in social relations. This means that specific technologies, for example online social networks, become productive cultural sites and elements of social relations. Users as well as other participants and nonparticipants of online networks can therefore be said to *construct* this type of technology by creating, consuming, criticizing, experiencing, arguing about, playing with, and working on it. Online social systems—in the way they function for the uses and conversations of those who are and are not involved in them—are the results of culturally produced understandings. How they function varies depending on the specific circumstances of participation and moments in time and space and includes material and immaterial entities and processes.

To narrow down the issues I want to explore, I rely on the idea of *performativity* of technology. Performativity implies that the phenomena have to be made and remade in the various ways and manners in which they are said to exist (Latour, 2005, p. 34). An online social network is created by specific material processes, such as the participation in activities associated with the network and the discussion of it. Law and Singleton (2000) claim that "particular and located enactment or performance of technological knowledge and practice" (p. 767) does particular kinds of work through telling the stories and acting out realities. Thus, what are the kinds of cultural work that are being produced on the electronic pages of social network systems? How do the activities and conversations off the Web pages—in other words off-line—create or construct meanings associated with the self?

One of the routes to take in responding to the first question is to explore the positions of individuals in the practices of everyday life: How is identity performed on the pages of a social network? Latour (2002) and Haraway (1991) suggest that nonhumans (e.g., cyborgs) have a human shape and, simultaneously, give shape to humans and human identities. Thus, this research examines the ways that humans (e)merge at the intersection of online and offline contexts. Contemporary theorists like Hall (2000) suggest that identification—the process of building the self—is always strategic and positional because institutional and historic sites allow or limit specific discursive formations and practices and invite specific enunciative strategies. In other words, identities are being multiplied and constructed across different, often competing and complementing, discourses and practices. When casually checking a friend's pages on a SNS, heavily engaged in a game on Facebook, or working on updating a profile, a participant makes a series of decisions that create and direct the process of identification. In addition to being heterogeneous and processual, identity is constituted through difference and depends on exclusion and otherness. Hall complicates the notion of identity, arguing that "it is only through the relation to the other, the relation to what it is not, to precisely what it lacks, to what has been called its *constitutive outside* that the positive meaning . . . of identity can be constructed" (p. 17). In online interactions, each participant of a social network brings some difference that defines communication choices within the possibilities allowed by the system.

Online identity has been often studied with emphasis on specific markers of difference: gender, sexuality, and race. Gender identity is often studied in relation to sex and sexuality as well as race (Kolko, Nakamura, & Rodman, 2000; Nakamura, 2002). These studies argue that race matters a great deal both offline and online, because when spending time online we are already shaped by racial cultural values and cannot help but bring our knowledge and experience when we log on. A social marker of class has also been the object of investigations by boyd (forthcoming) and Hargittai and Hinnant (2008). Following recent publications about social media (Barker, 2009; van Doorn, 2009; Gajjala, 2007) that create a nuanced examination of the practices in social network systems, I stress that race, gender, sexuality, and other markers of difference (ability, class, geography, etc.) intersect with each other and are made up of ongoing processes of definition, performance, and enactment. Considering cyberspace not as a physical place but as a locus around which hypertexts of texts, modes of social interaction, commercial interests, and other discursive and imaginative practices coalesce, I focus on contextual materialization of promises and limitations existing in cyberspace.

The analyses in this chapter emerge from a cyber-ethnographic engagement with Facebook and the system's participants (Rybas & Gajjala, 2007). In particular, the data come from such sources as self-observations and journals of the Facebook-ing process during the period of September 2006 to December 2007, in-depth interviews with 15 students of a medium-size Midwestern university (conducted in the spring of 2007), and content analyses of student-produced videos and essays for an upper-level undergraduate course focusing on digitally mediated identities (taught in the fall of 2006). The participants in the study were traditional students in their junior or senior year in college. Using these methods, I draw on the encounters of living both online and offline as a resource—on moments of interaction on the Facebook Web site and with other participants in the social network. Actively producing my cyber-self and observing others who are doing the same, I conduct an ethnography by immersing myself in the cyber-context. I collected the data from the second half of 2006 throughout 2007, during which time Facebook was becoming a phenomenon of cultural significance. I have categorized the experiences—interviews with Facebook members, analyses of videos produced by students, and self-narratives—into two themes: *producing authentic selves* and *creating/erasing differences at the interface*. These themes reflect the processes of living at the intersection of online and offline contexts and reveal the logic of constructing and managing the self at the digital interface.

Producing authentic selves?

The theme of producing authentic selves emerges from interviews I conducted with the study participants about their understanding of online interactions in the social network system. I focused a series of interview questions on asking what image the users project on their profiles. In general, the responses of the study participants have a common theme: *The profiles represent who the users are*. The major emphasis in the responses falls on the attempt to maintain the "trueness" of representation of the real self on the profile page. In the narrative below, I use pseudonyms to refer to the study participants.

The most evident assumption about profile content appears to be that real people create real profiles of themselves. This assumption becomes clear from what users include and accentuate in their profiles. For example, Jema says, "You know, you've got the basic information of who I am, where I am from, so, like, I am going to put myself out there and let people know who I am." The interviewees often stress that some of the information they share on the Web site does not change because that part of their personality remains constant. Other study participants describe their profiles through describing themselves, especially their tastes and preferences. The pattern *I like this, I post what I like on my profile, thus my profile is me* draws a syllogistic matrix for creating one's presence on Facebook.

In addition, the study participants discuss perceiving other users' profiles as more or less accurate representations of them. For example, Mishak, who is highly critical of the practices of social networking, doubts the overall usefulness of the online social-networking practices. However, he draws a strong parallel line between online and offline experience, in that individuals in both settings may make themselves vulnerable if they reveal too much information. In a similar way, another interviewee states, "If I am meeting a girl, like in class, or on campus, or restaurant what not, and I want to get to know her better . . . I'll make judgments about her based on her Facebook." Thus, the standard of true self-presentation guides the users when they work with profiles of others in the system.

The expectation that there will be a direct connection between the profile and the person is apparent from the account setup. This tendency for identifying the profile with the user stems from the literal symmetry in describing oneself and one's textual and digital embodiment in the profile. Taylor (2003) argues that virtual environments are designed with particular visions of identity, communication, and community in mind. The personal information that Facebook requires the user to fill in, such as "Current city," "Birthday," and "About me" in the "Basic Information" tab or "Employer" and "college/university" in the "Education" and "Work" tabs (as of July 2011) have been formulated to elicit details that will be interpreted by users as details that describe one's real self. Although these data are changeable, and, for the most part, voluntarily provided or not provided, they can serve as convenient anchors tightly connecting an account to a person. One of the masterminds of Facebook, Mark Zuckerberg, confirms that the site focuses on authenticity and representing one's *real* self and *real* connections in the *real* world (Locke, 2007). Having one account for each user indicates the drive for singularity in the social network system, in which, after creating a profile, users focus on developing it. Based on how the users understand the interface of the social network, the expectation that their profile is a true, or authentic, representation of the self is quite strongly embedded in the Facebook architecture.

Dubrosfsky (2011) suggests a different take on authenticity. In a recent publication about surveillance practices on Facebook, she argues that although heavier Facebook users focus on constantly inputting and circulating data about themselves, they may strive to appear consistent across contexts yet do not wish to achieve authenticity. Dubrofsky draws her understanding of authenticity from reality TV studies, suggesting that authenticity is based on a desire to appear natural and consistent across contingent contexts: on and off the television-surveilled space. Comparing the surveillance mechanisms of reality TV and social networks, Dubrofsky denies Facebook users' claims of authenticity on the basis that Facebook-ing does not involve a direct comparison of behaviors across spaces.

However, authenticity, if understood in terms of singular and stable trueness, as projected according to the network prompts, becomes an issue given the performative nature of identity. The process of identification, as explained by Hall (2000), places producing of the self at the center of activities in the network—a user becomes him or herself as he or she produces the multiplicity of data in the network. The major portion of engagement with Facebook focuses on maintaining one's presence in the network by updating and editing the details on one's profile and interacting with other profiles. Indeed, a Facebook user emerges as a data subject marked by the actions and movements in the network database (Dubrofsky, 2011). One of the study participants, Jordan, describes updating the movies he watches, changing profile pictures, adding more details to the section "About Me," posting pictures about his recent trips, or commenting on his friends' status updates. Thus, the self emerges from the work of producing one's presence in the network.

Users are involved in decision making about what information to include and within which blanks to enter it on their profiles. They interpret episodes of their life such as watching movies, playing sports, learning games, hanging out with friends, reading, and many others to make the narratives fit the architecture of the site. One of the study participants, Doug, explains negotiating the Web site structure in the following way: "I try to put everything that is true about me, but it is all the positive things." Doug's comments underscore the idea that profile work requires technical knowledge on how to post and make changes as well as involves making choices and accommodating, interpreting, and negotiating what the profile may imply.

Thus, authenticity, even though expected, becomes an illusion of the Facebook-ing process. I question the authenticity claim not because of the absence of comparison as Dubrofsky (2011) proposes but because users continuously interpret and tinker the expectations outlined by the network structure. This is why I placed a question mark in the title of the heading of this section. Users produce their own identities at the interface and interact with the identities of others by committing to the expected authenticity, while the layers of technological structure demand processual engagement that undermines the authenticity project.

Creating/erasing difference at the interface

Based on the expected similarity between Facebook profiles and the people who claim ownership of the Web space, profiles reveal details about them as if shedding light on their personality. Study participants often mention that they rely on Facebook to find out what is happening with their friends. Facebook users call this interaction "information exchange and sharing." However, in high-stakes situations, such as establishing and maintaining a relationship with a romantic partner, the profile plays into an assumption that when meeting new people, one needs to find more information "to follow up," to confirm initial impressions, and to search for common ground and differences. In this search, Internet users may discover some unexpected events or details about the other. These are the moments in which users who are striving for fleeing authenticity realize contextual configurations of differences.

In exploring the configurations of differences, I refer to videos created by undergraduate students for an upper level communications course. One of the clips makes an attempt to show a case in which, as described in a student's journal, "you might have met someone out at night and had one impression of them until you follow up and find out more information about them and you realize they are much different." The video plays out a script of two young people, male and female, meeting at a bar. After the first date, the male character, Aaron, checks the Facebook profile of his female counterpart, Noel, and finds that it is different from what he expected. Online, she presents herself as a fan of gothic mythology, whereas she seemed shy and less "edgy" when he had met her. This example speaks of multiple layers of embodied understanding of digital technology and cultural practices of everyday life.

This video suggests that some practices become easily normalized in the context of online networks. The characters rely on the ability of the online network to represent the user in the profile supposedly truly and authentically. Even though the video does not reveal Noel's reasoning, she offers Aaron an opportunity to see her profile, and Aaron does not hesitate to view it. This act of profile reading seems normal and is taken for granted by the personae acting in the video. The characters in the video perform the scripts of romantic relationship development. As potential romantic partners, Noel and Aaron engage in self-disclosure in which a profile-generating Web site becomes the locus of the disclosing process. Again, this practice is taken for granted, producing no unexpected reactions or doubts.

In addition to showing the network as a space for the practices of courtship, this video reveals the gender dynamics at the online–offline intersection. The role of Aaron, played out in the video and further explained in the student's journal, positions him as a heterosexual male who is actively looking for a heterosexual female partner. He considers romantic involvement and checks out Noel's profile to find more information about her. Following that, he makes decisions about whether Noel is the right person with whom to continue a relationship. Aaron ranks the two images, of Noel online and offline, in their desirability and tries to make Noel fit into an expectation of an ideal partner. The video portrays her as an object that is on display, as he is shopping around. He is shocked and offended at her two disparate presentations, taking her act of "being two different people" as pretense, but she has no opportunity to respond or explain, presuming these explanations are even needed. In general, this performance aligns with previous research (Kendall, 1998; Nakamura, 2002), arguing for augmented gender stereotyping at the intersections of online and off-line contexts.

The video represents how its creators understand identity formation online and off-line. Noel's character is written and performed in a double play: She is both a normal, good-looking girl at the bar or a different person—a fan of Dungeons & Dragons–style games. The two images oppose each other, especially in the eyes of Aaron, who interprets identity as necessarily consistent. He does not admit that a good-looking girl can also be involved in Dungeons & Dragons culture. Dora, a study participant, writes in her journal, "In our video, we have Noel acting in different roles changing from the hot girl that Aaron is very impressed with, to the dorky girl that he would rather never talk to again." Both the characters in the video and the video creators express concerns that the discrepancy between the off-line image and the online profile is taken as untrue. Thus, the male character feels confused, offended, and angry, tearing up the note Noel wrote to direct him to her profile. The video creators suggest that entering the digital world allows one to shed the baggage of his or her identity and become a completely opposite person, as if a network member takes off his or her theatrical makeup when stepping off the screen and into the physical world. This change suggests only two possibilities for communication choices: either to be considered a hot girl who will impress a guy or a dorky girl who will be never noticed again. These consequences emphasize that *the difference created at the interface* matters and has material effects. For one, the online presentation may define choices and influence decisions people make off-line. The articulation of difference at the interface reveals power dynamics and implicit privileges about who watches, who is being watched, who makes decisions, and who remains silent.

Other differences produced at the interface define communication practices. For example, study participant Jack draws a very clear distinction between joining Facebook as a community college student and as a university student. He shares that the community college at which he started his higher education career was not an appropriate place [or time] to join the online network because he would continue to be associated with a less prestigious (community college) institution. While Facebook programming did not make any value distinction between these two types of institutions, Jack explains that he did not know many people in that community college and did not keep in touch with his high school cohort. This explanation sounds plausible, yet a student in a community college interacts with other students and usually establishes at least some social relations. Peers from high school can still make a feasible circle of friends. This pattern of minute decisions about movements in online and off-line life, based on the image of "preferred Facebook users," contributes to the system of boundaries augmented by the computer interface.

The spillover of "offline zoning" into the realm of online interactions highlights the subtle character of social clustering , whereby users avoid crossing the boundaries of physical *and* digital neighborhoods. Facebook incrementally had users from colleges, adding high schools and then corporations. It is remarkable that at every stage the site preserved its exclusivity by claiming the specific audience and membership on the site. This simple condition has instilled a strong sense of elitism and social privilege. Current project participant Jack prefers not be associated with the context of a community college, just as undergraduate student researchers preferred to study Facebook rather than Myspace in the study described by Gajjala, Rybas, and Altman (2007). In this study, university students expressed apprehension about coming into contact with high school students on Myspace who lived in a poor neighborhood of a city and reported feeling more comfortable about interacting with their counterparts on Facebook. Another explanation is suggested by boyd (2006), who argues that on social network sites "people define their community egocentrically." This means that Facebook "friends" constitute the context for both members and audiences and signal the expected demarcations. Entering a certain context in online social networks happens through affiliation with other individuals who construct sets of interests and communication practices. boyd implies that online social networks collect users of similar walks of life to serve as a digital place where "birds of a feather flock together."

So far, I have focused on differences emerging at the interface. Further, I have observed that members of Facebook try to eliminate noticeable differences and smooth out the edges that make users stand out from the crowd. I have discussed how various categories of difference, such as race, class, gender, geography, and others intersect with specific place-based configurations of access to various literacies and ignorances constructed in the communities of production (Gajjala et al., 2007; Haraway, 1991). Acting on one's knowledge of the system and one's interpretation of personal biography and collective history allows a (re)drawing of the digital fabric of the self.

Study participant Doug refers to a video he created with his classmates to illustrate his claim that Facebook changes lives. The video narrates how a freshman becomes a part of the campus crowd after he opens a Facebook account and works the social network. This video illustrates that the Facebook interface is a template filled in with a variety of information. The standard organization of profiles presents all members in an identical format, making them look and feel more or less similar to each other. Participation in the same project of Facebook-ing, located on the same campus, at the same moment in time, imposes a feeling of belonging to the campus community. Ellison, Steinfield, and Lampe (2007), in their study connecting the use of Facebook and social capital, argue for the inclusive character of this social network system based on the online community not systematically differentiating between social groups and on the network striving to establish connections between anybody who desires to participate. However, the tendency to mold the participants into a certain form suggests a limiting function of Facebook. Being that a member voluntarily includes various details in his or her profile—by typing or posting oneself into online contexts—there seems to be a choice of what to say about oneself and how to say it. As an alternative, the user can invent the details to enter into the blanks, and, even in this case, the author is required to explore the past and imagine other ways it might be. This way, difference may emerge or disappear from the digital representation.

When working with a profile, a user is required to do some interpretive work about their biography and about the blanks suggested by the program. Applying fantasy and imagination helps reflexively understand oneself and the world aroundus. A critical scholar of gender, Judith Butler (2004), assigns fantasy the power of transformation. In her words, "Fantasy is part of the articulation of the possible; it moves us beyond what is merely actual and present into a realm of possibility, the not yet actualized or the not actualizable" (p. 28). She claims that fantasy establishes the limits of the real and allows looking elsewhere and connecting the possible and impossible. I explore the pattern identified by Butler in my personal journal entry in which I follow my choices in updating my profile. In May 2007, I wrote that I filled in the blank space for one's favorite music with "dance-trance-pance-mance—very exotic." This type of music does not exist, and the social network of Facebook lists no users with this example of music taste. This combination comes from a few areas of my life experience: First, "dance" and "trance" are music types I occasionally listen to; second, "pance" and "mance" are two words that rhyme with "dance" and "trance"; and third, "very exotic" is a phrase from my favorite movie *Moulin Rouge*. Thus, combining the three parts together, I drew from my life experiences in a creative way to produce humorous and sarcastic effects. Stating the style's exotic nature, I invoke two meanings at the same time—ordinary and extraordinary. These complex elaborations reflect the perceptions of my tastes in music and my anticipation of the audience's perceptions in the public atmosphere of Facebook. This way, I create an unusual feature on my profile.

If in the previous example the difference is deliberately created or emphasized in a certain way, another group of stories suggests that differences at the interface can be erased or de-emphasized. The following examples come from interview episodes that focus on discussing the future Facebook activities of participants in the study. These respondents believe that they will continue their engagement with the social network system to keep in touch and have fun with high school or college friends or family members. Even though the interviewees do not plan to exclude Facebook from their everyday experiences, they suggest that their routines will change. It is notable that some interviewees plan to considerably alter their profiles. For example, Kara projects that she will make her profile more professional "just to give them something to see that was positive, like, think of all the organizations [I am involved in]" and "have more pictures of me doing, I don't know, professional things or, like, school things." Neena targets future employers by eliminating people from her profile that might harm her reputation: "Before I actually graduate and start applying for jobs, [I plan] to weed out everybody on my friends' lists who I don't know because I don't want any spies." Kara and Neena are making an attempt to view their Web presence through the eyes of those who may have access to their profiles and who may have influence on their future life. To meet the expectations of these audiences—who are mostly identified as employers—the users strive to eliminate the differences that may possibly be interpreted as inappropriate or undesirable.

It is undoubtable that employers exercise the power of decision making regarding the future of current college students. As Facebook members and as potential employees, the study participants believe that the two images have to be reconciled and developed to fit each other. In her study of the dynamics of teenagers' participation in Myspace, formerly an SNS, boyd (2007) argues that certain groups, such as parents, schoolteachers and officials, and government administrators hold power over teens. The teens who interact with their friends on Myspace thus face the dilemma of how to be "cool" with their peers while also being appropriate with parents and other influential entities. Study participants who have accounts on Facebook are in a similar double bind. Unlike high school students, however, they contemplate changing their profiles and fitting the expectations of viewers.

While users of the SNS on Facebook actualize their identity by collecting a variety of details about themselves, they also strive to filter their Web presence on their profiles to fit the requirements of the program and to fit the imagined expectations of the audiences. From the point of view of program structure, the Web-based software may not have the functional possibilities to accommodate the full spectrum of identifying markers suggested by individual users. Even though the software has been constantly in flux, adding more features and lines in menus, individual users who do not fit the standards of the program have to choose the least conflicting variants, or delete the differences, in order to meet perceived audience expectations. In this way, participation in the online network creates and erases differences at the interface. Since participation in the network is more or less uniform, being a network member and engaging with the network creates the feeling of similarity and belonging.

Conclusion

This chapter pursued the objective of critically questioning networked media and culture and the way in which they rhetorically and pragmatically construct the self. What is at stake when users strive to present their authentic self yet fail to do so because of the interpretive work done about the self and about the network? What are the implications of creating and erasing differences at the interface of a software program? I started the chapter with the explication of techno-culture, suggesting that technologies saturate everyday life and create possibilities for social relations but also limit them. Seeing social network systems as an element of techno-culture, I argue that the self gets created *because, through,* and *with,* and maybe *despite,* the Web site (together with the systems of relationships and other interactions). Facebook is one of the places where users articulate themselves, similar to parks, offices, libraries, houses, and highways, by interacting with other members and using technology to make the context. The performative work of Facebook-ing occurs in the opportunities and processes of interpretation and articulation of one's relationships to the self and to imagined and actual others. By creating a profile, a member seems to freely choose what to say about the self and how to say it. However, this apparent freedom is limited by the options suggested by program menus and the experiences one has to rely upon. In order to construct oneself in an online system, a user has to find an optimal way to reflect his or her intentions, to respond to the prompts formulated by the Facebook team, and to meet the expectations of friends or other publics.

In other words, social network systems, similar to other technologically mediated environments, provide productive sites where cultural baggage gets created, recreated, consumed, and reiterated. The nuanced ways of "putting the self up" on Facebook—the intricate invisible negotiations of what is possible and not possible and what is required and what is expected within the program and the community—construct members (as well as nonmembers) in relationship to that particular discursive space and to each other. Difference is fluid in itself, as its specific articulations depend on the contexts and participants; thus the differentiation—a process embedded in identification (Hall, 2000)—may become a factor in the context of intercultural contact when people may have only Facebook-mediated options for communication. The consideration of the often-invisible and taken-for-granted meanings that are produced as technocultural practices emphasizes the need to be aware of naturalized approaches to digital contexts, in which users act on the basis of familiar ground rules and norms. This engagement is necessary for developing a critical praxis and critical attitude toward technology-mediated interaction, as well as for avoiding the acceptance of technical innovations as natural elements of contemporary culture that strengthen the status quo. Technologies carry meaning by providing a leverage for identity presentation and interpretation; thus, digital spaces like Facebook contribute to negotiating issues critical for life in society: power, authority, knowledge, and representation.

References

Barker, V. (2009). Older adolescents' motivations for social network site use: The influence of gender, group identity, and collective self-esteem. *Cyber Psychology & Behavior, 12*(2), 209–213.

Beer, D. (2008). Social network(ing) sites . . . Revisiting the story so far: A response to danah boyd & Nicole Ellison. *Journal of Computer-Mediated Communication, 13*(2), 516–529.

Bijker, W. (2006). The vulnerability of technological culture. In H. Nowotny (Ed.), *Cultures of technology and the quest for innovation* (pp. 52–69). New York, NY: Berghahn Books.

boyd, d. (2006). Friends, Friendsters, and Top 8: Writing community into being on social network sites. *First Monday, 11*(12). Retrieved from http://www.firstmonday.org/issues/issue11_12/boyd/

boyd, d. (2007). Why youth (heart) social network sites: The role of networked publics in teenage social life. In D. Buckingham (Ed.), *Youth, identity, and digital media* (pp. 119–142). Cambridge, MA: MIT Press.

boyd, d. (forthcoming).White flight in networked publics? How race and class shaped American teen engagement with Myspace and Facebook. In L. Nakamura & P. Chow-White (Eds.), *Digital race anthology*. New York, NY: Routledge.

Butler, J. (2004). *Undoing gender*. New York, NY: Routledge.

Doorn, N. van. (2009). The ties that bind: The networked performance of gender, sexuality, and friendship on Myspace. *New Media and Society, 12*(4), 583–602.

Dubrofsky, R. (2011). Surveillance on reality television and Facebook: From authenticity to flowing data. Communication Theory, 21, 111–129.

Ellison, N., Steinfield, C., & Lampe, C. (2007). The benefits of Facebook "Friends": Social capital and college students' use of online social network sites. *Journal of Computer-Mediated Communication, 12*(4). Retrieved from http://jcmc.indiana.edu/vol12/issue4/ellison.html

Facebook. (2011). *Facebook Factsheet*. Retrieved from http://www.facebook.com/press/info.php?statistics

Gajjala, R. (2007). Shifting frames: Race, ethnicity and intercultural communication in online social networking and virtual work. In M. B. Hinner (Ed.), *The role of communication in business transactions and relationships* (pp. 257–276). New York, NY: Peter Lang.

Gajjala, R., Rybas, N., & Altman, M. (2007). Epistemologies of doing: E-merging selves online. *Feminist Media Studies, 7*(2), 209–213.

Hall, S. (2000). Who needs "identity"? In P. du Guy, J. Evans & P. Redman (Eds.), *Identity: A reader* (pp. 15–30). London, United Kingdom: Sage.

Haraway, D. (1991). *Simians, cyborgs and women: The reinvention of nature*. New York, NY: Routledge.

Hargittai, E., & Hinnant, A. (2008). Digital inequality: Differences in young adults' use of the Internet.*Communication Research, 35*(5), 602–621.

Hargittai, E., & Hsieh, Y. P. (2010). From dabblers to omnivores: A typology of social network site usage. In Z. Papacharissi (Ed.), *A networked self* (pp. 146–168). London, United Kingdom: Routledge.

Ito, M. (2010). *Hanging out, messing around, and geeking out: Kids living and learning with new media*. Cambridge: Massachusetts Institute of Technology.

Kendall, L. (1998). Are you male or female? Gender performances on MUDs. In J. O'Brian & J. Howard (Eds.), *Everyday inequalities: Critical inquiries* (pp. 131–154). Malden, MA: Blackwell.

Kolko, B., Nakamura, L., & Rodman, G. (2000). Race in cyberspace: An introduction. In B., Kolko, L. Nakamura, & G. Rodman (Eds.), *Race in cyberspace* (pp. 1–14). New York, NY: Routledge.

Latour, B. (2002). Morality and technology: The end of the means. *Theory, Culture, and Society, 19*(5,6), 247–260.

Latour, B. (2005). *Reassembling the social: An introduction to actor-network theory*. Oxford, United Kingdom: Oxford University Press.

Law, J., & Singleton, V. (2000). Performing technology stories: On social constructivism, performance, and performativity. *Technology and Culture, 41*(4), 765–775.

Locke, L. (2007, July 17). The future of Facebook. *Time Magazine* (electronic edition). Retrieved from http://www.time.com/time/business/article/0,8599,1644040,00.html

Nakamura, L. (2002). *Cybertypes: Race, ethnicity, and identity on the Internet.* New York, NY: Routledge.

Paasonen, S. (2002). Gender, identity, and (the limits of) play on the Internet. In M. Consalvo & S. Paasonen (Eds.), *Women & everyday uses of the Internet: Agency & identity* (pp. 21–43). New York, NY: Peter Lang.

Rybas, N., & R. Gajjala (2007) Developing cyberethnographic research methods for understanding digitally mediated identities. *Forum: Qualitative Social Research Sozialforschung, 8*(3). Retrieved from http://www.qualitative-research.net/fqs-texte/3-07/07-3-35-e.htm

Slack, J. D., & Wise, J. M. (2005). *Culture + technology: A primer.* New York, NY: Peter Lang.

Taylor, T. L. (2003). Intentional bodies: Virtual environments and the designers who shape them. *International Journal of Engineering Education, 19*(1), 25–34.

7. *Who Am I in Virtual Space? A Dialectical Approach to Students' Online Identity Expression*

PING YANG, PHD

Introduction

The complexity of online identity issues presents new research challenges for intercultural communication scholars (St. Amant, 2002; Chester & Bretherton, 2007; Abbas & Dervin, 2009). Intercultural identity expression is a lived, situated experience. In a virtual community, the expression of cultural selves is related to cultural norms as well as the specific contexts in which it occurs. In the shrinking time and space relations of virtual environments, the lines between users' identities are becoming blurred. There exist fluidity and tensions between anonymity and honesty, visibility and invisibility, and temporality and permanency in cultural identities. Web users can now choose between a variety of media to negotiate their identities, such as e-mail, newsgroups, File Transfer Protocol (FTP), chat areas, blog, twitter, and social network sites (SNSs; McPhail, 2006; boyd & Ellison, 2008).

This chapter focuses on how American college students express their cultural selves through asynchronous time and space in text-based computer-mediated communication (CMC) contexts. It employs phenomenological and dialectical frameworks in examining the ways that students express and negotiate their identities via online communication. Phenomenology, according to Martinez (2006), enables the researcher to examine issues of cultural identity by fully engaging the complexities of race, gender, and ethnicity, as well as the questions of context and power in building knowledge. The dialectical approach, introduced by Martin and Nakayama (1999), emphasizes the processual, relational, and contradictory nature of intercultural interactions. The position taken in this chapter is that social and cultural identities are fluid and constantly being reconstructed, and related communication differences are relative, dynamic, and undergoing changes, always within structures of power relations. Therefore, the investigation of the expression of cultural identity requires an in-depth inquiry into the complex communication process online. This research investigates the realities of students' experience with online identity construction as what it is in the human lifeworld and interrogates the role of power dynamics in these virtual encounters. The following section reviews the existing literature focusing on identity issues in intercultural online communication.

Online identity construction

Prior studies have explored identity issues in CMC in various cultural contexts including the role of national backgrounds (Holmes & O'Halloran, 2003; Ye, 2006), gender issues (Baron, 2004), and linguistic and socioeconomic backgrounds (Cassell & Tversky, 2005). Recent research has emphasized the importance of a particular CMC context when examining identity formation and expression in online encounters (Abbas & Dervin, 2009; Chester & Bretherton, 2007).

The mode of computer-mediated encounters examined in this research context differs from face-to-face (FtF) interaction in two major ways: First, they involve the filtering of nonverbal cues by their lack of contextual clues. Second, they are asynchronous—mostly occurring in SNSs, online discussion boards, and e-mails. As St. Amant (2002) suggests, the characteristics of some CMC differentiate identity presentation online from off-line interactions—allowing for more fluid and even deceptive self-presentation (Turkle, 1995). The use of CMC has therefore increased the complexity of identity presentation and negotiation in intercultural interactions by creating a new context for constructing cultural selves. For example, St. Amant describes how the potential for plastic and changing identity in CMC may cause problems for cultural groups that require a stable and discernible identity for communication. For some cultural groups, it is important to know the social identity background of another interactant, such as gender and status, in order to communicate appropriately (St. Amant, 2002). Donath (1999) points out the ambiguity of identities and the possibilities of identity deception in virtual communities that result from anonymity. As Donath notes, the lack of basic cues about personality and social roles may result in deception about identities, making credibility one of the most significant issues in the mediated environment. The lack of identity cues (usually communicated through appearance and other nonverbal cues) can result in suspicion and deception in online encounters (Zywica & Danowski, 2008).

A number of studies in online identity development has also explored power issues. Gajjala (2004) points out that cyberspace today looks White because people of different races, genders, ethnicities, and nationalities are all connected within a "Westernized, Anglicized and white cyberspace" (p. 84). A related point that van Dijk (2004) focuses on is the digital divide—the gap in media access and use of computer and the Internet that results from differences in technology access and ownership, demographic variables, and deeper social, psychological, and cultural roots. On the other hand, Mehra, Merkel, and Bishop (2004) have focused on members of minority and marginalized groups and examined how the Internet and digital technologies can be used to empower these people in the process of their intercultural and interpersonal coconstruction of meaning in their everyday lives. These authors argue that the Internet can be used to close the digital divide and achieve social changes and to serve as an important site for intercultural communication, being that it is yet to create spaces for other "perspectives and voices" that are possible on a majority of sites online (Gajjala, 2004, p. 87).

The previously mentioned literature provides the foundation for the current investigation of cultural identity presentation via asynchronous text-based online communication. When students of diverse cultural backgrounds interact online, issues of race, gender, class, region, and education come into play. Because of the scarcity of identity cues, the fluidity and tensions between anonymity and honesty, visibility and invisibility, and temporality and permanency invite more in-depth explorations of the identity issues beyond the traditional research that has been done for statistical representation. These require interpretations that describe the unique process of expressing cultural selves online within the examination of context and power. With the growing cultural complexity of the new context of online interaction, the current research attempts to address the following research questions by interrogating students' experiences of online identity presentation: What modalities of consciousness and preconsciousness constrain and enable students' experiences? How do students' consciousness and preconsciousness of cultural self-presentation manifest in their discourses?

Methods

Phenomenology, an important theory and methodology, focuses on the life-world and lived experience of people (Martinez, 2000). It helps to explain the meaning of human phenomena and understand the lived structures that make particular social meanings possible (van Manen, 2003). With the research interest in understanding the lived experience of students' online identity expression, this study employs phenomenology to explore how students express their cultural selves online in a way that questions any particular utterance of fact or experience. For an in-depth understanding of the complexity and intricacies of students' lived experience online, this study takes a dialectical approach, which emphasizes the processual, relational, and contradictory nature of intercultural communication (Martin & Nakayama, 1999). According to Martin and Nakayama (1999, 2010), the dialectical approach helps researchers move beyond the limitations of research paradigms and opens up a new perspective for studying culture and communication. There are six intercultural dialectics proposed by Martin and Nakayama (1999): cultural–individual, personal/social–contextual, differences–similarities, static–dynamic, present/future–history/past, and privilege–disadvantage. These six dialectics are widely employed in studying intercultural communication (Cools, 2006; Orbe, 2008) and teaching intercultural communication (Fong & Chuang, 2004; Schmidt, Conaway, Easton, & Wardrope, 2007).

Phenomenology meets the criteria emphasized in the dialectical perspective by providing the technical specifications that allow the dialectical perspective to be fully implemented in practice. According to Martinez (2000), these interpretive and critical perspectives are inherently interconnected, that is, the interpretive descriptions produced by phenomenology quite easily allow for the critical interpretations of the dialectical perspective that focus on issues of power and context. Similarly, critical interpretations challenging existing structures can easily include interpretative descriptions.

This research took place in a large Southwestern university in the United States. Personal description, interviews, and online discussion were the primary strategies for capta (phenomenological data) collection. The researcher set up online forums by providing discussion questions using the Blackboard online learning management system. These forums invited students to share their intercultural experiences with the use of the Internet and how culture interacts with identity construction online. A total of 86 postings were collected from the forums. Personal descriptions were collected from students' written assignments about their intercultural online experiences with their intercultural conversation partners. Interview questions featured semistructured, open-ended questions, focusing on the students' experiences of cultural identity expression online. Each interpersonal interview lasted for about 60 minutes. All of the interviews were audiotaped and transcribed. Confidentiality was guaranteed and pseudonyms were used.

Phenomenological analysis was started after the entire set of capta was collected. The researcher followed the requisite three stages of description, reduction, and interpretation in analyzing the capta. These three phases were interdependent of each other and helped the researcher to generate richly nuanced descriptions about the students' intercultural online experiences.

A set of themes emerged from the phenomenological analysis as the capta were read and analyzed iteratively to generate research findings.

Modes of consciousness in expressing cultural self online

The research capta generated by the application of phenomenological methodology provide experiential descriptions of the phenomenon of interrogating the students' consciousness and preconsciousness in expressing their identities with CMC. A phenomenological analysis of the online experiences of the students support and extend the dialectical framework advocated by Martin and Nakayama (1999), generating new dialectics that characterize online identity expression: digital equality and marginalization, authenticity and anonymity, liberty and tensions, stability and fluidity, and trust and suspicion in online intercultural communication.

Digital equality and marginalization

The phenomenological interpretation of students' online experiences in this study reveals the dialectic between digital equality and marginalization in online identity expression based on the privilege–disadvantage dialectic advocated by Martin and Nakayama (1999). As revealed by the students' experiences, anyone who has access to the Internet may enjoy the convenience, effectiveness, and efficiency of expressing who they are to people anywhere in the world. This privilege allows them to write and express their status more as equals. Patricia is a student from the Czech Republic. She feels positively about new media in communicating with her friends:

> I sometimes enjoy the Facebook more than talking with my friends in person because if English is not my first language, I sometimes find it hard to be funny with it, to look funny and smart. . . . So I kind of enjoy Facebook and all the stuff there because I can joke there, I can think about it, I can write witty and humorous replies. I sometimes enjoy it more because I don't feel so dumb. People cannot hear my accent. I feel more confident there.

In the online context, the invisibility of nonverbal cues makes it possible to suspend prejudging others based on physical features such as race, ethnicity, gender, and age. Therefore, the fluidity and flexibility of online identities, as well as the suspension of prejudgment based on nonverbal cues, make online identity presentation an advantage to some marginalized group members. According to Claudia, "Interaction online is somewhat anonymous [and this] can be an advantage. Since people don't know who the person they are talking to looks like, it is harder to make a judgment based on their ethnicity or how they look." Intercultural communication using new media helps reduce the chances of stereotypes and discrimination and gives voices to the unheard and silenced people.

According to van Dijk (2004), there are four successive kinds of access in the context of digital technology: mental access, material access, skills access, and usage access (p. 234). Students' online experiences demonstrate differences in their access to the computer and online resources but also their different skills, usage, and mental access to the Internet. Although e-mails, computer, and the Internet play a vital role in the lives of many US students, those from other cultural groups or world regions may not enjoy the same opportunity and comfort levels using them. Molly traveled to Mozambique and experienced huge gaps in online technology, resources, and opportunities of expressing oneself:

> Online messaging is so prevalent [in the US], but where I traveled last summer, they actually don't have access to the Internet. It's sad because we don't have the opportunity to keep in touch, unless I travel back to Mozambique . . . Online identity can show a lot about one's culture and how important it is to you by the way you hold your values of being honest and private . . . The fact that [people] can use the Internet and have access shows that they must have at least low middle-class identity.

As these students' experiences reveal, to a certain extent communication technologies have enlarged the gaps among people of different cultural, racial/ethnic, and socioeconomic backgrounds. Various cultural groups may communicate in different ways, and technology only makes this more obvious. At the same time, digital equality becomes possible by the world being more connected electronically, allowing more perspectives and voices to be heard online.

Authenticity and anonymity: Presenting cultural selves in virtual spaces

Some scholars suggest that online identities on SNSs may be authentic (Chester & Bretherton, 2007), authentic, ironic, or "fakester" (Marwick, 2005), but others argue that self-profiles on SNSs can never be considered real (boyd, 2008). Students' experiences have revealed both the authentic and anonymous nature of identities on SNSs. Because of the lack of ascription in online identity negotiation, some communicators exercise the freedom to express their cultural identities the way they want. They manage and present who they are through self-description. Elian, one of the students in this study, shares how anonymity makes it possible to skew one's identity on Facebook:

> Your identity online can be whatever you want to make of it. There is a certain amount of personality that comes across when using online communication, but unlike face-to-face communication, you can distort who you really are. . . . For

> example, my friend made this fake girl on Facebook and became friends with most of my social group. He then started to ask some of my friends out and actually set my friend Brian up and we all showed up at the bar and had a good laugh when he realized the situation.

A phenomenological analysis of students' online experiences shows that the expression of online identity is greatly influenced by the specific site where the communication takes place (SNSs, discussion forums, e-mail, instant messages). The dialectic between authenticity and anonymity exists, depending on whether communication is conducted in a public virtual space or a private one and the particular medium it employs. Tim describes how authenticity of identity online is determined by the site and viewers:

> A few years ago when Myspace first came out, my parents found my little stepbrothers' pages. They were full of drug representations and alcohol pictures, although my brothers were not even in high school (and raised in a Mormon household). I think while it may give more information to determine who a person is, it cannot be used solely to judge, without taking into consideration who the page is meant to be viewed by.

In a more public virtual space, because of the lack of clear indications of who the viewers are, people have greater liberty in creating and experimenting with their own and others' identities. In a more person-to-person–based communication system such as e-mail or instant messaging, communicators tend to be more careful with the messages they compose, and the self-presentation is more authentic.

The analysis of these students' online experiences suggest that cultural identities online are real and anonymous, clear and ambiguous; they vary with the specific cultural space(s) that one visits. The dialectic of authenticity and anonymity reveals its presence in online identity expression through variations in the public or private space of a site, the potential viewers of a message, the specific technological tool one uses, and the cross-cultural differences for using the same medium.

Liberty and constrains: Experiencing a new space created by asynchronicity

A phenomenological analysis of the students' experiences reveals that the asynchronicity of online interactions creates a new space that allows them liberty in expressing, negotiating, and recreating their identities. Variations in whether the communication is conducted synchronously or simultaneously, between friends or strangers, or through the medium of a SNS or an e-mail all are associated with the dialectic of liberty and tensions in expressing one's cultural self online.

Asynchronicity seemed to allow these students to construct, edit, and change their identities within a larger time frame. For example, Harry thinks he has more control and freedom in presenting his cultural self in this online communication context:

> In online communication, I can control how much I can reveal of myself to my conversation partner. Since he cannot see me, he does not know my appearance, such as skin color, height, weight, gender, race, and ethnicity. . . . I can tell him that I am outgoing, nice, funny, but others might see me in a different way. . . . My conversation partner, before he e-mailed me, might have thought that I was an average white college student. In reality, I am of Latino/Hispanic origin.

At the same time that communication technologies make it easier to represent who one is online, they also make it easier to alter one's identities by editing, forwarding, and reposting messages. In fact, "Online identities are unstable and easy to alter by both original posters and by reposters of online messages" (St. Amant, 2002, p. 199). Internet users can separate messages from original posters or change what others said. The asynchronicity of interactions and invisibility of physical looking online blurs cultural differences to an extent.

The time difference in asynchronous online interactions not only allows communicators more opportunities to construct and edit their identities but can also create a comfortable space for interaction to occur. Another student, Aaron, prefers e-mail to FtF interaction with his parents because he finds that "text messaging or e-mailing my parents tends to keep them at a comfortable distance away from my life, while still being able to communicate in a somewhat effective way." However, the space created by time differences in online communication may also lead to confusion. Individuals in some cultural groups allow a longer time span for e-mail replies, whereas others may expect a prompt reply. Jessica is an international student from South Korea. Educated mostly in the US, Jessica is very prompt in her e-mail replies and not comfortable with her Korean friends' slower style. According to Jessica, "Most of my friends in Korea are very slow in their response. I expect them to reply within a week, sometimes two weeks or three weeks. That would be the quickest . . . [but] for American friends, a shorter time span." Differences in frequency and time spans of e-mail correspondence produce multiple interpretations: differing orientations to time, lack of skills on the computer, variable access to the Internet, and individual carelessness. This time difference could be confusing to communicators who are not from the same culture but is an important space for expressing cultural and individual identities.

The space created by the lapse of time reduces language barriers, minimizes cultural differences, makes it possible to experiment with sociocultural identities, and allows a comfortable distance for intercultural interactions.

At the same time, this space is likely to cause misunderstandings, as it is a culturally specific place to communicate who we are during online interactions.

Stability and fluidity: Moving in and out of postmodern online spaces

Online communicators have the freedom of moving in and out of the dynamic, fluid, tenuous, and temporal spaces created by asynchronicity. These virtual spaces are defined by cultural practices, languages spoken, and identities constructed. They only exist when we visit them. Online communicators experience tensions between authenticity and anonymity, stability and fluidity, and permanency and temporality because they are constantly moving in and out of these postmodern spaces in negotiating and expressing their cultural identities.

Students have illustrated how they express and learn about each others' cultural identities through e-mails, chat rooms, online forums, and SNSs. The specific site that students visit online has greatly influenced the stability and fluidity of their identities. According to Teresa and Jeffrey, most participants in chat rooms and online forums are anonymous; however, when using Facebook to communicate with acquaintances, they usually make a more specific and real presentation of themselves. As Jacob states,

> The creation of sites such as Myspace, Facebook, YouTube, etc., allows individuals to express their identity more specifically. While you can still learn a person's labels (i.e., age, religion, orientation, etc.), you can further learn about different aspects of that person's identity. For instance, pictures can communicate an individual's age identity, race identity, gender identity, family identity, religious identity, etc., all without any words.

The online identities of students vary with the site they visit, their relationship with other participants, and their purposes for communicating. For example, Patricia's friend in the Czech Republic has hundreds of friend connections on her Facebook account, demonstrating her popularity and social attractiveness. She uses Facebook to look for a boyfriend and gives permission to everyone to become her friend. Her identity management and friendship networks are different from Patricia's, who only uses Facebook to connect with those whom she has already met in person.

The space created by asynchronicity could be big, small, or even nonexistent. When people use online chat programs, such as MSN or Internet Relay Chat, to communicate, the interaction is almost spontaneous; there is little space for editing and organizing what one wants to say. Culture interacts with our construction of identities online as people move through these multiple cultural spaces. Our cultural selves are revealed by the way we speak and the behaviors we perform. How we represent our cultural selves is closely associated with who we are in real life. Many of the existing stereotypes associated with offline identities are reproduced in online context. Ahmad, an international student from Pakistan, shares his experience online:

> I probably have more of a unique perspective on how cultural and bias effect people online because I run an active technology blog. While yet to face an actual racial slur, at times I am the victim of name calling. Now I can't say this is because of my race or because my non-white name. But it does make me wonder if I published under a name like "John" if I would be treated the same way.

CMC provides a multiplicity of cultural spaces for participants to develop and express their cultural selves. The dynamic and fluid nature of these postmodern cultural spaces creates opportunities to construct, reconstruct, and experiment with one's identities. The unique context of cyberspace gives us more power but also causes tensions in expressing and negotiating identities on the Internet, making our online identities more contextual and fluid.

Trust and suspicion: Writing online identities with voices

CMC, in some cases, has reduced our interactions to the use of symbols. The use of linguistic symbols and special cues facilitate the expression of our cultural selves, but at the same time, it can also hinder the way we express who we are, thus leading to the dialectic between trust and suspicion toward people's online identities.

In online contexts, the language we use and the way we use it are largely related to our cultural upbringing. Native and nonnative English speakers, and speakers of different varieties of English, for example, all represent their cultural backgrounds through language. What is more, special cues, emoticons, and abbreviations are all closely associated with the culture of the communicators. Certain terms and special cues are used in online communication sometimes to include or exclude each other and negotiate group memberships and relationships, as illustrated in Lisa's experience:

> I believe that the amount of identities that show online is important when it comes to shared identities between people who communicate frequently online. For example, I usually use more Yiddish phrases only while conversing with Jewish peers online, which I never noticed until now. While I'm no closer to my Jewish friends than others, I realize that adding in an "oy vey" here and there allows us to share in that Jewish identity and create a bond that is difficult to maintain only in online situations.

Reliance on symbols in expressing cultural selves causes tensions between trust and suspicion. Symbols facilitate identity presentation but also cause problems when there is no agreement on the meanings they convey. In online conversations, labels are especially problematic because communicators tend to categorize people based on the labels. As Jacob shares:

> I learned that my partner was Muslim, but that's it. I don't have the nonverbal cues that would communicate to me information not obtained from what he told me orally. So it can present a problem, for instance, if Aziz told me he was Middle Eastern and I associated him as a terrorist. A person would be less likely to do that while communicating in person.

A White person, on the other hand, might be considered arrogant simply based on expressing his or her ethnicity online. According to Jade, "I've always found that it's a little difficult to express pride in your background, especially when you're from a majority group (White ethnicity, for example) without offending others who don't share your background. This is especially noticeable online."

Because online identities are communicated mainly through the use of symbols, most of the interviewees seem both suspicious and trusting toward others' online identities. As Lena illustrates, it is "difficult to fully trust that what they tell you about themselves to be true." Teresa experiences a lower level of trust in chat rooms. Because no one can verify the reality of the conversation partner in chat rooms, it is easy to polish, overstate, or fake one's identity in these places. Regarding Facebook and e-mails, Patricia has a high level of trust. Anyone can see her profile with her permission, view her pictures, converse with her, and learn her age and life experiences. E-mail is a place in which Patricia has a higher level of trust: "If I have not seen a person before, if it's on e-mail, I tend to trust him/her more."

A vast amount of information is communicated through the use of symbols in telling others who we are in CMC. Because we may not know the backgrounds of those culturally different others before we begin to interact with them, this may blur cultural differences but reinforce the stereotypes that have already existed in the off-line world. We experience both trust and suspicion when we travel through different online spaces and talk with different people in constructing our cultural identities.

Conclusion

This current study has recognized the fluidity and complexity of identity presentation and engaged questions of context and power in studying students' expression of their cultural selves online. The descriptive accounts of the students and themes that emerged from their lived experiences have revealed the significance of contexts in online identity presentation. The physical and social aspects of the situation impact how one presents one's cultural self and the degree to which it is connected to offline identities. Students tend to construct a more authentic profile when the site is comparatively private, such as on Facebook, but experience more liberty in experimenting with their identities in other online venues such as chat rooms. Cultural spaces have different meanings depending on their privacy/publicity, languages, communicators, viewers, and goals of communication. This leads to varying degrees of authenticity and anonymity, liberty and tensions, stability and fluidity, and trust and suspicion in online identity expression.

The Internet seems to create equal opportunities for identity expression, but cyberspace is not a utopia, devoid of social and political turbulence (Gajjala, 2004; Mehra et al., 2004). The interpretation of students' online experiences has revealed the resistance of both dominant and marginalized groups to certain cultural labels. The Internet can empower minority and marginalized users to overcome language barriers and may lead to prejudice reduction, but at the same time, students of marginalized groups encounter stereotypes because of the information they provide about their names, places of origin, religion, and pictures. Many demographic groups are excluded or underrepresented and experience bias in cyberspace due to their lack of skills, as well as lack of material, mental, and usage access to newer communication technologies.

Online identity expression is related to the historical, social, and political structures under which communication occurs. The historical conflicts between the dominant and marginalized, the West and East, the colonizer and colonized, and the developed and developing countries have led to different ways in which individuals negotiate and express their cultural selves online. For instance, recent events in the US related to fighting terrorism may cause students of the Muslim faith and Arab culture to encounter more name calling and negative stereotypes. Ideological differences between China and the West have led to heated discussions of the image of China and the Chinese online. How an individual expresses his or her cultural self is closely related to the past, present, and future, and also to the physical, social, and historical circumstances of the situation in which the communication takes place.

A growing corpus of the scholarship concerning SNSs has focused on impression management and friendship connections, self-profiles, and privacy issues, but identity presentation remains an area for more scholarly investigation. This study fills the void by employing the dialectical perspective in investigating the processual, relational, and contradictory nature of online identity expression. Phenomenological analysis of students' experiences has supported and extended the dialectical framework in identifying additional dialectics that are unique to the expression of identities in online communication. They are identified here as digital equality–marginalization, authenticity–anonymity, liberty–tensions, stability–fluidity, and trust–suspicion dialectics. These dialectics avoid essentializing sociocultural identities as fixed categories and regarding national culture as the most important organizing frame in intercultural online communication. They are interconnected with one another, helping us see the complexity of online identity expression as a dynamic and ongoing process from interpretive and critical perspectives.

References

Abbas, Y., & Dervin, F. (2009). Introduction. In Y. Abbas & F. Dervin (Eds.), *Digital technologies of the self* (pp. 1–11). Newcastle, United Kingdom: Cambridge Scholars.

Baron, N. (2004). See you online: Gender issues in college student use of instant messaging. *Journal of Language and Social Psychology, 23*(4), 397–423.

boyd, d. (2008). None of this is real. In J. Karaganis (Ed.), *Structures of participation in digital culture* (pp. 132–157). New York, NY: Social Science Research Council.

boyd, d., & Ellison, N. B. (2008). Social network sites: Definition, history, and scholarship. *Journal of Computer-Mediated Communication, 13,* 210–230.

Cassell, J., & Tversky, D. (2005). The language of online intercultural community formation. *Journal of Computer-Mediated Communication, 10*(2), article 2.

Cools, C. A. (2006). Relational communication in intercultural couples. *Language and Intercultural Communication, 6*(3,4), 262–274.

Chester, A., & Bretherton, D. (2007). Impression management and identity online. In A. N. Joinson, K. Y. A. McKenna, T. Postmes, & U.-D. Reips (Eds.), *The Oxford handbook of Internet psychology* (pp. 223–236). New York, NY: Oxford University Press.

Dijk, J. van. (2004). Divides in succession: Possession, skills, and use of new media for societal participation. In E. P. Bucy & J. E. Newhagen (Eds.), *Media access: Social and psychological dimensions of new technology use* (pp. 233–254). Mahwah, NJ: Lawrence Erlbaum Associates.

Donath, J. (1999). Identity and deception in the virtual community. In M. Smith & P. Kollock (Eds.), *Communities in cyberspace* (pp. 27–58). London, United Kingdom: Routledge.

Fong, M., & Chuang, R. (2004). *Communicating ethnic and cultural identity.* Lanham, MD: Rowman & Littlefield.

Gajjala, R. (2004). Negotiating cyberspace/Negotiating RL. In A. Gonzales, M. Houston, & V. Chen (Eds.) *Our voices* (pp. 82–91). Los Angeles, CA: Roxbury.

Holmes, P., & O'Halloran, S. (2003). Communicating across cultures on the Internet: Implications for global(ising) e-education. *Australian Journal of Communication, 30*(2), 65–84.

Manen, M. van. (2003). *Researching lived experience: Human science for an action sensitive pedagogy* (2nd ed.). Ontario, Canada: The Althouse Press.

Martin, J. N., & Nakayama, T. K. (1999). Thinking dialectically about culture and communication. *Communication Theory, 9,* 1–25.

Martin, J. N., & Nakayama, T. K. (2010). Intercultural communication dialectics revisited. In R. T. Halualani & T. K. Nakayama (Eds.), *The handbook of critical intercultural communication* (pp. 59–83). Malden, MA: Blackwell.

Martinez, J. (2000). *Phenomenology of Chicana experience and identity.* Lanham, MD: Rowman & Littlefield.

Martinez, J. (2006). Semiotic phenomenology and intercultural communication scholarship: Meeting the challenge of racial, ethnic, and cultural difference. *Western Journal of Communication, 70*(4), 292–310.

Marwick, A. (2005). "I am a lot more interesting than a Friendster profile": Identity presentation, authenticity, and power in social networking services. Paper presented at the Association of Internet Researchers 6.0 conference, Chicago, IL. Retrieved from http://microsoft.academia.edu/AliceMarwick/Papers/400480/IMa_Lot_More_Interesting_Than_a_Friendster_Profile_Identity_Presentation_Authenticity_and_Power_In_Social_Networking_Services

McPhail, T. (2006). *Global communication: Theories, stakeholders and trends* (2nd ed.). Malden, MA: Wiley-Blackwell.

Mehra, B., Merkel, C., & Bishop, A. P. (2004). The Internet for empowerment of minority and marginalized users. *New Media and Society, 6,* 781–802.

Orbe, M. P. (2008). Theorizing multidimensional identity negotiation: Reflections on the lived experiences of first-generation college students. In M. Azmitia, M. Syed, & K. Radmacher (Eds.), *The intersections of personal and social identities: New directions for child and adolescent development* (pp. 81–95). New York, NY: Jossey-Bass.

Schmidt, W. V., Conaway, R. N., Easton, S. S., & Wardrope, W. J. (2007). *Communicating globally: Intercultural communication and international business.* Thousand Oaks, CA: Sage.

St. Amant, K. (2002). When cultures and computers collide: Rethinking computer-mediated communication according to international and intercultural communication expectations. *Journal of Business and Technical Communication, 16*(2), 196–214.

Turkle, S. (1995) *Life on the screen: Identity in the age of the Internet.* New York, NY: Simon & Schuster.

Ye, J. (2006). An examination of acculturative stress, interpersonal social support, and use of online ethnic social groups among Chinese international students. *Howard Journal of Communications, 17*(1), 1–20.

Zywica, J. & Danowski, J. (2008). The faces of Facebookers: Investigating social enhancement and social compensation hypotheses: Predicting Facebook and offline popularity from sociability and self-esteem, and mapping the meaning of popularity with semantic networks. *Journal of Computer-Mediated Communication, 14,* 1–34.

8. New Media and Asymmetry in Cultural Identity Negotiation

GUO-MING CHEN, PHD
XIAODONG DAI, PHD

Introduction

The advent of a new era of telecommunications and human interconnection brings with it questions of identity, community, the place of the individual in a globalizing society, the connectedness that stems from the information superhighway, and the possibility that microcultures will soon be dissolved in a diffuse macroculture (Berger & Huntington, 2003; Pavlik & McIntosh, 2010). Among these trends, the impact of the emergence of global new media on cultural identity is an issue of utmost concern among intercultural communication scholars (Chiang, 2010; Collier, 2000; Halualani, 2008; Koc, 2006). This chapter examines the relationship between new media and asymmetry in cultural identity negotiation. The first section examines the fundamental relationship between new media and cultural identity, including how the integration of new media and globalization challenges the stability and autonomy of cultural identity by establishing a variety of new communities and social groups. The second section examines the relationship between new media and intercultural communication asymmetry, focusing on how new media and global power imbalances lead to asymmetry in cultural identity negotiation. In conclusion, strategies for managing this asymmetry are discussed.

New media and cultural identity

Integration of new media and globalization

In the last two decades, new media has gained formidable momentum in light of its functional innovations. New media integrates interpersonal media and mass media. Embedded in the context of globalization, it has inherited the strength of traditional media and developed its own features. New media lends great force to globalization, and globalization provides it with a huge platform to bring its potential into full play. As such, human communication has undergone significant changes that are in turn associated with changes in cultural identity.

The acceleration of globalization draws from the features of new media, including digitality, convergence, interactivity, hypertextuality, and virtuality (Flew, 2010; Guan, 2006). Digital media allows users to compress, retrieve, and manipulate a large quantity of data, giving rise to a different system of media production and marketing. New media converge the forms and functions of different media, which results in, for example, mergers in the media industries. Moreover, the interactivity and hypertextuality of new media provide users with resources within the interconnected nodes of the network and a space to develop virtual experience and reality.

These features impact human society in the global context, as the Internet has blurred the line between mass and interpersonal communication among people of different cultural groups. Through the transformation of physical and social settings, it has created a global town square in which people can project their self-image and construct new reality through the free expression of opinions and the redefinition of time and space. New communities for members of culturally diverse groups can also be established (Ellison, Steinfield, & Lampe, 2007; Murata, 2010). New media further fuels globalization's dynamism. Former boundaries of human societies, in terms of geography and cultural beliefs, have evolved into new patterns of interconnectedness because of the dialectical interaction between localization and globalization and between cultural identity and cultural diversity. Thus, the interrelationship and interreliance become the knots of this world's Web, not only in the global community but also in various independent but interconnected virtual communities. This shows how new media simultaneously shapes globalization into a process of differentiation and homogenization.

The freedom, flexibility, creativity, and innovation new media and globalization provide in physical- and cyberspace challenge the traditional emphasis on order, norms, organization and intellectual autocracy (Barry, 2001) as well as power distribution, human consciousness, and cultural patterns in human society (Castells, 2000). The integration of new media and globalization may improve human relationships (Zhao, Grasmuck, & Martin, 2008) or influence community identity, the establishment of civil society, and the enactment of global citizenship (Boulding, 1998; Jones, 1995). As Lister, Dovery, Giddings, Grant, and Kelly (2009) point out, the integration of new media and globalization brings people with diverse experiences together; provides new ways of representing the world in virtual space, new experiences, and new relationships between embodiment, identity, and community at both local and global levels; provides new challenges of differentiating reality from virtuality; and provides new patterns of organization and production. The next section further discusses the impact of the integration of new media and globalization on cultural identity and how this leads to asymmetry in intercultural communication—reflected in the inharmonious relationship between the West and the rest of the world (due to Western domination, cultural appropriation, and hybridization).

Impact on cultural identity

Cultural identity is the consequence of the self-internalization and the recognition of in group/out group members. In the new communities formed by the global new media, in which participants can flexibly adjust or remove individual and group boundaries, identity turns into a fluid, dynamic, and relativized phenomenon. As a result, it has transcended the traditional perception of identity and brought forth more opportunities for the construction, reconstruction, and negotiation of cultural identities (Tan, 2005).

Cultural identity is formed through the interaction between the self and the affiliated group. The intensity of cultural identity increases when one's beliefs are consistent with the experiences in a community. According to Collier (2000), cultural identity is presented through core symbols and norms and changes over time with multifaceted forms. It is enduring because of the involvement of affective, cognitive, and behavioral elements. It is changing because in different contexts the salience and intensity varies. Together, the nature of cultural identity demonstrates three characteristics embedded in the elements of coherence and consistency: temporality, territoriality, and interactionality (Belay, 1996; Preston, 1997).

Temporality stipulates that cultural identity is a product of historical development sustained by time, lifestyle, shared memory, and tradition. The continuity of cultural identity can only be achieved through the geographical or territorial boundary that distinguishes the in-group from the out-group members or from the "we" and the "they" (Morley & Robins, 1995). Moreover, the formation and construction of cultural identity is attained through the process of social interaction, which is conducted on the basis of individual or local perspective and can only be observed at the behavioral level. These characteristics of cultural identity construct the accepted notions of community. However, this traditional meaning of community and collective identity has been shaken by the integration of new media and globalization (Kraidy, 2002; Tenenboim-Weinblatt, 2010).

The global connectedness of new media due to *time–space compression* has changed the way we represent the world to ourselves and therefore the way we construct and reconstruct cultural identity (Harvey, 1989). Global new media have created a new *time–space relation*. Through the compression of time and space, these media produce a virtual space in which the costs of communication are significantly reduced and people from all the corners of the world are able to join the interaction, as long as they are digitally connected. The accessibility of this new mediated global interaction provides "numerous opportunities for people to form a virtual community, [and] a new cultural identity is then emerging and the old cultural identity is challenged" (Chen & Zhang, 2010, p. 810). Slater (1998) claims that this cyberspace develops a virtual community in which people practically deconstruct essentialism, because all things, including body and identity, lose their territorial connection and new opportunities are expanded.

Although the emergence of new collective identities at different levels of cyberspace—including local, global, and those in between—still require a sense of belonging to the electronic community, the community lacks territory or physical place, social system, and face-to-face (FtF) interaction (Maheswaran, Ali, Ozguven, & Lord, 2010). Beniger (1987) labels this electronic community a *pseudocommunity*, in contrast to the traditional *organic community*. The pseudocommunity is based on impersonal association in a space of physical and geographical separation, whereas the organic community is based on FtF interpersonal interaction in a confined physical place.

By changing the concept of territorially confined cultural identity, the new identity formed in a virtual community simultaneously reinterprets the past and the future development of the community. The *socially produced space* online is not only different in terms of the interpretation and narrative of virtual experiences but also in terms of mobility in cyberspace (Soja, 1989). Mobility refers to the movement without real physical travel that is possible in a virtual community and the constant change of character, role, status, and class (Jones, 1995; Wood & Smith, 2005). In other words, the integration of new media and globalization facilitates a de-territorialization that diminishes the traditional relationship between culture and geographical/social territories and gives us a sense that we can start over from the beginning through the interaction in the cyberspace (García-Canclini, 1995; Tomlinson, 1999). Bell (2001) contends that the freedom to construct, reconstruct, and negotiate one's identity in virtual community may lead to the decline of real-world sociality and in turn the demise of the real life and the loss of cultural identity. Bell's argument seems to exaggerate the power of new media and underestimates people's ability to maintain their sense of space. Moreover, online and offline identities might not be as different as predicted (Chua, Madej, & Wellman, in press; Huffaker & Calvert, 2006; Slater, 2002; Wang, Walther, & Hancock, 2009), but her argument reminds us of the significant implications of the virtual world for our sense of cultural belonging.

Although virtual groups do not constitute real communities, the virtual community provides a space where people can meet to share information and carry out various transactions in a different way. This mutual learning process facilitates the adoption of a flexible viewpoint, through which people can see themselves with new eyes through other participants' responses. Thus, people can potentially cultivate a collaborative and participatory culture by fostering a socially constructed and coconstructed identity founded on the creative production in the digital world (Weber & Mitchell, 2008). Whereas some scholars hold a positive view about the impact of digital community shaped by the integration of new media and globalization, other scholars are critical and pessimistic of the new virtual world (Nederveen Pieterse, 2009; Shaw, 1997). Another macrolevel concern is the problem of intercultural communication asymmetry caused by the integration of new media and globalization.

The push of new media and globalization has allowed cultural appropriation and hybridization to become mundane and routinized in the global context, which induces asymmetry in intercultural interaction (Kraidy & Murphy, 2008; Rogers, 2006). In general, Western culture penetrates more into non-Western cultures, and non-Western cultures have to deal with more challenges in identity reconstruction and self-transformation. In most cases, non-Western cultures are struggling for recognition and self-definition (Friedman, 1994; Tomlinson, 1999). This situation often makes identity negotiation problematic and difficult or even leads to intercultural conflict. The next section analyzes how asymmetry in intercultural communication, accelerated by the integration of new media and globalization, affects the cultural identity negotiation process and proposes strategies for managing the tensions caused by the asymmetry in cultural identity negotiation.

New media and asymmetry in intercultural communication

Identity negotiation is "a transactional process whereby individuals in an intercultural situation attempt to assert, define, modify, challenge, and/or support their own and others' desired self-images" (Ting-Toomey, 1999, p. 40). This process unfolds in two dialectical directions: convergence and divergence (Brewer, 1999). In intercultural communication, people want to be recognized and included, but they also want to be autonomous and differentiated, so that both self and mutual identification can be achieved (Dougherty, Mobley, & Smith, 2010). If people form a close attachment to similarity and adopt a convergence strategy without differentiating themselves from cultural others, the result will be either overaccommodation or assimilation. Neither overaccommodation nor assimilation will produce reciprocal relationships, because the former makes people feel that they are treated as inferior or incompetent, and the latter makes people feel that they are suppressed (Dai, 2010; Gallois,

Ogay, & Giles, 2005). New media help us transcend physical barriers to intercultural communication, but this by no means implies that cultural identity negotiation is free from geopolitics. There is a distinct *power geometry* to globalization in which some people are more advantaged than others and the dominant party always determines the topic, the orientation, and the form of social interaction (Hermans, 2001; Massey, 1994). This asymmetrical power structure in cultural identity negotiation is typically reflected in the domination of the West over the non-West.

Asymmetry in cultural identity negotiation

New media seem to offer both the West and the non-West equal opportunities to represent their cultures and facilitate mutual influences, but in reality mediated interactions work more to the West's advantage. With the help of new media, cultural events and images of the non-Western world become better known to Westerners, yet the wide dissemination of Western ideas and institutions continues to strengthen the West's dominance (Giddens, 1990). More specific, Western domination as a global reality inevitably leads to power asymmetry in cultural identity negotiation, which is reflected in four aspects: (a) the domination of Western modernity; (b) the existence of a digital divide, linguistic hegemony, and imbalanced representation; (c) the advantageous position of the West in cultural appropriation and hybridization; and (d) more challenges to non–Westerners in terms of self-transformation and identity reconstructions.

Although there has been progress with the introduction of new media to the non-West, there are several historical reasons for the West's continued dominance: First, since the Enlightenment, modernity has developed rapidly in the West and spread to other parts of the world with the expansion of the capitalist system. The core values of modernity, such as the cult of reason and science, the faith in individuality, and human freedom and democracy have been widely embraced. Although modernity is resisted by some cultures due to its threat to their traditions and its alienating and suppressive forces on humanity, it has long been legitimized and constituted as the metacontext in which all cultural values are assessed and re-evaluated. Modernity exerts so pervasive an influence on cultures and societies that it has become one of the most essential driving forces of globalization (Robertson, 1992).

Second, new media first appeared in the West as a direct product of the telecommunications revolution. Western countries not only have a technological edge over non-Western nations but also have greater access to digital resources. Of late, new media has spread quickly to the non-Western world, but a digital divide continues to produce unequal intercultural communication. In addition, linguistic hegemony lends force to Western domination.

European languages, especially English, are more used than non-European languages in the new media. According to Graddol (2006), English is the most used Internet language, followed by Chinese, Japanese, German, Spanish, French, and Italian, and 32% of Web pages use English as the lingua franca. Chinese and Japanese sites have a large readership, but these readers are mostly users in China and Japan and have far fewer international users than German, French, or Italian sites. Thus, the digital divide and linguistic hegemony lead to imbalanced communications. Non-Western events, values, and images are less represented, and when they are, it is done in a more negative way. As Kraidy (2002) points out, Western technology is often depicted as an instrument that undermines censorship in non-Western countries and also as a "fetish of Western slickness, modernity, and creativity to which foreign audiences aspire" (p. 325).

Third, due to the domination of modernity, Western countries are in a relatively more advantageous position in relation to cultural appropriation and hybridization. Cultural appropriation refers to the use of culture's symbols, artifacts, genres, rituals, or technologies by members of another culture, which is inescapable when cultures come into contact either in physical or in virtual space. It takes diverse forms such as reciprocal exchange, unidirectional domination, unjustified exploitation, or productive transculturation; it is a two-way borrowing and lending between cultures (Rogers, 2006). Reciprocal, balanced, and voluntary exchanges usually take place between cultures with evenly distributed powers. The process involves the two-way flow of cultural ideas, symbols and practices, resulting in mutual influence and transformation. Nevertheless, when powers are unevenly distributed, two undesirable situations may appear: The powerful culture may impose its values on the less powerful and produce cultural domination, or the dominant culture may intentionally or unconsciously use the dominated to serve its own interests without any financial compensation through the process of cultural exploitation. Power manifests itself in politics and everyday communication practices that express the interests of individual and groups (Mumby, 2001). The hegemonic influence of Western media in intercultural communication demonstrates that the "appeal of Western media products is already structured in power" and that "the political implication of acts of appropriation are not entirely determined by the intentions, motivations, and interests of the subordinated cultures doing the appropriating or by the dominant cultures that are imposing their media products" (Rogers, 2006, p. 483).

In cultural hybridization, Western countries are also typically better off than non–Western states. *Cultural hybridization* refers to the mixing, fusion, intermingling, and combination of the cultural elements of various cultures, which can be intentional and unintentional. It reflects global cultural conditions and highlights the growing intercultural connectivity. However, hybridity is not a power-free space where communicators equally interact with each other and negotiate their cultural identities. Whereas Westerners apply hybridization to enrich, innovate their culture, and establish hegemony, non-Westerners employ hybridization as a way to survive, develop, and gain world recognition. One of the most serious problems arising from hybridization is that the core values of the West penetrate into non-Western cultures and pose a formidable threat to their traditions and cultural identities. In contrast, though Asian technological products like video games are rising in popularity, non-Western cultures rarely affect the West or constitute any substantial threat to Western value orientation and cultural identities. Huntington (1996) claims that Islam and Confucianism might become a serious threat to Western civilization, but Islamic or Confucian values have only slowly penetrated into the mainstream culture of Western societies. On the other hand, the core values of Western culture such as rationality, liberty, individuality, and democracy, have gained foothold in China, Arabic nations, and many non-Western countries, threatening their cultural traditions. In this sense, hybridity is "an old functionalist notion of what a dominant culture permits in the interest of maintaining its own equilibrium" (Chow, 1993, p. 35).

Last, the asymmetry of intercultural communication produces more challenges to non-Westerners when transforming their traditions and reconstructing their cultural identities. Identity construction starts from self-definition and is accomplished in mutual recognition. It is unfortunate that non-Western countries have fewer chances to articulate their opinions and define themselves than their Western counterparts. Moreover, in transforming their traditions, non-Westerners face a perplexing dilemma: to incorporate modern Western ideas denotes both empowerment *and* marginalization. The domination of the West breeds Eurocentrism based on the attitude of Western triumphalism, according to which Eurocentrists tend to celebrate individuality, assume that only Europeans have the right to define the reality, and promote the European idea as the only valid form of human societies (Asante, 2006).The problem of Eurocentrism has been intensively criticized by scholars from different continents and backgrounds (Chen, 2006; Gordon, 2007; Ishii, 2007; Miike, 2006, 2007). With the spread of modernity the power and presence of some non-Western countries have risen dramatically but most still have difficulty in convincing Westerners of the legitimacy of their culture. More often than not, non-Western countries struggle for recognition, through adapting to Western standards or joining the West. They either overaccommodate or assimilate, leading to troubles in differentiation in the process of identity negotiation. Non-Western cultural traits are often tolerated and used to exemplify a token Western effort to preserve cultural diversity or are seen as alien to the modernized West.

The non-Western dilemma in self-definition and cultural transformation, along with Westerners' insensitivity to their difficulties, make intercultural identity negotiation problematic. In order to combat the domination of Western media and cultural hegemony, non-Westerners tend to adopt a strategy of cultural resistance. State censorship, Web site supervision, and ideological and moral attacks therefore become the most frequently used instruments for maintaining their own cultural particularities. Because human society has entered into the global age, the search for an effective way of managing the tension caused by the asymmetry of cultural identity negotiation becomes indispensable for developing a reciprocal relationship in order to coexist peacefully and productively. The next section proposes three strategies for reaching this goal.

Management of asymmetry in cultural identity negotiation

Constructive intercultural identity negotiation is embedded in identity validation, intercultural transformation, and identity extension (Dai, 2009). The dialectics of identity security and identity threat make up the basis of identity negotiation (Ting-Toomey, 2005). Only after identities are recognized and confirmed can interactants have a meaningful dialogue and negotiate intercultural agreements. An effective way to reach this consensus and construct mutually desired identities is through the process of intercultural adaptation. As an integral part of identity negotiation, intercultural transformation based on intercultural adaptation is a process through which participants change their cultural orientations, redefine their self-images, and enrich their cultural scripts. Moreover, in re-establishing cultural membership, communicators synthesize diverse cultural elements and extend their identities. When the two parties extend their cultural inventory and share more values and behaviors in common, it becomes easier for them to reach mutual understanding and negotiate intercultural agreements.

Although power asymmetry provides the powerful a better position and more choices in identity negotiation, it does not always guarantee their domination. By the same token, uneven power distribution places great pressure on the less powerful, but it does not rule out their agency. The less powerful can manage resources in such a way as to exert control over the more powerful in established power relationships (Giddens, 1984). Thus, the key to reciprocal communication lies in the appropriate management of the tension and turning it into creative dynamics. Here, we propose three strategies for managing the tension between the West and the non-West.

The first strategy is to develop communicative rationality to remove cultural prejudices. Culture originates from a specific physical and social space, where identity is maintained by shared values, experiences, memories, and patterned ways of life. Cultural identity not only expresses the sense of belonging but also functions to develop collective cohesion and differentiates in-groups from out-groups; hence, ethnocentrism arises easily in the process of intercultural communication. Ethnocentrism becomes a barrier when it prevents people from attempting to interpret others' points of view from their perspective. When power asymmetry emerges, ethnocentrism tends to nurture cultural prejudices. Prejudiced communicators often stigmatize others and underestimate the value of their cultures.

Although new media, especially social media such as Facebook, Twitter, and YouTube help people transcend physical distance for communicating and engaging with information easily, potentially promoting intercultural dialogue (Kaplan, 2010; Notafish, 2010), the historical and social gaps are far from being removed. Communicative rationality can help to remove cultural prejudices and facilitate two-way communication. Communicative rationality, advocated by Habermas (1984), emphasizes intersubjective/intercultural agreement and demands that communicators express themselves rationally—with truth, truthfulness, and rightness—and give reasons that others can accept when validity claims are challenged (Warnke, 2006). In other words, validity claims are open to criticism, and the effective way to reach intersubjective/intercultural consensus is not through imposition but through rational argumentation. Thus, in identity negotiation, both the West and the non-West should make true representations, through which cultural identities are treated as dynamic and open to criticism and transformation, whereby the intentions, interests or motivations of participants can be articulated.

The second strategy is to strengthen intercultural sensitivity and intercultural awareness. Intercultural sensitivity and intercultural awareness are conducive to identity recognition and validation. Intercultural sensitivity refers to a communicator's ability to develop positive emotional responses to cultural differences. It is a transformation process in which communicators' cultural orientation gradually changes from ethnocentrism to ethno-relativism (Bennett, 1986). When communicators achieve ethno-relativism, they are more open to cultural others and capable of appreciating diversity. Intercultural sensitivity motivates people to understand and acknowledge other people's needs, and translate emotions into actions. With the development of an intercultural orientation, they can truly engage in others' lives and cultivate intercultural awareness, which refers the understanding of cultural similarities and differences (Chen, 2010). Cultural awareness provides us with a cultural map that reduces situational ambiguity and uncertainty.

Strengthening intercultural sensitivity implies that both Westerners and non-Westerners have their old stereotypes removed and that they develop a positive attitude toward each other; increased intercultural awareness implies that both Westerners and non-Westerners learn more about the distinct cultural characteristics of their counterparts. After acquiring intercultural sensitivity and intercultural awareness, one's perspective is broadened, which leads to increased intercultural competence (Chen & Starosta, 1996). As Ting-Toomey (2005) indicates, the "competent identity negotiator emphasizes the importance of creatively integrating knowledge and positive attitudinal factors, and putting them into mindful practices in everyday intercultural interactions" (p. 226).

The last strategy is to develop interculturality to enhance reciprocal intercultural relations. Cultural identity negotiation takes place in physical or social spaces. Intersubjectivity constitutes the primary field in which people interact with each other and coproduce the meaning of life. Through a mutually shared language, self and other coordinate their actions, develop patterned behaviors and common values, and define their identities. When individuals venture into another culture and resocialize into a larger intercultural community, they negotiate intercultural agreement, produce interculturality, and develop intercultural identity (Kim, 2001). Whereas intersubjectivity refers to the interpersonal connection between individuals in the same society, interculturality refers to the complex connection between and among cultures whose members negotiate intercultural agreements and mutually acceptable identities. In the intercultural space, self and other are existentially separate and relationally asymmetrical, but they are assumed to be relatively independent partners who live in different worlds trying to develop mutuality and reciprocity (Dai, 2010). Interculturality is then a ground that lies between cultural universals and cultural particularities, in which intercultural communicators can meet each other in the full range of human relations. The space between self and other remains open in the sense that "each party recognizes his or her dependence upon the other, and each can allow the judgment of the respective other to be valid as an objection against oneself" (Honneth, 2003, p. 12).

Conclusion

As a space in which communicators from different cultures coexist, interculturality helps people harness intercultural tensions. Differences are legitimized, appreciated, and treated as dialogue promoters. Communicators with diverse cultural backgrounds can freely articulate their opinions and negotiate shared meanings and desired identities. With shared life experiences and the ability to assume the role of the other, interactants are capable of reaching mutual understandings, locating where problems lie, and negotiating workable solutions. Thus, intercultural tensions can be constructively managed and changed into creative dynamics for the development of reciprocal intercultural relationships.

References

Asante, M. (2006). The rhetoric of globalization: The Europeanisation of human ideas. *Journal of Multicultural Discourses, 1*(2), 152–158.

Barry, A. (2001). *Political machines: Governing a technological society.* London, United Kingdom: Athlone.

Belay, G. (1996). The (re)construction and negotiation of cultural identities in the age of globalization. In H. B. Mokros (Ed.), *Interaction & identity* (pp. 319–346). New Brunswick, NJ: Transaction.

Bell, D. (2001). *An introduction to cybercultures.* New York, NY: Routledge.

Beniger, J. (1987). Personalization of mass media and the growth of pseudo-community. *Communication Research, 14*(3), 352–371.

Bennett, M. J. (1986). A developmental approach to training for intercultural sensitivity. *International Journal of Intercultural Relations, 10*, 179–196.

Berger, P. L., & Huntington, S. P. (Eds.). (2003). *Many globalizations: Cultural diversity in the contemporary world.* New York, NY: Oxford University Press.

Boulding, E. (1988). *Building a global civic culture.* Syracuse, NY: Syracuse University Press.

Brewer, M. B. (1999). Multiple identities and identity transition: Implications for Hong Kong. *International Journal of Intercultural Relations, 23*, 187–197.

Castells, M. (2000). Materials for an exploratory theory of the network society. *British Journal of Sociology, 51*(1), 5–24.

Chen, G. M. (2006). Asian communication studies: What and where to now. *The Review of Communication, 6*(4), 295–311.

Chen, G. M. (2010). *A study of intercultural communication competence.* Hong Kong: China Review Academic.

Chen, G. M., & Starosta, W. J. (1996). Intercultural communication competence: A synthesis. *Communication Yearbook, 19*, 353–384.

Chen, G. M., & Zhang, K. (2010). New media and cultural identity in the global society. In R. Taiwo (Ed.), *Handbook of research on discourse behavior and digital communication: Language structures and social interaction* (pp. 801–815). Hershey, PA: Idea Group.

Chiang, C.-Y. (2010). Diasporic theorizing paradigm on cultural identity. *Intercultural Communication Studies, 19*(1), 29–46.

Chow, R. (1993). *Writing diaspora: Tactics of intervention in contemporary cultural studies.* Indianapolis: Indiana University Press.

Chua, V., Madej, J., & Wellman, B. (2011). Personal communities: The world according to me. In J. Scott & P. J. Carrington (Eds.), *The Sage handbook of social network analysis* (pp. 101–115). Thousand Oaks, CA: Sage.

Collier, M. J. (2000). Reconstructing cultural diversity in global relationships: Negotiating the borderlands. In G. M. Chen & W. J. Starosta (Eds.), *Communication and global society* (pp. 215–236). New York, NY: Peter Lang.

Dai, X.-D. (2009). Intercultural personhood and identity negotiation, *China Media Research, 5*(2), 1–12.

Dai, X.-D. (2010). Intersubjectivity and interculturality: A conceptual link. *China Media Research, 6*(1), 12–19.

Dougherty, D. S., Mobley, S. K., & Smith, S. E. (2010). Language convergence and meaning divergence: A theory of intercultural communication. *Journal of International and Intercultural Communication, 3*(2), 164–186.

Ellison, N. B., Steinfield, C., Lampe, C. (2007). The benefits of Facebook "Friends": Social capital and college students' use of online social network sites. *Journal of Computer-Mediated Communication, 12*, 1143–1168.

Flew, T. (2010). *New media*. New York, NY: Oxford University Press.

Friedman, J. (1994). *Cultural identity and global process*. London, United Kingdom: Sage.

Gallois, C., Ogay, T., & Giles, H. (2005). Communication accommodation theory: A look back and a look forward. In W. B. Gudykunst (Ed.), *Theorizing about intercultural communication* (pp. 121–148). Thousand Oaks, CA: Sage.

García-Canclini, N. (1995). *Hybrid cultures: Strategies for entering and leaving modernity*. Minneapolis: University of Minnesota Press.

Giddens, A. (1984). *The constitution of society*. Cambridge, United Kingdom: Polity.

Giddens, A. (1990). *The consequences of modernity*. Cambridge, United Kingdom: Polity.

Gordon, R. D. (2007). Beyond the failures of Western communication theory. *Journal of Multicultural Discourses, 2*(2), 89–107.

Graddol, D. (2006). *English next*. London, United Kingdom: British Council.

Guan, S. J. (2006). *International communication*. China: Beijing University Press.

Habermas, J. (1984). *The theory of communicative action* (Vol. 1). *Reason and rationalization of society*. Boston, MA: Beacon Press.

Halualani, R. T. (2008). "Where exactly is the Pacific?": Global migrations, diasporic movements, and intercultural communication. *Journal of International and Intercultural Communication, 1*, 3–22.

Harvey, D. (1989). *The condition of postmodernity*. Oxford, United Kingdom: Blackwell.

Hermans, H. J. M. (2001). The dialogical self: Toward a theory of personal and cultural positioning. *Culture & Psychology, 7*(3), 243–281.

Honneth, A. (2003). On the destructive power of the third: Gadamer and Heidegger's doctrine of intersubjectivity. *Philosophy & Social Criticism, 29*, 5–21.

Huffaker, D. A., & Calvert, S. L. (2006). Gender, identity, and language use in teenage blogs. *Journal of Computer-Mediated Communication, 10*(2), article 1.

Huntington, S. (1996). *The clash of civilizations and the remaking of the world order*. New York, NY: Simon and Schuster.

Ishii, S. (2007). A Western contention of Asia-Centred communication scholarship paradigms: A commentary on Gordon's paper. *Journal of Multicultural Discourses, 2*(2), 108–114.

Jones, S. G. (1995). Understanding community in the information age. In S. G. Jones (Ed.), *Cybersociety: Computer-mediated communication and community* (pp. 10–35). Thousand Oaks, CA: Sage.

Kaplan, A. M. (2010). Users of the world, unite! The challenges and opportunities of social media. *Business Horizons, 53*(1), 59–68.

Kim, Y. Y. (2001). *Becoming intercultural: An integrative theory of communication and cross-cultural adaptation*. Thousand Oaks, CA: Sage.

Koc, M. (2006). Cultural identity crisis in the age of globalization and technology. *The Turkish Online Journal of Educational Technology, 5*(1), 37–43.

Kraidy, M. M. (2002). Hybridity in cultural globalization. *Communication Theory, 12*, 316–339.

Kraidy, M. M., & Murphy, P. D. (2008). Shifting Geertz: Toward a theory of translocalism in global communication studies. *Communication Theory, 18*(3), 335–355.

Lister, N., Dovery, J., Giddings, S., Grant, I., & Kelly, K. (2009). *New media: A critical introduction*. New York, NY: Routledge.

Maheswaran, M., Ali, B., Ozguven, H., & Lord, J. (2010). Online identities and social networking. In B. Furht (Ed.), *Handbook of social network technologies and applications* (pp. 241–268). New York, NY: Springer.

Massey, D. (1994). *Space, place, and gender*. Cambridge, United Kingdom: Polity.

Miike, Y. (2006). Non-Western theory in Western research? An Asiacentric agenda for Asian communication studies. *Review of Communication, 6*(1,2), 4–31.

Miike, Y. (2007). An Asiacentric reflection on Eurocentric bias in communication theory. *Communication Monographs, 74*(2), 272–278.

Morley, D., & Robins, K. (1995). *Spaces of identity: Global media, electronic landscapes and cultural boundaries*. New York, NY: Routledge.

Mumby, D. K. (2001). Power and politics. In F. Jablin & L. Putnam (Eds.), *New handbook of organizational communication* (pp. 585–623). Thousand Oaks, CA: Sage.

Murata, T. (2010). Detecting communities in social networks. In B. Furht (Ed.), *Handbook of social network technologies and applications* (pp. 269–280). New York, NY: Springer.

Nederveen Pieterse, J. (2009). *Globalization and culture: Global mélange*. New York, NY: Rowman & Littlefield.

Notafish. (2010, October 16). How intercultural is social media? [Web log post]. Retrieved from http://blog.notanendive.org/post/2010/09/19/how-intercultural-is-social-media

Pavlik, J. V., & McIntosh, S. (2010). *Converging media: A new introduction to mass communication*. New York, NY: Oxford University Press.

Preston, P. W. (1997). *Political/cultural identity: Citizens and nations in a global era*. Thousand Oaks, CA: Sage.

Robertson, R. (1992). *Globalization: Social theory and global culture*. London, United Kingdom: Sage.

Rogers, A. R. (2006). From cultural exchange to transculturation: A review and reconceptualizaton of cultural appropriation. *Communication Theory, 16*, 474–503.

Shaw, M. (1997). The theoretical challenge of global society. In A. Sreberny-Mohammedi, D. Winseck, J. McKenna, & O. Boyd-Barrett (Eds.), *Media in global context: A reader* (pp. 27–36). London, United Kingdom: Arnold.

Slater, D. (1998). Trading sexpics on IRC: Embodiment and authenticity on the Internet. *Body & Society, 4*, 91–117.

Slater, D. (2002). Social relationships and identity online and offline. In L. A. Lievrouw & S. Livingstone (Eds.), *Handbook of new media: Social shaping and consequences of ICTs* (pp. 533–546). London, United Kingdom: Sage.

Soja, E. (1989). *Postmodern geographies: The reassertion of space in critical social theory.* London, United Kingdom: Verso.

Tan, S-H. (Ed.). (2005). *Challenging citizenship: Group membership and cultural identity in a global age.* Burlington, VT: Ashgate.

Tenenboim-Weinblatt, K. (2010). Producing identity: Introduction. In M. M. Karidy & K. Sender (Eds.), *The politics of reality television: Global perspectives* (pp. 15–17). New York, NY: Routledge.

Ting-Toomey, S. (1999). *Communicating across cultures.* New York, NY: Guilford Press.

Ting-Toomey, S. (2005). Identity negotiation theory. In W. B. Gudykunst (Ed.), *Theorizing about intercultural communication* (pp. 211–234). Thousand Oaks, CA: Sage.

Tomlinson, J. (1999). *Globalization and culture.* Chicago, IL: The University of Chicago Press.

Wang, Z., Walther, J. B., & Hancock, J. T. (2009). Social identification and interpersonal communication in computer-mediated communication: What you do versus who you are in virtual groups. *Human Communication Research, 35*(1), 59–85.

Warnke, G. (2006). Communicative rationality. In S. K. White (Ed.), *The Cambridge companion to Habermas* (pp.120–142). Beijing, China: SXD.

Weber, S., & Mitchell, C. (2008). Imaging, keyboarding, and posting identities: Young people and new media technologies. In D. Buckingham (Ed.), *Youth, identity, and digital media* (pp. 25–47). Boston, MA: The MIT Press.

Wood, A. F., & Smith, M. J. (2005). *Online communication: Linking technology, identity and culture.* Mahwah, NJ: Lawrence Erlbaum Associates.

Zhao, S., Grasmuck, S., & Martin, J. (2008). Identity construction on Facebook: Digital empowerment in anchored relationships. *Computers in Human Behavior, 24,* 1816–1836.

9. *Negotiating a New Identity Online and Off-Line: The HeartNET Experience*

DEBBIE RODAN, PHD
LYNSEY URIDGE, PHD
LELIA GREEN, PHD

Introduction

Once individuals are diagnosed by medical practioners with heart disease, they are given the identity of a "heart patient." Imposing the identity of heart patient challenges the individual's notion of self. The challenge comes from the threat to an identity that until that point in their lives might have appeared whole and stable. As result of having a heart event, heart patients are named and positioned in particular ways by others, and in this respect a heart patient identity operates in ways similar to other identities such as gender, race, and class. One reason for this positioning is the media's very narrow conception of a heart patient's identity, which is constructed as either being part of the tragically ill medical model or the superhuman recovered tri-athlete. HeartNET (www.heartnet.com.au/) is one of the few online communities that offers a range of differing identities to heart patients and the potential for a renewed idea of the self. As we reveal in this chapter, forming an identity as a new heart patient can often be reflected in a person's online interactions.

HeartNET is a community-based Web site that supports people living with the effects of heart disease: both the patients themselves and the families and friends who might wish to be involved. Operating since 2005, HeartNET is funded jointly by the National Heart Foundation of Australia (Western Australian Division) and the Australian Research Council. For new heart patients seeking out positive inputs and contacts, HeartNET can offer an opportunity to forge an acceptable and positive identity as a heart patient, at the same time as developing a digital identity as an online community participant. Often it is hard for "nonpatients" to provide the reassurance that is craved by a person who feels that their next heart event may be their last. For many newly diagnosed heart patients, an online community is the most accessible means of meeting people who understand their own particular situation—that is, people who have experienced real heart health challenges but who are still alive and, in many cases, thriving. This is particularly the case for remote area residents.

This chapter investigates how members of HeartNET negotiate their own "heart patient" identity and identify with the online identities of others. We begin with a brief discussion of the methodology. We then review the literature about online identity and explicate how HeartNET members form an online identity. To conclude, we offer some reflections on the global implications and applications—present and future—for identity formation in online community groups. This Inquiry will include the voices of heart patients who are members of or who have accessed the HeartNET online community.

Methodology

HeartNET was established in 2005 as a research-driven online community that would allow an evaluation of the benefits, if any, of peer-based online support for recovering heart patients. At that point, it also allowed investigation of the formation, development, and operation of an online community, as well as some exploration of what constitutes "community" online. By the time that project was completed in 2008, it had raised a number of questions around the formation of a heart patient identity in the context of a supportive online community. Although the full project will compare online identity formation with the development of a heart patient identity in everyday face-to-face (FtF) contexts, it is the online identity component of this agenda that forms the focus of this chapter.

This research was established based on the principle that, postdiagnosis, heart patients in Australia and other Western democracies constitute a cultural grouping. This notion is adapted from anthropology and from Henrich's (2004) work on acculturation. Henrich sees cultural groupings as comprising (a) *conformist transmission*, or *copying the majority* behavior; (b) *prestige-based transmission*, in which people with prestige are identified and disproportionately influential upon the behaviors of other, and newer, members of a group; (c) *punishment of non-conformists or norm violators*; and (d) *normative conformity*, which is behavior that is adopted more to avoid punishment than because it is embraced, as is the case in the first example. In the case of heart patient acculturation, the grouping is constructed through diagnosis and identification with other people who model what it is to behave like a "successful" heart patient.

Table 1. Descriptions of study interviewees quoted in this chapter

Brian	About 40 years old, living with partner. He has been involved in HeartNET since its inception, although he has not been actively involved in the last few months due to other commitments.
Laura	Age 50, separated from partner following her heart event, lives on her own and very active on HeartNET.
Sally	Around 30 years old, chronically ill, and a potential candidate for major heart surgery; has great support from family and friends; discovered HeartNET about 12 months ago but found it to be too supportive and loving of an environment and could not cope with all these new people who cared for her. She appears on the site about every 4 weeks or so to tell people how she is but is not actively involved.
Henry	Around 55 years old, lives in country, very involved on HeartNET, and has found poetry to be a way of dealing with his depressive thoughts and feelings following a major heart attack. Has withdrawn from the site after being very active but keeps in contact with other members off-line. He remains a member of HeartNET.
Bronwyn	In her early 60s. Initially found that HeartNET met her needs but then found that being honest and being open to complain seemed not to be allowed by some of the members of the site; so she chose to leave HeartNET.
Clarissa	Another woman in her 60s who felt she had to be positive all the time and decided to leave HeartNET when she found that if she was outspoken about her illness and came across as unhappy, nobody would try to support her.

The HeartNET research project, currently including about 800 members, is positioned within the overall context of the field of media and communications studies and focuses upon audience and reception studies approaches, although it takes its definition of cultural group from anthropology. It is qualitative research. The first stage of the research involved 16 in-depth one-on-one interviews with HeartNET members (*the online group*; table 1) and 11 separate interviews with heart patients from the wider community who do not participate in an online forum (*the off-line group*). Initial interviews were timed to be early in the journey of heart patients, ideally within the first three months of initial diagnosis, so that the process of acculturation within the heart patient community was just beginning and more accessible to the research process. This time frame had to be extended because it proved difficult to recruit sufficient numbers of participants so soon after the first diagnosis.

Information from the first set of interviews was supported by data from FtF meetings over coffee in the morning, telephone calls to members, and personal communications principally by e-mail. The research also draws upon posts from the message board and chat room discussions that take place on HeartNET. A final set of one-on-one in-depth interviews was scheduled for six to twelve months after the first stage of data gathering: Every participant was thus interviewed twice. By this time, it was supposed that the identity of the heart patient would have stabilized somewhat, and it would be possible to deduce factors influencing people who form their heart patient identity as part of a supportive online community and compare them with factors influencing people whose heart patient identity is not supported by online interactions.

Online identity: Review of literature

In the 1990s, Internet enthusiasts believed that cyberspace would allow people to be their "true selves"; writers often drew upon Rogers's 1951 conception of the self[1] (cited in Bargh, McKenna, & Fitzsimons, 2002, p. 33). Other enthusiasts concurrently believed individuals on the Internet could be anyone they wanted to be. Turkle's (1995) investigations of college students' online gaming in multiuser domains (MUDs) exemplifies the latter perspective in her view that online identities can be multiple, fluid, and fragmented. Turkle (1995, p. 164) argues that "when we live through our electronic self-representations we have unlimited possibilities to be many" (cited in Chester & Bretherton, 2007, p. 234). For Turkle, "People become masters of self-presentation and self-creation," and for this reason the "very notion of an inner, 'true self' is called into question" (cited in Chester & Bretherton, 2007, p. 234). These two beliefs about identity in cyberspace continue today.

It was inevitable that early speculation and research about identity production[2] on the Internet tended to make general observations rather than context-specific ones. Few ethnographic studies—or what is now referred to as *netnography*[3]—had been performed in situ. Later research shows that the context in which identity production takes place does need to be taken into account before proffering any conclusions about online identity formation (boyd, 2008; Hine, 2001).

Some 20 years after the launch of the World Wide Web, research about online identity reveals that individuals online are neither a true self nor multiple or fragmented identities. Individuals who participate in online forums in the form of online dating sites (Ellison, Heino, & Gibbs, 2006), online communities (Rodan, Uridge, & Green, 2010; Palandri & Green, 2000), and Usenet (Donath, 1998), as well as Web page producers (Kennedy, 2006), consistently render personality traits in their use of language, writing styles, signatures, avatars, images, and hyperlinks. These markers of identity signify aspects and traces of the individual's off-line identity. Internet researchers currently point to other aspects that mobilize identity construction: *the distinction between being and feeling in Internet identities*, *identification* with particular kinds of identities, *affect* in terms of belonging, and *as-if* and *becoming* in which identity production is a continual process (Kennedy, 2006, p. 872; see Ferreday on belonging, 2009).

How the self is formed online is the focus of some recent studies. Although the concept of the self has a somewhat different meaning than that of identity, theoretical work exploring how the self is conceptualized nonetheless contributes to the discussion in this chapter of how online identity works. Studies on selfhood, such as Suler's (2004), have revealed that there is a *constellation of selves* (also see Palandri & Green, 2000, p. 634; de Vries, 2009, p. 18; Dervin & Abbas 2009, p. 6). This constellation of selves emerges in specific online contexts (boyd, 2008; Larsen, 2007). Suler claims that the self is mobilized through *clusters of affect* and *cognition* through the individual participating in different online contexts (p. 321). An assemblage[4] of selves is produced in context, but parts of the assemblage may only emerge within specific online communities or forums. For this reason, Suler claims that the "self does not exist separate from the environment in which that self is expressed" (p. 325).

Individuals may fully express certain aspects of themselves in environments under certain conditions—for Suler these aspects are neither fragmented nor new. He explains: "If someone contains his aggression in face-to-face living, but expresses that aggression online, both behaviors reflect aspects of self: the self that acts non-aggressively under certain conditions, the self that acts aggressively under other conditions" (2004, p. 325). Each self is one of "two dimensions of that person." But each self is only "revealed within [a] different situational context" (p. 325). The "dynamics" of an individual's personality from this perspective "involves" the multipart interactions between the dimensions of the self and the specific milieu (p. 325).

Suler's (2004) constellations model can be used to explain certain behaviors on the Internet. For example, the model offers an explanation for "the disinhibition effect as well as other online phenomena, like identity experimentation, role playing, multitasking, and other more subtle shifts in personality expression as someone moves from one online environment to another" (Suler, 2004, p. 325). What online behavior shows, then, is an assemblage of somewhat "different constellations" of the self. The constellations are made up of "affect, memory, and thought that surface in and interact with different types of online environments" (p. 325).

Other studies explain online identity production slightly differently than does Suler, claiming that the individual's off-line and online self "is blurred"—that is, online identity is not necessarily wholly a person's off-line identity (Larsen, 2007, p. 15; boyd, 2007). Some aspects of identity can be expressed online without social sanction, but these aspects are not divorced from an individual's off-line self (Palandri & Green, 2000). This perspective supports Suler's (2004) constellations of selves model in the sense that there are traces of oneself in both off-line and online contexts. Dervin and Abbas (2009) clarify:

> So whatever actions, discourses or emotions that one discovers on the web cannot be separated from the "real world" or discarded as having no influence whatever on the individual. In the complex identification processes in which we are all involved (with more or less freedom), our presence "out there" plays an increasingly important role. (p. 6)

In other words, one's presence off-line seeps into one's online presence. Here, Dervin and Abbas suggest that it is not possible to contain our online identifications—that is, our online identity categories or markers are also evident in the "real world." Reviewing online identity from the perspective of an individual's identification processes allows a more nuanced conception of how online identity works (also see Kennedy, 2006).

Yet, what can be stated about online identity is that in certain contexts individuals can more fully express aspects of their self that cannot be expressed to the same extent off-line. In her study of teens online, boyd (2008) explains that "each profile is unique, but there are network effects in terms of tone, genre, and style suggesting that teens are positioning their digital identities in relation to those around them" (p. 136). Her study reveals that identity production by teens is often very specific to the audience:

> Their decisions about what to say are deeply connected to their sense of audience. That audience, by and large, is their friends and what teens say is best understood in that context. Out of context, some of what appears is not quite what it seems. (p. 143)

Online identity can enable individuals to experience and express a form of *self-transformation*, as de Vries (2009, p.17) claims in her study. This self-transformation may only occur in an aspect of the self such as self-esteem, in support of identity development (de Vries, 2009), or, in the case of HeartNET, in positive affirmation as a heart patient. Self-transformation (see Foucault, 1988) is enabled through the *techniques of self*[5] that the individual employs online that affect one's offline conception of the self. Online techniques can include the *self-disclosure* that happens through writing about oneself and/or responding to others' posts and the *self-revelation* that happens through choices of signatures, avatars, and photographs of one's family (Donath, 1998; Rodan,

Uridge, & Green, 2010). Off-line techniques can include the *self-examination* (Foucault, 1988, p. 44) that happens in the reflection on one's online identities, which may well enable self-transformation in a limited realm (as illustrated by Palandri & Green, 2000).

Online Identity: HeartNET

Ultimately, we think that Suler (2004) offers the best explanation of how online identity works in his idea of constellations of the self. He explains how in many ways participants in an online community feel they can be less inhibited—for example, by throwing out reasonable self-impression management—and express feelings, and views, that they would not in a FtF situation. As discussed earlier, Suler points out that participants are expressing only one aspect of themselves online and that this aspect is no "truer" than other parts of themselves they express offline. Online communities like HeartNET can and do allow a social space for this expression of uninhibited self-disclosure that is not always available off-line.

Members of HeartNET often comment on how their online community provides them with the support they want: the support that loved ones are often not able to give. Although loved ones can support and encourage heart patients, it is the encouragement and support of someone who has been in a similar situation that many people are looking for when they visit online communities. To identify with another person's situation often means finding other people who have an identity as a heart patient—that is, people who have been given a medical diagnosis of cardiovascular disease (CVD). For this reason, being part of an online community can be a positive experience. It can be positive in that it allows individuals to express an aspect of themselves that they may feel inhibited about expressing offline. For Laura, this was particularly significant:

> And then I saw this thing for HeartNet so I went in there and I looked around for two or three days reading posts and all that. And because I couldn't talk to [my husband] about it and couldn't really talk to my kids as I wanted to protect them, so I had nobody who I could talk to about how bad I was feeling. I didn't want to talk to my friends about it because "gung ho Laura: she can do anything." And so I saw HeartNet and it's been my savior in many ways. . . . At times when I've been really crushed and down and I've poured my heart out lots of support from so many people . . . I feel like I've really made some very good friends . . . No matter what you say no one judges you, no one says, "Oh you big la la, get your act together," nobody says that.

In the initial period after a heart event, Brian looked for help through the usual channels of doctor's surgery and the local hospital. He was given a telephone number and reports, "When I contacted this support group, and I got an answering machine . . . that was the last thing I wanted to hear was a voice recorded message of, 'and leave a name and number' . . . so I didn't talk on that." Brian was already aware of how helpful online communities were (his partner was actively involved in several), so he decided to look for a similar group. Brian found he could fully express himself in the context of HeartNET:

> I always found HeartNET to be there when I needed it. It's just one of those things, you know it's there, and there's always someone online at some point, or you can just post a message and there's going to be someone you know or are familiar with who's going to respond.

Finding techniques to express aspects of the self allows for the kind of self-transformation to which de Vries (2009) refers. Brian's technique was to post online at the moment he most needed to, which allowed him to be less inhibited and act in a manner generally less available to him with family and friends in the off-line world. Being uninhibited about one's feelings in the context of an online community such as HeartNET is generally accepted, but out-of-context comments may appear to be peculiar or alarming behavior (see boyd's 2008 study of teens online discussed earlier).

Yet, not all participants who accessed HeartNET felt supported in the same way as Brian. Sally coped for so long without a supportive environment that when she found others in similar situations who offered ongoing support, it felt "great" but, at the same time, "overwhelming." When asked about the support available on HeartNET, she responded:

> I . . . think I haven't been in a very supportive environment until now so I feel a bit overwhelmed on one hand and really great on the other. I have felt so alone outside of this as well. Even before this with all my other medical problems, I felt like I've had to deal with everyone on my own, so much with the exception of my husband that I feel a little claustrophobic I guess, I don't quite know how to deal with people being there, being nice to me. So, yeah, I think that's why I haven't been on very much the last couple of weeks.

HeartNET provided a context for Sally to express an aspect of herself that she was often constrained in revealing in an off-line environment. She discloses in her comments her perception that an online environment can allow for the emergence of "subtle shifts in personality expression" (Suler, 2004, p. 325); shifts that HeartNET members have suggested are more often regulated in an off-line environment (except in specific contexts such as one-on-one counseling, and so forth).

HeartNET members can more fully express their identity as heart patients in an online context because of the virtual space of the community. Given that this virtual space is available, there is the possibility of self-transformation of the individual as a heart patient. For HeartNET members, this possibility of transformation can be a beneficial experience that allows them to feel positive about themselves as heart patients, which is not how they often experience themselves off-line when interacting with family members or when dealing with the medical profession.

The main technique that HeartNET members use to express their identities online is self-disclosure, in the form of writing about oneself and/or responding to others' posts.[6] Self-disclosure through the process of writing can enable self-transformation of one's identity as a heart patient. Some members write poetry to express feelings. Uridge, Rodan, and Green (2008) found HeartNET to be an "ideal locale for this kind of communication" (p. 102): ideal in the sense that members can express creative aspects of themselves that are suppressed in other environments (also see Braithwaite cited in Sullivan, 2003). Some male HeartNET members have expressed their fears and concerns using poetry: Most of their poems were posted to the site in the early hours of the morning when people are known to be at their most vulnerable and do not wish to concern their families with their fears. Writing and posting poetry online can be a release, as well as a form of venting; it can also be a form of purging, and cleansing, as can be seen in this poem created by Henry:

"A Walk on the Dark Side"

I took a walk on the dark side
and looked into the dark Abyss . . .
was a dark haunting ride . . .
thort[7] of the loved ones I'd miss.

I took a walk on the dark side
saw the monsters and demons down there..
sat there frightened, eyes open wide..
heart filled with terror, dread an despair.

I took a walk on the dark side
shadowy figures reach out from their den . . .
always waiting, at the end of the slide . . .
and if not today . . . then when.

I took a walk on the dark side
felt the battle raging within . . .
tearing at my heart, destroying my pride . . .
this is a battle, they could so easily win.

I took a walk on the dark side
so people place your bet . . .
for the demons no not, of those allied . . .
they take on, not just I . . .
but friends from HeartNET.

The responses to Henry's poetry from HeartNET members varied. At one point, the moderator was contacted as several members were concerned that Henry might be suicidal. When the moderator telephoned him, Henry assured her he was fine and commented that using the site to write his thoughts and feelings gave him an outlet he had never had before.[8] In total, Henry published around seven poems on HeartNET. Initially his poetry was quite "dark and depressing"; however, over time he wrote lighter and more joyful verse. Henry's self-reflection on the telephone about his poetry (see Endnote 8) enabled some personal insight into his online identity as a heart patient (similar to Palandri, cited in Palandri & Green, 2000). His poetry reveals a form of self-transformation over time to a more positive conception of his identity as a heart patient.

A few members of HeartNET expressed concern that there was *too much* support on the site for them. These members had posted on the site about how sad or how disappointed they felt about their health: They wanted to be allowed to be, and feel, miserable at this time. These particular members commented that they received too much support and encouragement and because of this decided to stop accessing the site. Bronwyn explains:

> MY name is Bronwyn I was a member of heartnet, but they failed to understand where I was coming from, they never allowed me to voice how i truely felt it always had to be positive, lets face it we have a right to feel S—. I never read your postings I just came on for a sticky beack[9] and saw you had deleted your posts, and I read one other post that would have pissed me off. I would love to email and chat to you, mabey we have something in common and tell the truth of how we feel. :)

And Clarissa writes:

> And yes, when I was describing some of the stuff happening to me as events, they were taking it that I wasn't being positive enough. I won't repeat some of the comments made to me. In fact, I do feel that I am quite positive about what is going on and can actually laugh at some of the stuff that has happened. I was thinking that maybe because I never had an actual heart attack that they just weren't able to understand what I was going through.

When the HeartNET site decided to set up a dedicated "whinging"[10] page, members did whinge: but ultimately one of the members would find themselves writing words of encouragement. Whether Bronwyn and Clarissa found the comfort they needed outside HeartNET is unknown. What is known is how supportive HeartNET members are of each other. Bronwyn and Clarissa were made very welcome, and any communication they made either on the site or in the chat room was met with compassion and support. However,

HeartNET, like other online communities, may not be the right place for all people, even in the case where people have experienced a heart event.

Clarissa's comments indicate something about the nature of HeartNET members and the milieu of the site that focuses on giving support and encouragement. Clarissa and Bronwyn reveal that even in online communities, members may not feel they can fully express all parts of the self and be recognized in authentic ways. As Suler (2004) points out, individuals may fully express certain aspects of themselves in environments under certain conditions—for Suler these aspects of the self are neither fragmented nor new. In his model of the constellations of the self, each self is dimensional but is only "revealed within [a] different situational context" (p. 325). The "dynamics" of an individual's personality from this perspective "involve" the multipart interactions between the dimensions of the self and the specific milieu (p. 325), and HeartNET offers a further milieu in which suppressed dimensions of the self may arise.

Conclusion

The online community is an environment that supports the formation of a new and unwelcome identity: that of the heart patient. This research was designed to allow the benefits derived from developing a new identity as a heart patient in the context of an online community to be evaluated. The HeartNET community offers benefits for some people who are coming to terms with a CVD diagnosis. Further, although most research into cultural groupings concerns groupings into which people are born, such as gender and race, or groupings with which people choose to identify, such as religious identity, attention also needs to be paid to cultural groupings that arise as a result of unexpected and or unwelcome life events, such as cardiovascular disease.

Some members engage with HeartNET only at arm's length: They claim that the attention and support on offer in the community is overwhelming. This indicates that the site is operating as a context for acculturation, and a proportion of potential members are unable to respond appropriately to the requirement for "conformist transmission": They identify a person on the site who is accorded prestige and whom they wish to emulate. Rather than negotiating a minority identity as a heart patient, for example as a whinger,[11] a number of "occasional" HeartNET-ers resist the pressure to acculturate by withdrawing.

It would be oversimplifying matters, however, to suggest that individuals have a fixed notion of self: an unchanging true self, whinger or not. Instead, people exhibit a constellation of selves, and such a perspective is reinforced in the HeartNET findings in which people express a range of emotions and responses in authentic ways and as a progression over time. Indeed, it is through expression on the site that many different aspects of the members' selves become visible, conditional upon the situational context. When people sought to express negative thoughts and feelings without triggering positive responses from other members, the site found it necessary to establish a whinging page; the context had to be created, before the behavior could be embraced.

Developing an identity in the context of an online community offers positive support for some heart patients and also an opportunity for self-reflection. The fact that online community is generally expressed in written form means that communications in this context are amenable to a process starting with self-disclosure, moving to reflection, and resulting in transformation over time. Other research, particularly into creativity, indicates the value of writing as a means of discovering what the writer feels to be true: "The only time I know that something is true is the moment I discover it in the act of writing" (Malaquais, cited in Avieson, 2008, opening quote). Richardson (1994) makes a similar point, commenting that writing is "a way of 'knowing'—a method of discovery and analysis. By writing in different ways, we discover new aspects of our topic and our relationship to it" (p. 516). The implication is that HeartNET members can achieve a deeper realization of what it is to have acquired the identity of heart patient as a result of writing their heart patient selves into existence.

The global implication of this research is that minority experience and group membership can be harnessed and expressed within the context of online community, regardless of how rarely a situation or circumstance occurs on the ground. Indeed, this is the experience outlined in an earlier publication about a bondage, domination, and sadomasochist subculture community by Palandri and Green (2000). Although not all communities will suit all eligible online members, it is more likely that self-expression, self-disclosure, reflection, and transformation can occur online than in FtF contexts, in which the experience can be particularly challenging to assimilate for individuals or their likely audience. This is especially the case when the on-the-ground situation fails to offer appropriate acculturation cues. Jackson, Peters, O'Brien, & East (2008) have discussed this dynamic in terms of evaluating the usefulness of computer-mediated-communication in nursing research, noting that online communication may reduce "fears and [negative] perceptions through enhancing anonymity and privacy and increasing comfort, which may yield greater disclosure and more information from participants" (p. 85). If this self-disclosure is combined with the "writing into truth" dynamic mentioned above, then the experience of feeling affirmed as a member of a community of like-minded people can be very powerful.

The potential of online communities to combine the passion and commitment of those who might find themselves in a social, cultural, or health minority has also been harnessed in recent years by a range of radical political movements. Although at odds with media perspectives circulating in the public sphere, online organizations such as MoveOn.org and the Tea Party (Phillips, 2010), and the global campaigning organization Avaaz (Green, 2010, p. 140) seek to recruit participants for digital activism. Like the 2008 campaign to elect President Barack Obama to the White House, they all mobilize individuals and groups who may feel sidelined by mainstream political discussions, including the disadvantaged and the poor. The Barack Obama Web site allowed American citizens who had previously been ignored by other political parties and campaigns to discover a sense of agency that could effect significant change (McQueen & Green, 2010).

Context, culture, authenticity, and self-expression have been demonstrated to be important aspects of identity production. These observations affirm the present uses of online community in supporting the production of new identities and indicating the possible global benefit for future applications of such communities, as populations increasingly face a range of age- and other health-related challenges. What has been demonstrated here with regard to heart disease can be replicated in terms of diabetes and other chronic life-changing conditions. Where such communities include people who are empowered to respond to their disease with lifestyle changes and symptom mitigation strategies, these approaches can inspire significant cohorts of fellow online community members. This is one of the many positive benefits that can be gained from negotiating a new identity as a heart patient in the context of a supportive online community.

Notes

1. The *true self* for Rogers exists "psychologically (i.e., a present, not a future version of self) but not fully expressed in social life (i.e., not the actual self)" (cited in Bargh, McKenna & Fitzsimons, 2002, p. 34). Suler (2004, p. 324) suggests that accounts of the true self are influenced by "Freud's archaeological model of the mind." Freud's archetype assumes that "personality structure is constructed in layers, that a core, true self exists beneath various layers of defenses" as well as "the more superficial roles of everyday social interactions" (Suler, 2004, p. 324).
2. We are drawing on Paul du Gay's (1997) notion of identity formation as production.
3. Robert Kozinets (2002) is often cited as the first to use the term *netnography*.
4. Here we are drawing on Deleuze and Guattari's (1988) concept of *assemblage*. Their discussion of the *proper name* explicates the idea that individuals are constituted by selves that come to the fore in specific contexts (pp. 263–264).

5. In his chapter "Technologies of the Self," Foucault (1988, p. 34) uses the term *techniques of self* as a catchall phrase for the rules and conduct individuals practice and apply to transform themselves. For instance, three techniques used by the Stoics—*letters to friends*, *disclosure of self*, and *examination of self and conscience* (pp. 34–35)—are similar techniques used online and offline by individuals today.
6. There is evidence that males and females utilize support sites differently with women accessing sites for support and men for information (Anderson, Jeneson, & Ruland, 2007).
7. "Thort" in the context of this poem means thought.
8. When contacted for permission to print his poems, Henry responded: "I read some of the poems I posted, and ones I didn't . . . and realised the pain I was feeling bk then in July 2007 still linger very close to the surface to this day (Rattled me badly reading them again)."
9. The usual spelling of the word is "stickybeak"; it is an Australian and New Zealand colloquial expression for someone who is inquisitive, nosy, or interfering (*The new shorter Oxford English dictionary*, 1993, p. 3058).
10. To "whinge" is an English colloquial expression that means to whine, "complain peevishly, grumble" (*The new shorter Oxford English dictionary*, 1993, p. 3669). In the context of HeartNET, one of the members, Sam, created a thread "Just wanna [want to] whinge," which he defined as: "You know those days when you feel that you just wanna whinge, & that no advice is wanted. YOU JUST WANNA WHINGE. Feel free to WHINGE here." In this thread, members can complain without other members feeling that they need to make the person feel better.
11. A "whinger" is an English colloquial expression that means "a person who whines or grumbles" (*The new shorter Oxford English dictionary*, 1993, p. 3669).

References

Anderson, T., Jeneson, A., & Ruland, C. M. (2007). Gender differences in online messages among cancer patients. MedInfo 2007: *Proceedings of the 12th World Congress on Health (Medical). Informatics: Building Sustainable Health Systems*, Amsterdam, Netherlands. Amsterdam, Netherlands: IOS Press.

Avieson, B. (2008). Writing—A methodology for the new millennium. *TEXT*, *12*(2). Retrieved from http://www.textjournal.com.au/oct08/avieson.htm

Bargh, J. A, McKenna, K. Y. A., & Fitzsimons, G. M. (2002). Can you see the real me? Activation and expression of the "true self" on the Internet. *Journal of Social Issues*, *58*(1), 33–48.

boyd, d. (2006). Why youth (heart) social network sites: The role of networked publics in teenage social life. In D. Buckingham (ed.), *Youth, identity, and digital media* (MacArthur Foundation Series on Digital Learning) (pp. 119–142). Cambridge, MA: MIT Press.

boyd, d. (2008). Taken out of context: American teen sociality in networked publics (Doctoral dissertation, University of California, Berkeley). Retrieved from http://www.danah.org/ papers/TakenOutOfContext.pdf

Chester, A., & Bretherton, D. (2007). Impression management and identity online. In A. N. Joinson, K. Y. A. McKenna, T. Postmes, & U.-D. Reips (Eds.), *The Oxford handbook of Internet psychology* (pp. 223–236). Oxford, United Kingdom: Oxford University Press.

Deleuze, G., & Guattari, F. (1988). *A thousand plateaus: Capitalism and schizophrenia* (B. Massumi, Trans.). London, United Kingdom: The Athlone Press.

Dervin, F. & Abbas, Y. (2009). Introduction. In Y. Abbas & F. Dervin (Eds.), *Digital technologies of the self* (pp. 1–11). Newcastle, United Kingdom: Cambridge Scholars.

Donath, J. (1998). Identity and deception in the virtual community. In M. Smith & P. Kollock, (Eds.), Communities in cyberspace. London, United Kingdom: Routledge. Retrieved from http://smg.media.mit.edu/people/Judith/Identity/IdentityDeception.html

Ellison, N., Heino, R., & Gibbs, J. (2006). Managing impressions online: Self-presentation processes in the online dating environment. *Journal of Computer-Mediated Communication, 11*, 415–441.

Ferreday, D. (2009). *Online belongings: Fantasy, affect and web communities.* Bern, Switzerland: Peter Lang.

Foucault, M. (1988). Technologies of the self. In L. H. Martin, H. Gutman, & P. H. Hutton (Trans. and Eds.), *Technologies of the self: A seminar with Michel Foucault* (pp. 16–49). Amherst: Massachusetts University Press.

Gay, P. du. (Ed.). (1997). *Production of culture/Cultures of production.* London, United Kingdom: Open University and Sage.

Green, L. (2010). *The Internet: An introduction to new media*, Oxford, United Kingdom: Berg.

Henrich, J. (2004). Cultural group selection, coevolutionary processes and large-scale cooperation. *Journal of Economic Behavior & Organization, 53*(1), 3–35.

Hine, C. (2001). Web pages, authors and audiences: The meaning of a mouse click. *Information, Communication and Society, 4*(2), 182–198.

Jackson, D., Peters, K., O'Brien, L., & East, L. (2008). The benefits of computer-mediated communication in nursing research. *Contemporary Nurse: A Journal for the Australian Nursing Profession, 30*(1), 83–88.

Kennedy, H. (2006). Beyond anonymity, or future directions for Internet identity research. *New Media & Society, 8*(6), 859–876.

Kozinets, R. (2002). The field behind the screen: Using netnography for marketing research in online communities. *Journal of Marketing Research, 39*(1), 61–72.

Larsen, M. C. (2007, October). Understanding social networking: On young people's construction and co-construction of identity online. Paper presented at the Internet Research 8.0: Let's Play, Association of Internet Researchers, Vancouver, Canada. Retrieved from http://www.mendeley.com/research/understanding-social-networking-on-young-peoples-construction-and-coconstruction-of-identity-online/

McQueen, T., & Green, L.(2010). Obama's election campaign and the integrated use of social media. In F. Sudweeks, H. Hrachovec, & C. Ess (Eds.), *Cultural attitudes towards technology and communication 2010, Proceedings 2010* (pp. 315–330). Retrieved from http://blogs.ubc.ca/catac/proceedings/proceedings-2010/

Palandri, M., & Green, L. (2000). Image management in a bondage, discipline, sadomasochist subculture: A cyber-ethnographic study. *Cyber Psychology & Behavior, 3*(4), 631–642.

Phillips, C. (2010). A good coalition. *M/C Journal, 13*(6). Retrieved from http://journal.media-culture.org.au/index.php/mcjournal/article/viewArticle/316

Richardson, L. (1994). Writing: A method of inquiry. In N. Denzin & Y. Lincoln (Eds.), *Handbook of qualitative research* (pp. 516–529). Thousand Oaks, CA: Sage.

Rodan, D., Uridge, L., & Green, L. (2010). Using nicknames, pseudonyms and avatars on HeartNET: A snapshot of an online health support community. In K. McCallum (Ed.), *Proceedings of the Australian and New Zealand Communication Association on Media Democracy and Change Conference.* Canberra, Australia: Old Parliament House. Retrieved from http://www.canberra.edu.au/anzca2010/attachments/pdf/Rodan,-Uridge-and-Green_ANZCA2010.pdf

Suler, J. (2004). The online disinhibition effect. *CyberPsychology & Behavior, 7*(3), 321–326.

Sullivan, C. F. (2003). Gendered cybersupport: A thematic analysis of two online cancer support groups. *Journal of Health Psychology, 8*(1), 83–103.

The new shorter Oxford English dictionary (Vol. 2, 3rd ed.). (1993). Oxford, United Kingdom: Clarendon Press.

Turkle, S. (1995). *Life on the screen: Identity in the age of the Internet.* New York, NY: Simon & Schuster.

Uridge, L., Rodan, D., & Green, L. (2008). HeartNET: Moving towards a transformative space? *Proceedings, Transforming Information and Learning Conference on Transformers: People, Technologies and Spaces.* Perth, WA: Edith Cowan University. Retrieved from http://ro.ecu.edu.au/ecuworks/919/

Vries, K. de. (2009). Identity in a world of ambient intelligence. In Y. Abbas & F. Dervin (Eds.), *Digital technologies of the self* (pp. 15–36). Newcastle, United Kingdom: Cambridge Scholars.

10. *Inoculating against Invisibility: The Friendly Circle of Cancer Patients, Chinese Blog*

WEI SUN, PHD
ANDREW JARED CRITCHFIELD, PHD

Introduction

This chapter focuses on identity and social support within a Chinese virtual blog community about cancer and how cancer patients find the group effective in their personal fight against the disease. The Internet has become increasingly important in helping people get information about health care and make decisions concerning their care. According to the Internet World Stats (2011), China has 485 million Internet users as of June 2011, up from 420 million the previous year. This makes China the largest national online population in the world. Of Chinese Internet users, 79.7% are between the ages of 18 and 50, and 82.9% of them have at least a high school diploma (China Internet Network Information Center, 2010).

This research performs a qualitative textual analysis of the influential Friendly Circle of Cancer Patients' blog community, and it evaluates new media as a vehicle for transmitting health information regarding narrative medicine. The increasing industrialization of China over the last two decades and the subsequent increase in industrial waste and environmental pollution is often cited as a cause of increasing cancer rates across all age groups. The Chinese Health Ministry (Li, 2008) announced that in the past 30 years, cancer deaths have increased by 80%. For the Chinese, cancer is the number one killer of urban residents and the second highest killer of rural residents.

Blogs and intercultural health communication in China

The concepts of *health communication* and *health campaigns* are still new in China. Health campaigns, which encourage healthy behaviors and discourage unhealthy ones, were first introduced in China in the late 1990s. Modeled on American health campaigns, these campaigns had little impact on behavior of the Chinese, due to differing cultural contexts. Cultural communication scholars conclude that people of high context and collectivistic cultures (such as China) take *other oriented*, *other concerned*, and *third-party* approaches in interpersonal communication settings. This reserved communication style can be associated with delivering and receiving health messages delivered through media outlets. To have the greatest positive impact, health campaigns must therefore be culture focused and sensitive to those who will benefit from the message (Liu & Chen, 2010).

In Chinese health communication, breaking the news of a cancer diagnosis has always been a cultural taboo among health providers, patients, and caregivers. Stigmatization (Meisenbach, 2010; Smith, 2007; Goffman, 1963) has served as a deterrent in discussing cancer and other health ailments openly. Communication between doctors and patients also demonstrate social and cultural chasms (Charon, 2009), leading to further isolation of the patient and distrust of medical professionals. As a result, cancer survivors in China have started to form virtual support groups, maintain diaries, and use blogs to communicate their emotional experiences concerning cancer diagnosis and treatment in order to build a community of support.

The community support, visibility (Orgad, 2005), and implicit acceptance found on blogs is important, because "as human beings hear one another out, donating serious and generous attention, listening from the teller's side, they are able to perceive one another, bestowing and accepting recognition as a result" (Charon, 2009, p. 119). The barriers between the ill and the "healthy" and the shame of diagnosis and treatment in China are collapsing through the use of blogs and other online support systems. This may help to empower cultural minorities. Sun (2007) and Sun and Starosta (2006) suggest that members of various cultural groups are rendered invisible within mainstream culture. Invisible group members, such as those with cancer, whose experiences differ from the larger population, usually experience their concerns being downplayed, ignored, or simplified. Invisible group members may also be stereotyped and stigmatized, either in conscious or unconscious and deliberate or nondeliberate ways. The effect of being rendered invisible and ignored is typically negative, resulting in a lower self-esteem and a demeaned existence for marginalized members of various cultural groups.

It is traditional for the Chinese to seek organizational help when they encounter hardship. There are labor unions and women's associations in most formal organizations in China. In government-owned organizations, there is a guarantee of job security for employees, and medical insurance is usually fully covered. But that benefit is limited solely to the employee and does not extend to family members.

Since the 1990s, the increase in the number of privately owned businesses and foreign joint ventures in China has significantly changed how health care benefits are administered, typically resulting in worse health care coverage for nongovernment workers. The transformation of health care continues to move toward privatization. The expense and physical challenges of seeking cancer treatment in China create additional health and financial problems for the patients. Not everyone has medical insurance, and cancer treatments are very costly and involve restrictions on treatment and care. Rural agricultural workers (noncity residents) often incur large medical expenses as cancer patients.

Research such as Mirivel and Thombre's (2010) demonstrates the importance of blogs for community connection when confronted with a health crisis. Weblogs, or blogs, have been used worldwide since 1999, mainly for "democratic self-expression and networking" (Kahn & Kellner, 2004, p. 91). Blogs promote visibility, inspire motivation to accomplish tasks or learn a new skill, and encourage participation in many activities. According to Yan (2006), blogs were initially introduced to China in 2002; there were fewer than 10,000 bloggers in the first year. By 2003, the number of bloggers increased to 200,000. The following year, blog sites were commercialized, and blogging became popular among Internet users. Yang, Zhao, Yu, and Zheng (2008) also indicate that in China, blogs have become an important source for information seekers, and people use blogs for interactive social networking by connecting and befriending others. The online connections concerning a particular experience or issue, hobby, or event help facilitate cyber-collectivism among the participating blog members (Kim, 2007). Although blogs provide a forum for intimate self-expression, the accuracy of content in blogs has been often questioned (Herring, Scheidt, Bonus, & Wright, 2004), especially when blogs are focused on health or technical information.

Cancer patients can use their blogging and Web connections to follow their paths of diagnosis. Gruman (2007, pp. 6–7) lists the stages of a patient being diagnosed with cancer and other fatal diseases, as follows: First, the person responds to the shock of diagnosis. She then learns about the condition and available treatments, while deciding whether or not to involve others, finds the right doctors and hospitals, makes timely medical appointments, seeks other opinions about what is wrong and what to do, manages work lives, pays for care, finds relief, and takes next steps, and so forth. Stein (2007) states that patients usually experience emotional stress such as betrayal, terror, loss, and loneliness after receiving a diagnosis of cancer, suggesting that "illness involves a specific loneliness, a set of limits and invisible walls surrounding the sudden and incomprehensible crime and betrayal that has occurred to one's body" (p. 169). And, "for many patients, there is the fear of the unknown, the fear of making wrong turns and bad choices, of being separated from companions and of traveling alone, of humiliation, of getting into something they can't escape" (p. 71). Building connections and a community online can decrease the loneliness and invisibility of cancer patients. As Charon (2006) suggests, "The healing process begins when patients tell of symptoms or even fears of illness—first to themselves, then to loved ones, and finally to health professionals" (p. 31).

Research context and cancer blogs

In September 2007, *sina.com (a popular Chinese website similar to yahoo.com) People's Daily* news column published an article entitled "A Cancer Girl and Her Blog Circle," highlighting cancer patient blogger Xu Liping and her blog Friendly Circle of Cancer Patients (blog.sina.com.cn/xuliping). Xu began blogging in 2005 after her cancer diagnosis and soon formed an online community of some 20 bloggers, which is essentially a database of cancer patients and their blog addresses. By 2007, 249 people had joined the circle of bloggers and over 200,000 people had visited the site (Chen, 2007). Xu passed away in September 2008 after battling cancer for three years. Her family updates her personal blog on the Friendly Circle site now on a bimonthly basis, as a virtual memorial to her. By August 2010, four million unique visitors visited her blog. The Friendly Circle of Cancer Patients community is now managed by another long-term cancer survivor, whose online ID name is Liqiao. The Friendly Circle of Cancer Patients' blog is a closed, online community, requiring membership. The first author requested membership on the site for research purposes and was granted access. The membership list was created by a member with the ID name "Cancer." There are currently 354 members of the Circle community. Of the blog's members, 79 who were members of the blog circle prior to January 2009 have been designated as senior members and given deference because of their status within the community.

The research questions of this study are as follows:

1. What are the cultural barriers in communicating about cancer among patients, family members, and health care providers in China?
2. How has the Internet as a new medium affected the public in relation to cancer prevention, diagnosis, and treatment?
3. How do the above findings add to our understanding of minority invisibility as theorized by Sun (2007) and Sun and Starosta (2006)? This is an important topic of study, for as Charon (2006) states,

> Without the narrative acts of telling and being heard, the patient cannot convey to anyone else—or to self—what he or she is going through. More radically and perhaps equally true, without these narrative acts, the patient cannot himself or herself grasp what the events of illness mean. (p. 66)

Content analysis and critical discourse analysis

This study combines content analysis and critical discourse analysis to describe and interpret the Web entries of selected cancer patient bloggers. Starosta (1988) and Sun and Starosta (2008, 2001a, 2001b) offer qualitative content analysis as an approach to signifying recurring themes in a text. The requirements of this type of analysis include that:

1. The unit selected for analysis should be rich in meanings.
2. The researchers should offer groupings of references and discuss their formation and meanings.
3. The research should offer examples from the themes and illustrate how they are significant in cultural, historical, or contextual terms. (Sun & Starosta, 2001a, p. 64)

Critical discourse analysis creates an integrated understanding of a text in social interaction, of mediated discourse analysis, and of the problem of social action (Scollon, 1999). Wodak & Meyer (2009) offer a four-step strategy for critical discourse analysis, suggesting that researchers should establish the contents of topics of a specific discourse with strong ideological ingredients before examining argumentation, linguistic meanings, and contextual meanings of the texts.

Research procedures

At the time of this writing, The Friendly Circle of Cancer Patients had 354 members, including 79 senior members. We selected 21 senior members' blogs to examine regarding the bloggers' influence within the community and length of time blogging (table 1). For example, Fragrance (Xu Liping) is the founding Circle blog host and Liqiao is the current blog host, so they are frequently cited within other members' blogs. Cancer's blog was chosen because he developed the directory of Circle members. The blog entries of Ziyou, who died at age 19 and is frequently mentioned within the blog, and Sage, a popular fashion magazine editor whose book was published before she died of cancer, were both chosen because of their popularity within the Circle. Some of the chosen blogs are now maintained by family members of the deceased bloggers. The blogs are all written in Mandarin Chinese, the first author's native language. The first author read the blog entries under analysis and translated key themes into English. The second author reviewed the themes and double checked groupings for accuracy.

Table 1. List of cancer patient bloggers and blogs examined in this study

Blog ID	Gender	Age	#Entries	Blog
1. Yaran	F	20	162	http://blog.sina.com.cn/chilianyuanran
2. Mercy	M	59	230	http://blog.sina.com.cn/cbxh
3. Desire	F	30s	138	http://blog.sina.com.cn/sos3332002
4. An's Mom	F	39	77	http://blog.sina.com.cn/ananmademeiliresheng
5. Angela	F	30s	297	http://blog.sina.com.cn/yangela
6. Liqiao	F	50s	394	http://blog.sina.com.cn/u/1273176952
7. Peony	F	40s	32	http://blog.sina.com.cn/u/1658864607
8. Greenwater	F	40s	104	http://blog.sina.com.cn/u/1604607510
9. Zhang Lirong	M	75	252	http://blog.sina.com.cn/u/1653071604
10. Walk in Cloud	F	40s	40	http://blog.sina.com.cn/u/1720998647
11. Lonely Fish	F	40	137	http://blog.sina.com.cn/deshun887
12. Fragrance*	F	28	288	http://blog.sina.com.cn/xuliping
13. Cancer	M	45	282	http://blog.sina.com.cn/jeson510
14. Zhaojun	F	44	454	http://blog.sina.com.cn/hegemuqin
15. On Road	M	50s	82	http://blog.sina.com.cn/wussn
16. Xin'er*	F	30s	232	http://blog.sina.com.cn/u/1445969723
17. Phoenix	M	41	142	http://blog.sina.com.cn/lifeofphoenix
18. Sage*	F	30s	447	http://blog.sina.com.cn/u/1183581574
19. Ziyou*	M	19	315	http://blog.sina.com.cn/ziyou
20. Danqing (for 8-yr-old son)			130	http://blog.sina.com.cn/danqingfanke
21. Emmanuel (for mother)			460	http://blog.sina.com.cn/51qiji

*Indicates that this blogger passed away, and their blogs have been maintained by family members. Not all entries reference cancer; some document bloggers' daily activities or thoughts.

Findings

Theme 1. Fears about cancer, treatment, and isolation of patients

Fears concerning cancer and mortality are common blog entry themes among cancer patients. With their lives in jeopardy, cancer bloggers express their feelings and experiences, hoping the blog posts will help calm them down and find solutions. They believe that in sharing their experiences and best practices, they will help someone else and that another's blog post will provide them with solutions. Almost every member of Friendly Circle of Cancer Patients experienced and described the feeling that by getting cancer, they became "disabled" and "distanced." "Darkness" seemed to be the universal word used among patients to describe their depression. Lonely Fish wrote 37 entries of "Black Diaries," in which "black" symbolized death. Yayan, a young girl diagnosed cancer when she was a high school junior, described her fear and uneasy feelings toward the chemotherapy's negative effects:

> The day before yesterday, I was injected with Salmon Calcitonin to treat my renal osteodystrophy bone disease (caused by chemotherapy treatment). Before the injection, Mom told me the instructions said only 0.1% people would feel nausea. That night, the side effect emerged and I threw up feeling just like I was going through chemo treatment. It's been two years now since I last had chemo; I thought that kind of pain had left me forever. . . . When I was engulfed by the pain that night, I was defeated by the drug and fear once again occupied my heart. . . . I was alone kneeling down in the bathroom, seized by the darkness.

Lonely Fish recalled "the most threatening day" on February 3, 2010, during the Chinese New Year holiday, when she started to lose her hair due to chemotherapy:

> I suddenly became so depressed. I kept myself inside my room, keeping my grey and cold soul inside. The clock ticked, and a life is lost by each second. Outside the building, children are laughing and playing with fireworks; their happiness and my darkness coexist. But their joyfulness and my greyness and cold are approaching death at the same frequency.

Within the Friendly Circle of Cancer Patients' blog, most bloggers write of their personal experiences, sharing in others' sorrows and joys. A blog about a death or a recurrence of cancer can spread worry and fear across the blog membership. After the health of several senior members worsened, Liqiao reassured the Circle members on March 16, 2007:

> This is a special crowd, recurrence is almost inevitable; and this is also a fragile crowd, storms outside the Circle will stir up our emotions. Keep calm. From Fragrance and Sage's articles, you should not only read the word "spread," but also read of their underlying bravery and calm. Let's live with grace and calmness.

With encouragement and comfort found within the Circle, patients found solace among the Circle members to overcome fears and isolation.

Theme 2. Patient relationships and expectations

According to Circle bloggers, the support of loved ones is important for cancer patients' rehabilitation. Cancer patients and their families deal with cancer within their existing responsibilities and commitments (Weiss & Lorenzi, 2005). It is unfortunate that many of the relationships of Circle members are strained and many marriages have ended due to cancer. Fragrance's husband was initially supportive when her treatment began and blogged as "The Purple Crystal," reporting her treatment and recovery. But, after a year of ineffective treatments, Fragrance and her husband divorced in 2006. In many blog entries afterward, Fragrance recalled her marriage bitterly, explaining how her husband had been impatient with her while she was sick and how he left for his own "bright future" when she needed him the most, causing the following reaction:

> My love left me forever. It walked away, leaving me with endless sorrows. How do I mourn you, my love?

Most Friendly Circle of Cancer Patients members supported Fragrance through their blog postings, blaming the husband for their divorce. The Purple Crystal blog was shut down for several months because Fragrance's supporters posted increasingly negative comments to Fragrance's ex-husband and about him after their divorce and her death four months later. More than 500 people visited the blog, where they read 26 damning comments posted there suggesting that he should be stricken with cancer or that he would have to eternally repay the moral debt of leaving her. This public flaming of the husband demonstrates the importance of insider status, or membership within a group, versus outsider status. It is interesting to see how some members of the Circle, a group who feel ostracized due to their cancer status, ostracized the husband of a peer until the blog was shut down.

A diagnosis of cancer can change the vows "in sickness and in health." Angela and Desire were both abandoned by their husbands due to cancer. Desire's husband, who described himself as "a traditional Chinese man," wanted to have a son who could bear his name and care for him in his old age and left Desire after she was diagnosed with cancer to find a new wife "who could bear him a child." Desire writes:

> After eight years of a romantic relationship, I thought we would live happily forever, no matter what happened. Now I know all vows are temporarily beautiful; one cannot resist the cruel reality . . . I don't want to force him to stay with me in sympathy. I'd rather take the pain by myself. I know he is conflicted at the moment because we had great years together. But he admires and desires more of his next fifty years than I can offer.

Not all marriages are affected by cancer end in divorce. Liqiao, who was married for 20 years, wrote about frustrations in her marriage, thoughts of giving up the relationship, and "setting my husband free" in her early entries. But her later blogs express a desire for reconciliation:

> True love solves everything. Maintaining a marriage requires true love and proper devotion. I adjusted my mood and strategies. Then I won him back!

A patient can become physically and emotionally fragile due to cancer treatments. This emotional fragility is compounded by others' lack of knowledge in how to communicate with the patient and provide support. Peony's July 10, 2010, blog posting demonstrates her resolve not to let the changing nature of her relationships after her diagnosis affect her negatively. One of her friends promised repeatedly to visit her but did not and stopped sending routine holiday greetings. She reminded her Circle friends, "If being sick is destined to let us lose something, then lose this shallow friendship. Compared to those who lost their family and marriage, it's nothing!" ("Sentiment")

Xin'er considered closing down her blog and deleting all entries because she was upset that her colleagues and friends gossiped about her situation. She blogs:

> Classmates, please do not request any of my information and do not gossip about me. I beg you to keep it to yourself. I also beg my colleagues to give me some privacy. I know all of you have good intentions, but do you understand my feelings?

She decided to keep the blog open for the next four months—until her death—after many Circle friends left comments encouraging her to keep the blog open.

Cancer is not contagious. However, the word cancer itself can make people nervous and cause superstitious imaginings. There are taboos about communicating about the disease, such as the karmic belief that discussing bad news will cause further bad luck. Others avoid getting mammograms for fear that the test or discussion about cancer will result in a diagnosis. Zhaojun recalls in one blog entry that her husband defaced the "anticancer" label on her prescription bottle. She added on September 10, 2010, that when she put her checkup appointment reminder on the family calendar, her husband told her, "You are damned to mention the word 'cancer'; you are going to bring bad luck to yourself and our family again!" Her husband also asked Zhaojun to reschedule her appointment for a later date, so it would not interfere with their son leaving for college.

It is noticeable that there are more female patients than male patients in this blog circle, and the narrative styles between men and women are quite different. Women tend to be sentimental in terms of their perception about the disease, life, and relationships, whereas men appear to be more "rational" and "goal oriented" in an attempt to "overcome" the disease. Among the 21 bloggers chosen for this study, only 6 are male. Mercy categorizes his entries as all types of "treatments," such as spiritual treatment, intimate relationship treatment, selflessness treatment, fresh air treatment, and so on. In one entry written on February 22, 2008, in response to a worried Circle member whose parent was recently diagnosed with cancer, Mercy writes:

> You should be calm and be open minded . . . because the best doctor is you. To survive, you need to inspire your spirit and believe that you can conquer the cancer. Spiritual energy is a powerful atom bomb; it could destroy all the diseases.

On the Circle blog, women seem to be more emotional and easily inspired by others than men. An's Mom wrote of attending a surprise "fighting cancer" party. The party was in her honor, celebrating what was declared her third birthday of survival after surgery. The party was hosted and organized by Aunt Wan, a 75-year-old lady and 15-year cancer survivor. She writes:

> Thanks to Aunt Wan for giving me a special birthday. The evil disease can torture our bodies, but it cannot take away love and warmth from people. Where there is love, there is hope. Aunt Wan teaches me with her 15 years of experience fighting cancer that cancer is not horrible, and with spirit, we can make the cancer withdraw. I'm fortunate to fight cancer within the Circle group.

Theme 3. Friendly Circle of Cancer Patients as community support

Danqing keeps blog diaries for her son Shan'er, an 8-year-old boy with leukemia, the youngest cancer patient in this study. Danqing is a teacher in the remote northwestern Chinese province of Xinjiang, where the local hospitals do not have expertise in leukemia. As a single mother, she is the sole caregiver for her son. Her insurance plan—which is typical of most in China—does not cover children, so she is deeply in debt from her son's treatments in Beijing. Danqing was worried about spending six months in Beijing without financial or emotional support. However, she and her son have found comfort on the Circle blogs, where Danqing is highly respected for her devotion to her son. She writes:

> On this thorny journey of my son's rehabilitation, people who offered us help are cancer patient friends, relatives, classmates, colleagues, social workers, even passengers. Your loving encouragement in text messages, phone calls, comforting written words, or monetary donations are appreciated. No matter how small those kindnesses are, for us, they are warm sunshine in winter and the coal we need on snowy days. We'll appreciate your kindness for a lifetime. I will certainly pass down the generosity I received in the Circle to others in need.

Within the Chinese health care system, many cancer patients find the suggestions of fellow patients more credible than the admonitions of hospital authorities. Many bloggers within the Circle explain the cause of their cancer in nonmedical terms. Sage concludes on her blog that being overworked and having a lack of sleep, not watching her diet carefully, and a high-pressure job caused her cancer and cautions others against overworking (July 8, 2007).

Cancer, a six-year stomach cancer survivor, deduced that working on deadlines as a journalist and his carelessness about his health was the cause of his cancer. His "50 Benefits of Having Cancer," entry lists the importance of keeping a healthy lifestyle, spending more time with family members, and establishing a rapport with doctors and nurses. Some items on his list are negative, though, such as those commenting on the unfairness of China's health insurance policies.

Among the friends of Circle, exchanging dietary suggestions and the effectiveness of prescriptions are important benefits to members. Bloggers wholeheartedly believe that correcting dietary habits is an important step in overcoming cancer. Traditional Chinese medical treatment, such as herbal medicine, acupuncture, mediation, Qigong energy, and Tai Chi exercises are also widely discussed and encouraged. Patients and family members actively seek information and transmit it to others. The credibility of a person who has survived five years or more of cancer is important to Circle members, and his or her diet and rehabilitation procedures are often faithfully copied and shared with others. Most bloggers keep a diary about their treatment and rehabilitation, hoping someday it can be a guidebook for others who are diagnosed with cancer. Because the bloggers' entries are constantly available after their recovery or their death, the blogs continue to offer information and inspiration to other Circle members.

Emmanuel, who kept a blog for his mother who was a lung cancer patient to publicize his mother's prognosis, prescriptions, diet, and treatments for other patients and wrote about his plan to establish a nonprofit cancer charity community, to provide assistance to final-stage cancer patients, especially those who can't afford medical treatment or can't afford hospice care. One year after the death of Emmanuel's mother, he established the I Want a Miracle (www.51qiji.com) Web site, which is a platform for cancer patients, family members, health care professionals, and volunteers to communicate information about fighting cancer and donating prescription drugs to others.

The oldest blogger in this study, 75-year-old Zhang Linrong, has survived lung cancer for 20 years. He created a workshop to help others through their cancer diagnosis and treatment, sharing his own story and those of heroes who have fought cancer through living and eating healthily as a way to promote a natural cure. Zhang Linrong refused his second chemotherapy treatment, believing it would kill him, in 1990. Since then, he has utilized

Chinese herbal medicine, dietary treatment, and consistent exercise to fight cancer. His natural and holistic approach has been recommended to many cancer patients as a complement to surgical treatments and chemotherapy, although there are some who appeared to use Zhang's suggestions as their sole cancer treatment.

Discussion

Schiavo (2007) notes that health campaign messages about cancer research focus on progress, benefits, and hope, though intended audiences receive these messages differently. It is still very common in China that medical authorities give patients ambiguous information about their conditions and let the family members make decisions for them, especially if the afflicted are elderly or in advanced stages of disease. The belief is that the patient is too weak to receive the diagnosis and will not have the spirit to fight the disease. Several of the bloggers in this study did not know initially that they had cancer, being that doctors and family members kept the information from them. The first author of this study had relatives and family members of friends who were not informed of their cancer until the day before they died. However, in the age of the Internet, and access to information about living with cancer, the ambiguity of not telling the patient he or she has cancer can create additional complexity, especially for patients whose symptoms are very similar to cancer, who are told they are suffering from another illness. There is an absence of a voice of authority when doctors and family members choose not to tell the patient that they have cancer. Most bloggers in the Friendly Circle of Cancer Patients received second or third opinions from out-of-network doctors and experts because they believed their doctors lack credibility. When a blogger is deemed cured of cancer, his or her prescriptions and methods are quickly copied by others, which further fuels the distrust of medical experts and hospitals among patients. The Chinese proverb "A small recipe can cure a major disease" provides insight into the importance of following a Circle blogger's regimen, as some people hope—and believe—that by following the current correct diet or treatment, the cancer cells will disappear. Medical professionals, however, caution patients and family members about various "myths" concerning cancer treatment that are spread through word of mouth and on the Internet. Lustig and Koester (2006) conclude that in an intercultural communication context, there exist different health care approaches: Westerners believe in biomedical data and treatment, whereas Africans and Asians believe in magico-religious and holistic approaches. Therefore, communicating with Asian family members and the patient him or herself is important.

Conclusion

In summary, and to address the original research questions:

1. There are several cultural barriers in communicating about cancer between patients, family members, and health care providers in China. As mentioned above, there is often a misunderstanding between doctors and patients. In Western health communication, building a relationship of trust and a rapport between doctors and patients is important. But, in China, it is difficult for cancer patients to receive individualized attention. In many hospitals, it is unfortunate that there is the unwritten rule that patients and family members need to bribe doctors and nurses in order to guarantee that they get effective treatment. Even though cancer is not contagious, many members of the public still fear and stigmatize cancer patients and are apprehensive about interacting with them. The first author experienced this fear several years ago, when a friend's husband was diagnosed with lung cancer. She found herself unable to communicate directly with the patient and would change the topic when he mentioned he had a severely fatal disease. Overprotection and oversensitivity may make cancer patients feel even more stressed or isolated.
2. Digital media use has influenced the public's understanding of cancer prevention, diagnosis, and treatment. The growth in the numbers of young and educated Internet users is reflected in the population of Friendly Circle of Cancer Patients. Most of the bloggers in this study are in their 20s to 50s and have college degrees. This group is active in seeking health care information, and they feel an obligation to share the information they find with others, especially through their blogs and the Circle site. Many describe the intention of their blogging as keeping others informed of their own condition but also supplying health information to others. Thus, the Friendly Circle of Cancer Patients has served as a health campaign, promoting health messages among cancer patients and to the wider society.
3. The above findings add to our understanding of minority invisibility. Sun (2007) and Sun and Starosta (2006) argue that members of various cultural groups have always been invisible in mainstream cultures. Those invisible group members face the reality that their concerns are downplayed, ignored, or simplified in a discriminatory way. The interactions toward those unrepresented and unprivileged are sometimes conscious or unconscious and deliberate or nondeliberate. However, the damage is often the same, resulting in a demeaned existence and lowered self-esteem. Cancer patients in China can be considered a

minority group, as their voices are unheard and their needs unsatisfied. In China, this group of people, along with the unemployed, uninsured, and less educated, are generally called a "disadvantaged population," being that they have less economic and political power than others. Although the Circle blog is helpful to many who suffer from cancer, a large percentage of cancer patients do not have Internet access and access to online resources. They are absent and invisible from society, and their life stories and battles with the disease and financial challenges are not acknowledged or considered by the general population. Medical professionals should be especially aware of this and seek to provide them with as much information as possible.

References

Charon, R. (2006). *Narrative medicine: Honoring the stories of illness.* New York, NY: Oxford University Press.

Charon, R. (2009). Narrative medicine as witness for the self-telling body. *Journal of Applied Communication Research, 37*(2), 118–131.

Chen, J. (2007, September 12). A cancer girl and her blog circle. *People's Daily.* Retrieved from http://news.sina.com.cn/s/2007-09-12/045513870620.shtml

China Internet Network Information Center. (2010, July). *Statistical Report on Internet Development in China.* Retrieved from http://www.cnnic.cn/uploadfiles/pdf/2010/8/24/93145.pdf

Goffman, E. (1963). *Stigma: Notes on the management of spoiled identity.* Englewood Cliffs, NJ: Prentice-Hall.

Gruman, J. (2007). *AfterShock: What to do when the doctor gives you—or someone you love—a devastating diagnosis.* New York, NY: Walker & Company.

Herring, S. C., Scheidt, L. A., Bonus, S., & Wright, E. (2004). Bridging the gap: A genre analysis of weblogs. *Proceedings of the Thirty-Seventh Hawaii International Conference on System Sciences*, HICSS-37. Los Alamitos, CA: IEEE Press.

Internet World Stats. (2011). China Internet statistics June 2011. Retrieved from http://www.internetworldstats.com/asia.htm

Kahn, R., & Keller, D. (2004). New media and Internet activism: From the "Battle of Seattle" to blogging. *New Media & Society, 16*(1), 87–95.

Kim, E. (2007). Blog power: Examining the effects of practitioner blog use on power in public relations. *Public Relations Review, 33*(1), 92–95.

Li, H. J. (2008, April 29). Chinese cancer deaths increased more than 80% in the past 30 years.*Caijing Newspaper.* Retrieved from http://health.sohu.com/20080429/n256589557.shtml

Liu, S., & Chen, G.M. (2010). Communicating health: People, culture and context. *China Media Research, 6*(4), 1–2.

Lustig, M., & Koester, J. (2006). *Intercultural competence: Interpersonal communication across cultures.* Boston, MA: Allyn & Bacon.

Meisenbach, R. J. (2010). Stigma management communication: A theory and agenda for applied research on how individuals manage moments of stigmatized identity. *Journal of Applied Communication Research, 38*(3), 268–292.

Mirivel, J. C., & Thombre, A. (2010). Surviving online: An analysis of how burn survivors recover from life crises. *Southern Communication Journal, 75*(3), 232–254.

Orgad, S. (2005). The transformative potential of online communication: The case of breast cancer patients' Internet spaces. *Feminist Media Studies 5*(2), 141–161.

Schiavo, R. (2007). *Health communication: From theory to practice.* San Francisco, CA: John Wiley & Sons.

Scollon, R. (1999). Mediated discourse and social interaction. *Research on Language and Social Interaction, 32*(1,2), 149–154.

Smith, R.A. (2007). Language of the lost: An explication of stigma communication. *Communication Theory, 17*, 462–485.

Starosta, W.J. (1988). A national holiday for Dr. King?: Qualitative content analysis of arguments carried in the *Washington Post* and *New York Times. Journal of Black Studies, 18*(3), 358–378.

Stein, M. (2007). *The lonely patient: How we experience illness.* New York, NY: William Morrow.

Sun, W. (2007). *Minority invisibility: An Asian American experience.* Lanham, MD: The University Press of America.

Sun, W., & Starosta, W. J. (2001a). As heavy as Mount Taishan: A thematic analysis of Wang Wei's memorial website. *World Communication, 30*(3,4), 61–79.

Sun, W., & Starosta, W. J. (2001b). A thematical analysis of 20th century classical Chinese fairy-tales collection, an implication of conflict management. In G. M. Chen & R. Ma (Eds.), *Chinese conflict management and resolution* (pp. 73–84). Westport, CT: Greenwood.

Sun, W., & Starosta, W. J. (2006). Perceptions of minority invisibility among Asian American professionals. *Howard Journal of Communications, 17*(2), 119–142.

Sun, W., & Starosta, W. J. (2008). Covering "The Lord of the World": Chinese sense-making of the U.S. 2004 Presidential election. *Human Communications, 11*(1), 1–16.

Weiss, J. B., & Lorenzi, N. M. (2005). Online communication and support for cancer patients: A relationship-centric design framework. *AMIA 2005 Symposium Proceedings* (pp.799–803). Retrieved from http://www.ncbi.nlm.nih.gov/pmc/articles/PMC1560480/pdf/amia2005_0799.pdf

Wodak, R., & Meyer, M. (2009). *Methods of critical analysis discourse.* Thousand Oaks, CA: Sage.

Yan, K. (2006). A chronicle of blog development in China. *People Online.* [Web log post.] Retrieved from http://media.people.com.cn/GB/40606/4170838.html

Yang, Y. H., Zhao, T. J., Yu, H., & Zheng, D. Q. (2008). Research on blog. *Journal of Software, 19*(4), 912–924.

Section Three: Negotiating Community

What happens in praxis when new digital media are implemented across cultures is contested and negotiated within complex local and political conditions. The five chapters in Section 3 focus on the taboos and constraints, as well as the opportunities and empowerments afforded by new media. Contributions to this section are united by an exploration of oppositional and dialectical negotiation of media use in each community context: by gendered, generational, ethnic, and national cultural groups. The increasing global importance of mobile media in the current decade is illustrated by the fact that three of the five chapters in this section examine community use of mobile technologies.

In the first chapter, Joonseong Lee examines the social trends and political meanings surrounding the recent rise of "cyber-memorial zones" in the Republic of Korea, within its contemporary neoliberal economic and political context. Drawing on interviews with funeral service professionals in Seoul, Lee describes how traditional rites of death have changed—cremation is on the rise, and funerals are increasingly commercialized and regulated. Web-based cyber-memorial zones have been created—and even promoted— by government policy in order to allow mourners to create and post memorial letters or media files of the deceased and to ostensibly ease their sense of loss and alienation, particularly under the conditions of cremation. Lee argues that cyber-memorial zones are in fact primarily an economic ploy implemented by the funeral industry, in collusion with the government. Through the combined theoretical lenses of Deleuzian and"culture as governmentality" perspectives, he argues that the current commercialization of Korean funeral culture subordinates the traditional Confucian cultural foundation to the deterritorializing power of capital flows. In this scenario, new media, such as cyber-memorial zones, are argued to both empower individuals and permit whole-scale cultural shifts by simultaneously offering solace for the bereaved and facilitating a neoliberal erosion of traditional cultural practices.

In the next chapter, Gado Alzouma explores the ways in which mobile phones offer the largely illiterate population of Nigerien migrant workers a blend of advantages and disadvantages in navigating the changing context of globalizing work and worker mobility. Based on interviews with a group of migrants from Niger now living in Nigeria, Alzouma reveals how mobile telephony is facilitating a new negotiation of relationships with and membership in their extended "transnational" families. Mobile phones offer new means of overcoming family fragmentation, keeping kin connected in reliable, intimate, and immediate ways not previously available. These same affordances, however, bring with them a suite of new challenges: unwanted contacts and obligations, such as requests for money, from extended family members. Alzouma rejects a purely instrumental or deterministic view of these communication technologies as having a "transformative" influence on societies and argues against clear-cut notions of *before* and *after* concerning these migrants' family relationships. Instead, he proposes that new media has not fundamentally changed the character of their family ties, the people involved, and the general configuration of the family but has only allowed the migrants and their families to adapt to new economic and social contexts.

The last three chapters of this section focus on the intersections between gender and new media, exploring how new technologies in some cases trouble the existing balance of gender relations within a community and in others are employed to maintain the status quo. First, Robert Shuter explores how the interpersonal "rules" of text messaging reflect the preexisting power inequities and societal disciplining of Indian women. In interviews, Shuter discovered that Indian women (but not men) experience strongly negative reactions from parents, extended family members, husbands, and male friends when sending or reading text messages in their presence. In addition, "eve teasing"—face-to-face harassment of women by men in their own community—now appears to extend to mediated contexts, including text messaging. Differing "textiquettes" for women and men in India, Shuter argues, both reflect the existing gender disparities in India and highlight the degree to which text messaging is viewed as a potential threat to traditional Indian patriarchy. Agreeing with other scholars who assert that the technology itself is gendered, Shuter offers this case study as an illustration of the theoretical proposition that the language of science and technology is conflated with masculinity (in both traditional and postmodern societies)— posing unique challenges for women who aspire to education and greater autonomy.

From Israel, Azi Lev-On and Rivka Neriya-Ben Shahar report the results of their quantitative study examining the practices of ultra-Orthodox Jewish women who use modern technology (in particular, the Internet) for purposes that are considered illegitimate in their community. Unlike the majority of the ultra-Orthodox community who still do not use the Internet, this cohort of women occupy a privileged position at the crossroads between modernity and Orthodoxy. The authors analyze the womens' perception of the impact of the Internet on themselves and others and describe what they call a *third-person effect*—the perception that even though the Internet may be harmful to the ultra-Orthodox lifestyle, and to their own religiosity, it is more dangerous to others than to themselves. The authors speculate that these views may be a strategic means of rationalizing the continuing dissonance between rabbinical decrees regarding the dangers of new technology as well as the opportunities that new technologies afford them. It will be interesting, they speculate, to discover whether this third party effect is strategically adopted by other groups in Israel and around the world as an identifiable stage in the process of adaptation of and to new media.

In the final chapter of this section, Carla Ganito and Cátia Ferreira explore how the cultural rituals of telling and preserving family stories through family photo albums—traditionally the responsibility of women—are changing, as a result of mobile phone photography. Employing in-depth interviews with Portuguese women representing seven different age groups, the authors investigate the uses and meanings of mobile phone cameras and conclude that roles are indeed changing. Using Foucault's and de Lauretis's notions of *technologies of gender*, they argue that even though women remain the primary keepers of family photographs and memories, the primary rituals of the new "mobile family album" are organized around the *production and collection* of photos and not the *organization and sharing* of them. Because mobile phones are personal devices (which are not easily shared) and digital photos are rarely printed, mobile phone photography is bringing about a shift in traditional forms of storytelling and memory production. Ganito and Ferreira argue that these new mobile technologies are increasingly being used as collaborative and creative tools, and as digital family archives, but that family photos are now becoming fragile and fragmented objects without much order or sequence.

11. Rite of Death as a Popular Commodity: Neoliberalism, Media, and New Korean Funeral Culture

JOONSEONG LEE, PHD

Introduction

South Korean funeral traditions, heavily influenced by Confucianism and little changed for centuries, are now experiencing great change. Small-scale mortuaries have dissipated, while large commercialized funeral parlors, mostly run by large hospitals, have emerged. Hospitals that previously subcontracted services to the funeral parlors residing on their premises now operate the parlors directly. Independent funeral parlors have also flourished, with changes in government policy aimed at promoting cremation to resolve a shortage of burial grounds. During the last two decades, the cremation rate has risen sharply, along with the number of large-sized columbaria.

Among these trends, cyber-memorial zones are considered the commercial vanguard for attracting the attention of mourners, a promotional tool of the funeral industry. In cyber-memorial zones, people memorialize their loved ones by writing memorial letters and posting photos or movies and setting up cyber-altars for remembering the deceased. During my fieldwork in 2005 in South Korea, I found that cyber-memorial zones had been developed to comfort mourners whose loved ones had been cremated. Cremating a body leaves the bereaved with only a handful of ashes, whereas a grave becomes the tangible connection between the bereaved and the deceased. For this reason, the bereaved may have a greater sense of loss after cremation than in ground burial, and memorial letters or media files of the deceased in cyber-memorial zones have served as a way to ease this sense of loss.

This chapter describes how traditional rites of death have changed into a popular commodity related to the neoliberal changes occurring in Korea, resulting in new Korean funeral cultural traditions—commercialized funeral parlors run by hospitals, drastic hikes in cremation rates, and cyber-memorial zones.

Exploring South Korea's new funeral culture: Two perspectives

Neoliberalism focuses on "the 'self-limiting' state, unregulated investment capital and the 'free-trading' open global economy" (Peters, 2001, pp. 207–208). A question arises, then, as to why Korea adopted neoliberal policies. Hundt (2005) points out that the introduction of neoliberal economic policy was a mandatory requirement for the International Monetary Fund (IMF) loan package offered to South Korea (p. 242), as the most pertinent remedy for overcoming the financial crisis of 1997. As a result, the neoliberal paradigm provided the Korean government options with which to handle the crisis.

The wave of democratization in the 1990s allowed industrial conglomerates, or *chaebols*, to be geared toward liberalized marketing without the direct state intervention, reflecting the weakening of state power in controlling the policy-making agenda (Hundt, 2005, pp. 242–245). However, the 1997 economic crisis provided the momentum for the government to regain controlling power of policy making. By carrying out neoliberal policy, the government was able to gain support from the people who were outraged by the poor management of *chaebols* by the large, conglomerate, family-controlled firms of South Korea, which were characterized by strong ties with government agencies, as well as from the US government and international financial institutions (Hundt, 2005, p. 257). However, the effect of the paradigm has been a shake-up of the Korean socioeconomic basis for possible liberalization of the medical sector and public education. In the following sections, two perspectives are explored that offer a better understanding of the new Korean funeral culture in the neoliberal turn.

A Deleuzian perspective

Deleuze and Guattari (1983) insist that the main characteristic of capitalism is "[t]he decoding of flows and the deterritorialization of the socius" (p. 34). They claim (1987) that capitalism's power of deterritorialization "consists in taking as its object, not the earth, but 'materialized labor,' the commodity" (p. 454). With the power of deterritorialization, capital as "convertible abstract rights" (p. 454) constantly flows beyond social realms such as nation-state, institution, or neighborhood. In this sense, the characterization of "[t]he neoliberal paradigm as globalization" (Peters, 2001, p. 208) appears reasonable. With the deterritorializing power of capitalism, neoliberalism can be seen as "one of the most pervasive, if not, dangerous ideologies of the 21st century" (Giroux, 2004). Giroux's argument suggests important points for understanding the commercialization trend in the Korean funeral culture in its highlighting of the capacity of capital to de-historicize society. The deterritorializing power allows capital flow to move beyond the territorial boundaries, including the permeation of the perceptual boundaries of human existence: life and death.

In addition to causing capital to flow beyond social realms through deterritorialization, capitalism also has a tendency to restore the territorialized social realms to facilitate capital flows. Considering the contradictory characteristics of capitalism, Deleuze and Guattari (1983) state that capitalism reestablishes "all sorts of residual and artificial, imaginary, or symbolic territorialities" (p. 34). The deterritorializing power of capitalism allows the flow of capital to penetrate multiple realms, but capitalism is also inclined to restore all the territories, though not in exactly their original form. In a paradoxical way, whereas capitalism needs the process of deterritorialization to enhance capital flow, it also calls for the process of reterritorialization to reestablish territories that can smooth capital flow.

The process of reterritorialization does not, however, aim to tighten the control of the state, being that excessive state power cannot make the flow of capital efficient. In capitalism, the schizophrenic processes of de-and reterritorializing social realms occur for maximizing capital gains. Although these two territorializing processes are contradictory, they share the same destination. The commercialization trend in Korean funeral culture in the neoliberal paradigm has the schizophrenic traits of capitalism. Deterritorializing capital flow has the effect of taking apart the meaning of the rite of death and changing it into a commodity. In contrast, government funeral policy reterritorializes the realm of the funeral culture. Such schizophrenic traits of capitalism that are embedded in Korea's new funeral culture call for closer attention to Foucault's notion of *governmentality.*

Culture as governmentality

Governmentality, which Foucault (1988) suggests to be the "contact between the technologies of domination of others and those of the self" (p. 19), is the result of contemplating the role of the "schizophrenic" and "immanent" flow of capital in the technologies of domination. In articulating the art of government in the modern states, Foucault (2000) addresses the concept of *economy* in maintaining the continuity between the rulers and the ruled (pp. 206–207). Because economy for Foucault suggests the flows of capital within a state, the governing techniques in modern states center on how to control "the sort of complex composed of men and things" (p. 208), constructed through capital flows. The term "things" includes "means of subsistence," "ways of acting and thinking," and "accidents and misfortunes" (p. 209). By "complex," Foucault insinuates that the abstract and immanent yet constant flow of capital functions as the technology of domination and influences the relations between men and things.

This is the basis on which Bennett (1992) proposes that culture is governmentality, insisting that culture as governmentality provides the "means of a governmental intervention in and regulation of culture" (p. 26). The object of government, such as "morals, manners, codes of conduct," becomes the domain of controlling culture (p. 26). Bennett's view of culture as governmentality is pertinent for understanding the proliferation of funeral parlors at medical centers in Korea and the government intervention. Above all, the discourse on hygiene and sanitation that has been formulated in Korea since colonial modernity consolidated the technique of *noso-politics* (Foucault, 1980) to regulate culture. Noso-politics expanded its influence because the discourse of hygiene created a culture in which the universe is ordered through "selective perception and labeling" (Bradbury, 1999, p. 119). In this cultural discourse, living is considered to be an order; death a disorder. Moreover, the decomposition of the body causes pollution, which becomes a threat to life (Bradbury, 1999, pp. 119–120). According to the clear-cut logic of this division, the discourse of hygiene becomes a dominant "regime of health for a population" (Foucault, 1980, p. 175), justifying the government's and the hospital industry's establishment funeral parlors at medical centers.

As governmentality, culture is equated with "a whole way of life, material, intellectual, and spiritual" (Williams, 1983, p. xvi). From Deleuze and Guattari's perspective, culture as a way of life becomes an immanent realm where de- and reterritorializing social realms constantly occurs, enhancing capital flow. With culture as governmentality, the technologies of domination are internally exercised on the relations of humans and things, and technologies of the self are constructed corresponding to the application of technologies of domination. Tension always exists at the threshold between the technologies of domination and those of the self.

Such technologies of dominance are associated with the process of de- and reterritorialization in capitalism. In this power structure, technologies of domination are closely interconnected with the flow of capital, both of which exercise immanent power on the relation between men and things. The exercise of power becomes immanent and rarely perceivable because capital functions as "convertible abstract rights" (Deleuze and Guattari, 1987, p. 454) and the flow of capital in the territorializing process is obscured.

When the flows of capital de- or reterritorialize social realms, technologies of domination use the power immanent in de- or reterritorializing the complex constructed between men and things. Thus, as capitalism progresses to the abstract and immanent, so do governmental techniques. Culture as governmentality becomes less discernible in the neoliberal paradigm, in which the flow of capital is extremely profit oriented. It becomes an attractive technique of governing because capital flow maximizes profit creation, and capital flow is increased without losing the efficiency of governing power. The commercialization trend in the Korean funeral culture can be considered in this context.

New Korean funeral culture in the neoliberal paradigm

Funeral parlors at medical centers

How, then, is the neoliberalization of the medical sector relevant to the emergence of the new funeral culture? Most Korean medical centers currently run or subcontract funeral parlors, meaning that hospitals have become the main places for funeral ceremonies. When patients die, their bodies are sent to mortuaries that belong to the funeral parlors run by these hospitals. Even when someone dies at home, the body is usually moved into a funeral home located in a hospital. In this context, if medical centers, which were previously regarded as existing for the benefit of the public, turn toward profits under the direct influence of the neoliberal paradigm, then the commercialization of the funeral parlors in these medical centers will undoubtedly be accelerated.

In addition, when commercial specialists conduct the funeral process, the bereaved become alienated. The specialists and the bereaved have a commercially oriented relationship. Song (2003) points out that funeral parlors at medical centers attempt to simplify the funeral ceremony; however, they preserve *yomsup* and *ipkwan*, which refer to "dressing and placing the dead body in a coffin" (Lee, 1996, p. 53). As in the case of the viewing ceremony in the typical commercialized funeral of the United States, the funeral parlors at medical centers make the most of yomsup and ipkwan for their profits while preserving the traditional funeral ceremony (Song, 2003, p. 309). Korea has not fully opened its medical sector yet, but the current trends allow us to predict the commercialization of its funeral culture.

Before the 1980s, according to Lee (1996), people considered dying away from home as a "bad death" (Bradbury, 1999) that should be avoided. For this reason, mortuaries in medical centers were mostly used for handling cases of "accidental deaths, deaths at hospitals, and deaths on the road" (Lee, 1996, p. 55). However, the trend of medical centers running funeral parlors has proliferated since the mid-1980s (Lee, 1996), when the process of state-driven industrialization came to a climax. Industrialization changed the traditional Korean family structure from consisting of extended to nuclear families; therefore, when a death occurred, the traditional communal way of the family helping with the funeral process could not be expected.

Furthermore, urbanization forced people to adapt themselves to small, high-density housing, which entails the "restriction of spatiality" (Song, 2003, p. 299) for performing funerals. As a result, funeral specialists replaced the work of the community, and dying at the hospital came to be regarded as convenient and hygienic.

In 1982, the Korean government made it mandatory that medical centers with more than 100 patients establish mortuaries within the center (Song, 2003, p. 302). According to Song (2003), funerals for those who died in medical centers were frequently held in these mortuaries, although this was illegal at the time. Since 1993, based on the revision of the Standing Rule of Family Ritual, enacted under the rein of the former president Park Jung Hee, mortuaries could legally provide funeral ceremonies. Song argues that active government interventions in collaboration with the hospital industry enabled the proliferation of funeral parlors run by medical centers. These changes demanded more educated funeral professionals, such as morticians and funeral directors, prompting some colleges to establish departments of mortuary sciences. The funeral professionals that I interviewed were graduates from these colleges.

Most of the professionals showed positive attitudes to the industrialization of the funeral culture and the opening of the medical and funeral sector to foreign investments. They pointed out that in the past, the funeral culture operated in an unsanitary environment; now the industry is competitive and the quality of service is very high. A funeral professional, Jung-tae, testified to his attempts to market the services of a provincial funeral parlor as its managing director in a region where there are 65 average monthly deaths. After learning that another funeral parlor was planning to open in the same town, his company invested money to make his parlor more luxurious, including purchasing a Cadillac hearse. This marketing strategy proved effective, and the new parlor did not open. Jung-tae's experience reflects the market-driven and commercialization environment of the new Korean funeral culture.

A dramatic increase in cremation

A dramatic increase in cremation is another characteristic of the new Korean funeral culture. In the 1980s, the cremation rate was less than 15%, which constitutes only a 10% increase from the preceding three decades (Song, 2003, p. 294). It seems that industrialization had not influenced a change from ground burial culture or the traditional way of disposing of bodies. However, since the late 1990s, the cremation rate has increased sharply: almost half of the deceased across the country were cremated in 2003 (Song, 2003, p. 294). In large cities such as Seoul or Busan, the annual cremation rate was more than 60% that year, which is much higher than in rural areas.

Furthermore, nearly 90% of Korean people now prefer cremation to ground burials (Kim, 2010). The ostensible reason for this sudden change can be attributed to a shortage of burial space, which would make it difficult to consign a body to the grave. The habitual inclination of hygiene discourse can be ascribed as another reason behind the increase in cremation. The modern habitus of hygiene and sanitization encouraged Koreans to stay away from the traditional funeral culture and adapt themselves to a simplified and practical approach (Lee, 1996, p. 54). Based on these practical considerations, Koreans presently do not have as many meaningful connections to the traditional funeral process (including ground burials) as they did in the past.

In this context, the traditional shamanic perspective that views the body as an important medium for maintaining balance and harmony between this and the other world becomes less popular, as reflected in the changing funeral tradition of *imjong*, which involves "keeping watch with the dying person" (Lee, 1996, p. 53). In the traditional funeral culture influenced by Confucianism, it was considered a filial duty that descendants perform imjong for a dying person's last moment. Performing imjong means caring for a body in the threshold between order and disorder, representing hope for a safe journey to the other world and the recovery of the broken balance caused by death. But imjong as carried out by funeral parlors diminishes its symbolism thereby making light of the meaning of the body (Lee, 1996). The government-driven funeral policy, therefore, also becomes a catalyst for the increase of cremations (Song, 2003; Kim, 2003; Chun, 2003).

In the 1990s, the Korean government attempted to fully revise its funeral policy, which was established during the Japanese Occupation (1910–1945). At this time, the government eased its policy on cremation and the establishment of charnel houses but made it difficult to consign the body to the ground. Whereas the approval of the government is now required for ground burials, cremation only needs to be reported. Government attempts to implement this policy faced strong opposition initially, but the revised policy was enacted in 2001, encouraging cremation and expanding the establishment of public charnel houses as well as family vaults in order to maximize land use efficiency.

An interesting point in this process is the role of the media. Most print and electronic media reported the problem of burial and the benefits of cremation from one perspective (Chun, 2003, p. 152), supporting the government-driven policy of cremation. The main subject of the reports was the problem of graveyards. The cremation discourse gained power as time progressed. The media tendency faces the criticism that the preference of the discourse on cremation prevented people from recognizing the more fundamental problems of ground burial than the claim that it encroaches on lands. Chun argues that the bigger problem with ground burials is the illegal graveyards that account for 70% of the total burials (graveyards for the unknown comprise 40% of these) and the graveyards for the elites. If these problems could be resolved, the burial system would not be a problem. Kim (2003) concurs that although the media condemned the graveyard problem, it remained silent on the issue of golf courses—more problematic in terms of encroachment of land space. Thus, media discourse functioned as an instrument of governing techniques, leading people to form a preference for cremation.

Cyber-memorial zones: An instrument of governance

Among these trends, cyber-memorial zones, another component of the new Korean funeral culture, are in the commercial vanguard as a means of attracting the attention of mourners. However, cyber-memorial zones began as an instrument of governance. According to one government official, the attempt to computerize the outdated funeral management system motivated the creation of cyber-memorial zones because burial and cremation information had been incorrectly recorded in many cases. In the process of computerizing the system, the government recognized that computerization could not only be used to organize the outmoded internal system but could also be the most appropriate instrument with which to disseminate a discourse of cremation preference. This situation, representing a combination of computerization and government-driven information policy, is another example of a top-down strong government intervention.

The Seoul Metropolitan Facilities Management Corporation (SMFMC) is a public corporation in charge of the management of public facilities in Seoul, including various columbaria and cemeteries in the Seoul area, as well as cyber-memorial zones. According to SMFMC, cremation has become a common practice in Korean society, and different ways to revere the dead have been introduced. SMFMC have provided hardcover memorial books since July 1997, consisting of memorial letters written by the bereaved. Some memorials that had been written in these books were later published in the series called *Tearful Letters*.

In light of the growing popularity of the Internet, SMFMC recognized the need for a digital version of such memorial books and thus started operating a cyber-funeral home, Post Office in Heaven, which also served as a memorial-writing Web site. Post Office in Heaven, one of the oldest cyber-memorial sites in Korea, began as a substitute for memorial book services to promote cremation but led to the huge popularity of the cyber-memorial zone alongside the blossoming of cyber-cultures in Korea.

Figure 1 shows the Web site of Post Office in Heaven in the cyber-memorial zone. As of January 2011 (the date shown in figure 1), more than 62,000 visitors have posted memorial letters to the deceased on the site. The

Figure 1. Post Office in Heaven established by the SMFMC

** In all figures, to protect users' privacy, some contents have been blurred, including the names of the bereaved who posted memorial letters in cyber-memorial zones, the names of the funeral service companies, and the obituary information of the bereaved.*

following was posted to Post Office in Heaven by a bereaved daughter to her father who had passed away not long before:

> Dear Dad,
> How are you? We have heavy snow today. All the places of the city were laden with the snow. Feeling the snow falling down on my shoulders, I remembered the time when you and I built a snowman together . . . But I still cannot believe you live in another realm absolutely different from where mom and I live. Isn't the place you live now cold? I hope not. When can I see you again? I also want to be with you. As you know, you were all in my life.

Mourning her father's death, a young female looked back on the memories that she shared with her father. Although it had been two years since her father died, she shared the same bodily and spatial recollection with her father in this memorial letter. In expressing that her father was "all in [her] life," she attempted communicate with her dad beyond the border between this world and other world.

A slightly different approach is evident in the following:

> Daddy, I hate you. Aunt told me that you endured the horrible pain but did not let us know of your illness. Why did you do that . . . Aunt said to me, "Your dad called me one day and asked me to watch out for you and your brother in the future." Do you know how much I sobbed to hear that? Are you happy to leave us like that without saying anything?

In contrast to the former letter, here the daughter's sharing of memories with her late father does not lead to comfort. The opening sentence, "Daddy, I hate you," insinuates her emotions toward her father. The letter allows her to purge her heartbreaking feeling toward her father in the other realm. As the above memorial letters from Post Office in Heaven demonstrate, for the bereaved, the memorial writing zones have become a significant place to communicate with their late family members or loved ones.

Thus, although cyber-memorial zones began as an instrument of governance, the successful cyber-memorial zones of today are not what the government intended when they created these zones. Funeral professionals point out that cyber-memorial zones are a byproduct of changing funeral trends that combine information technology and the new funeral culture. A comment by a funeral industry professional that I met during my fieldwork implies that the advance of information technology has facilitated the adoption of the cyber-memorial zones within the funeral industry:

> At first, cyber funeral culture was difficult for Koreans to accept. But as computers and the Internet are popularized, people better understand it and it's settling down in the Korean society. They might feel empty about cremation and charnel, but the cyber funeral will be a good way to soothe their hearts.

This comment can also be associated with a government official's acknowledgment that cyber-memorial zones were developed to appease the mourner's distress at having to cremate their loved ones. According to one government official, the bereaved are more likely to experience a greater sense of loss in instances of cremation than in ground burial. For this reason, memorial letters or media files of the deceased in cyber-memorial zones may be a good way to ease this sense of loss.

Chul-min, a salesperson working in a charnel house, suggests that there is a correlation between the new trends of cremation and such memorial writings:

> Koreans tend to have something tangible for the worship of forebears. In burials, graves were regarded as the space where the relationship between ancestors and descendants were constructed and preserved. But cremation leaves nothing except a handful of ashes. The government needed something tangible in

> introducing the scattering model so that people could accept the new model without resistance. . . . Memorial writings can be considered in this context.

It can be inferred that SMFMC was well aware that, as Aries (1974) suggests, cremation is a method that completely eradicates the connection between the past and the present, and that cremation and scattering are easier ways than ground burial for the bereaved to stay away from the deceased. However, SMFMC was also concerned that there should be something tangible for the bereaved to relieve themselves from the sorrow. Cyber-memorial zones and memorial books were created and promoted to encourage people to adopt cremation.

Cyber memorial zones: The commercial vanguard

Because of their popularity, the main concept of memorial writing zones—cyberspace used as a medium to communicate with the departed—was adopted and developed into commercialized cyber-memorial zones. Figure 2 shows a cyber-memorial letter Web site called Letters to Heaven hosted by a funeral service company that has led the funeral service industry since cremation

Figure 2. A cyber-memorial writing zone at a charnel house

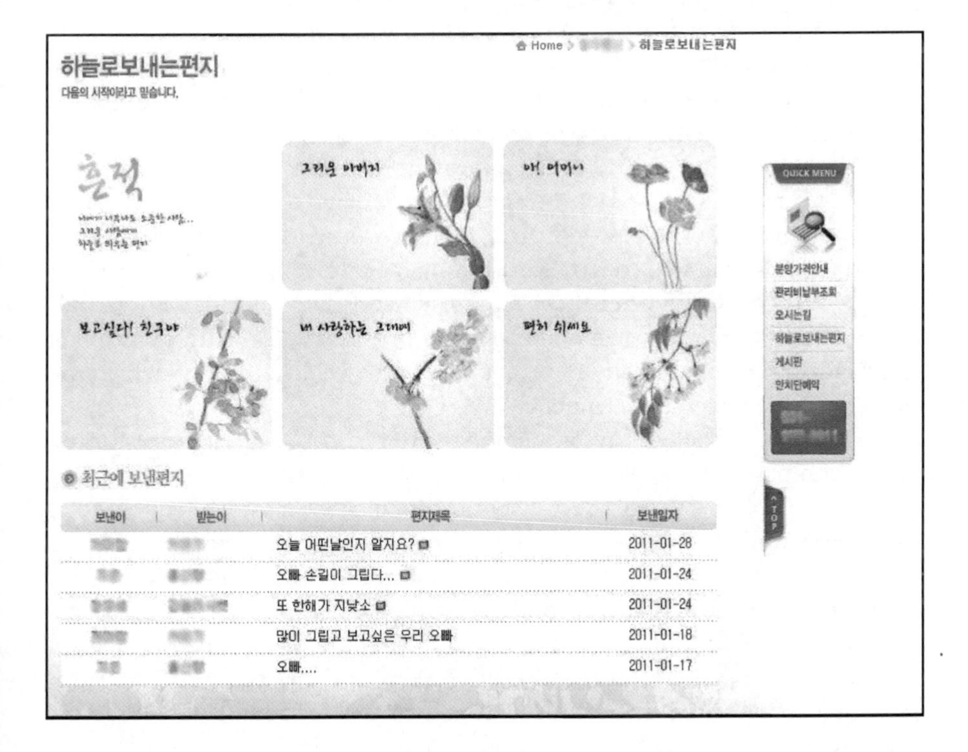

became common in Korea. Well known for its active marketing strategies targeting the bereaved, the company's home page is directly linked to the memorial writing Web site, Letters to Heaven.

As shown in figure 2, Letters to Heaven includes different memorial writing sections in which the bereaved can write letters to the deceased, whether they are their fathers, their mothers, or their friends. These letters do not look very different from the memorial letters posted on Post Office in Heaven, the difference is the way in which cyber-memorial writing zones are promoted. On this site, memorial writings are considered an event. I interviewed Ki-sung, who worked for this funeral service company as a funeral professional. While working for the company in the early years of Letters to Heaven, he promoted the notion of memorial writing as an "event" to promote a private cinerarium, as follows:

> For the private cinerarium business, how to increase the numbers of charnel house reservations is the key to success. When I suggested the event, I considered the usages of memorial writing zones as the potential source for promoting the company as well as generating profit. In other words, the main purpose of creating cyber memorial zones is to gain profit. The prospect and efficiency were considered at the same time.

The event of writing memorial letters became successful, and many bereaved family members participated in the event by posting memorial letters to Letters to Heaven within the designated days. It was an event only accessible by the bereaved family members, whose loved one's remains were laid to rest in the charnel house. The event was successful, particularly because the idea of writing a memorial letter and posting it online was new. Ki-sung was a wedding photographer and sometimes produced movies of funeral ceremonies before he joined the funeral industry. Having worked in the industry since the early 1990s, he realized that Korean funeral culture had changed into an event-driven culture. Under these circumstances, the memorial letters that the bereaved write in cyber-memorial zones in conjunction with the funeral home's main sites offered an important supportive vehicle through which the company could generate revenue.

How, then, is the writing of memorial letters a commercial boost for the funeral company? An approach from the point of view of event marketing allows us to explore this question. Event marketing, as an integrated and total marketing strategy, is a practice "adopted by a firm which sponsors or owns a sport team in order to galvanize its consumers into spending" (Lee, Lin, & Su, 2010). Although this marketing strategy is frequently used in sports marketing, it has become a popular trend in the business industry because it focuses on live and interactive relationships between the business and consumers. When a funeral service company uses such an event marketing strategy, emphasizing immediacy and interactivity, funeral processes, including cleaning and dressing of the deceased, placing a body in a coffin, or starting a funeral procession, become commodities. In this context, it is not surprising to discover that cyber-memorial zones and memorial letters become attractive products that are sold by the funeral service company.

Utilizing the concept of the "Five P's of event marketing" (Graham, Goldblat, & Neirotti, 2001, pp. 169–171), I further explain the components of cyber-memorial letters as a funeral commodity:

- **Participation:** For the bereaved, writing memorial letters enables a vivid and interactive relationship where they can participate in communication with the beloved deceased at any time.
- **Product and brand experience:** Whenever the bereaved writes the letter, they need to visit the memorial writing zone that is linked with the home page of the funeral service company. This allows the bereaved to experience a number of funeral products that are displayed on the home page.
- **Promotion:** The miscellaneous collection of the stories created by the bereaved are excellent promotion tools to attract first-time visitors to the cyber-memorial zones and to experience the products and brand of the company.
- **Probing:** The memorial writing zones are maintained on the basis of research done before, during, and after the memorial letters are posted. The position of the link of the memorial zone on the company's home page is changed if needed. The link of the memorial zone Letters to Heaven was calculated for maximum effect and located in the upper corner of the company's home page, which affirms that the cyber-memorial zone is an excellent promotional tool for the company.
- **Prospecting:** The funeral companies try to have a long-term relationship with the bereaved. For this purpose, in addition to cyber-memorial zones, companies have set up cyber-communities, such as Naver, Daum, or Nate for the bereaved via mainstream Korean Web portals. In the cyber-community, the bereaved get the information not only about the status of the remains of the deceased but also about various memorial services provided by the funeral company. The bereaved also feel a sense of kinship with other bereaved in the cyber-community. This is an effective afterservice marketing tool for the funeral company.

The use of the event marketing strategy elaborated above shows how the memorial writing event has turned out to be a popular funeral commodity that can boost sales of other funeral products.

Moreover, Ki-sung claims that following the initial success of memorial writing zones, many other funeral service companies benchmarked them and provided similar services for grieving family members. Examination of the cyber-memorial zones provided by another funeral service company reveals a number of new and different features beyond the memorial zones mentioned earlier. This company has created memorial zones that are more individualized. Whereas the memorial writing interfaces shown in figure 1 and figure 2 start with the collection of memorial letters and can move to the individual zone with a click, this company's more individualized model allows direct entry to a designated individual memorial zone. Moreover, this company's memorial zone offers a combined package of the memorial writing zone and the memorial photo/video zone, whereas in the products considered earlier, the memorial writing zone is separated from the memorial photo/video zone. In addition, this company's memorial zones show a photograph of the deceased on the left middle side of the zone—a further feature of customization and individualization. Last, whereas Post Office in Heaven, in figure 1, had received more than 62,000 postings of memorial letters by the bereaved at time of writing, examination of the individualized zones developed by this third company showed very few memorial letters, and those that were up were typically posted by the company rather than by the bereaved.

The above discussion of memorial writing zones has elaborated how cyber-memorial zones and memorial writings are increasingly geared toward commercialization and how the funeral service companies profit from them. Cyber-memorial zones do not exist in isolation from the commercially driven memorial zone. The examples of the cyber-memorial zones discussed suggest that the de- and reterritorialization of commercial forces in cyber-memorial zones reflect the increasing commercialization of Korean funeral culture under the influence of the neoliberal paradigm. The cases discussed here show that cyber-memorial zones have become the commercial vanguard, with a flow of capital to attract the attention of the mourners, which, from a Deleuzian perspective (Deleuze & Guattari, 1987), provides space for constantly de- and reterritorializing commercial forces.

Conclusion

Before the neoliberal paradigm had fully emerged in Korea, the rite of death was not considered a commodity but rather was mainly considered a space for contemplating the meaning of life. However, in the neoliberal paradigm, where the de-territorializing power of capital has reached its peak, even the rite of death is commoditized. What happened in the funeral industry of the United States more than a hundred years ago is now happening in Korea. The US funeral industry, under the influence of the de-territorializing power of capital in the late-19th century, established its modern shape (Mitford, 1963). In contrast, it was not until the late-20th century that the contemporary funeral industry began to take shape in Korea. Korean funeral culture has become rapidly commercialized in the past decade, and the funeral industry has demonstrated the power of capital in de-territorializing the meaning of the rite of death.

Although a characteristic of the neoliberal paradigm is the maximization of the flow of capital, I suggest that it is unconvincing that this paradigm alone has de-territorialized the inherent meaning of the rite of death and reterritorialized it with commercialized values in such a short period. How, then, did the sociopolitical and economic environment become so susceptible to the de-territorializing power of capital? The various technologies of domination in Korean modernity suggest a direction for exploring this question. Korean modernity took its form under Japanese Occupation, whose technologies of domination were internally exercised on the Korean people. The unique trait of the technologies of colonial power was to use Korean funeral culture as a tool for consolidating colonial control. By implementing the policy of cremation, the colonial power attempted to have Koreans construct their technologies of the self in detachment from their funeral tradition, which had been built on shamanism, the indigenous Korean belief system. The colonial power oppressed shamanism in the same way that they exercised technologies of colonial dominance on Koreans and their funeral culture.

Even after the Japanese Occupation had ended, the colonial mentality persisted through postcolonial modernity. Korean postcolonial modernity was established under the influence of American modernity. Americanized modernity was burned onto the body of Koreans during the Japanese Occupation and has exercised its influence with the continuation of Cold War ideology (Yoo, 2001); therefore, the de-territorializing power of capitalism was positively engrained on the body of Korean society. Illegitimate military regimes have continually stirred people's propensity for capitalism in the name of democracy. Technologies of the Korean self were constructed to identify democracy with American modernity and capitalism. Because of these situations, the Korean social body changed to the extent that many Koreans became susceptible to the de-territorializing power of the neoliberal paradigm. Korean society became vulnerable to the de-territorializing power of capital, and it is not surprising that the current funeral culture is under the influence of capitalism.

The commercialization of the rite of death in cyber-memorial zones can be considered in a similar context. As discussed, the meaning of life has been devalued to the status of a governmental tool for promoting the policy of cremation and that of a product in the funeral marketplace. Based on fostered nationalism, the "Korean neoliberal policy regime" shows the drastic conversion of the meaning of life and death into a governmental tool or commodity. Under these circumstances, the technologies of the Korean populace's self do not seem to involve the countertechnologies that respond to the power of capital flow and the technologies of government domination that one would so fervently hope for them to.

References

Aries, P. (1974). *Western attitudes toward death: From the Middle Ages to the present.* Baltimore, MD: Johns Hopkins University Press.

Bennett, T. (1992). Putting policy into cultural studies. In L. Grossberg, C. Nelson, & P. A. Treichler (Eds.), *Cultural studies* (pp. 23–37). New York, NY: Routledge.

Bradbury, M. (1999). *Representations of death: A social psychological perspective.* New York, NY: Routledge.

Chun, S. Y. (2003). *Maejangkwa hwajang munjereul dullussan sahoejuk damron bunsuk* [Analysis of social discourse on burial and cremation]. *YuksaMinsokhak, 16,* 129–152.

Deleuze, G., & Guattari, F. (1983). *Anti-Oedipus: Capitalism and schizophrenia* (R. Hurly, M. Seem, & H. Lane, Trans.). Minneapolis: University of Minnesota Press.

Deleuze, G., & Guattari, F. (1987). *A thousand plateaus: Capitalism and schizophrenia* (B. Massumi, Trans.). Minneapolis: University of Minnesota Press.

Foucault, M. (1980). *Power/knowledge: Selected interviews and other writings 1972– 1977.* New York, NY: Pantheon Books.

Foucault, M. (1988). Technology of the self. In L. H. Martin, H. Gutman, & P. H. Hutton (Eds.), *Technology of the self: A seminar with Michel Foucault* (pp. 16–49). Amherst: University of Massachusetts Press.

Foucault, M. (2000). Governmentality. In J. D. Foubion (Ed.), *Power: Essential works of Foucault* (pp. 201–222). New York, NY: New Press.

Giroux, H. A. (2004). Neoliberalism and the demise of democracy: Resurrecting hope in dark times. Retrieved from http://www.dissidentvoice.org/Aug04/ Giroux0807.htm

Graham, S., Goldblat, J., & Neirotti, L. (2001). *The ultimate guide to sports marketing.* New York, NY: McGraw-Hill

Hundt, D. (2005). A legitimate paradox: Neo-liberal reform and the return of the state in Korea. *The Journal of Development Studies, 41*(2), 242–260.

Kim, H.-C. (2010, January 31). 9 out of 10 people prefer cremation to burial. *The Korea Times.* Retrieved from http://www.koreatimes.co.kr/www/news/biz/2010/03/123_60006.html.

Kim, K. (2003). *Hankukui maejangmunhwawa hwajangmunhwa* [Ordinary burial and cremation practices: The meanings and prospects of those practices in cultural and environmental terms]. *YuksaMinsokhak, 16,* 101–128.

Lee, H. (1996). Change in funeral customs in contemporary Korea. *Korea Journal, 36,* 49–60.

Lee, J. (2004). Cyber-confession: Korean women and technologies of the gendered self. Unpublished paper presented at International Communication Association Convention, New Orleans, LA.

Lee, T., Lin, J., & Su, X. (2010). Retailing logistics and sale operation strategy for uncertainty demand due to the effect of consumers' motives in promotion of sport marketing events. *International Journal of Management and Enterprise Development, 8*(3), 209–227.

Mitford, J. (1963). *The American way of death.* New York, NY: Simon and Schuster.

Peters, M. (2001). Environmental education, neoliberalism and globalisation: The 'New Zealand experiment.' *Educational Philosophy and Theory, 33*(2), 203–216.

Song, H. D. (2003). *Hyundai hanguk jangreui byunwhawa gue sahoijuk uimi* [Change and the social meaning of modern Korean funeral ritual]. *Jongkyo Yunku* [Studies in Religion] *32*, 289–314.

Williams, R. (1983). *Culture and society, 1780–1950.* New York: Columbia University Press.

Yoo, S. Y. (2001). Embodiment of American modernity in colonial Korea. *Inter-Asia Cultural Studies, 2*(3), 423–441.

12. *Far Away from Home . . . With a Mobile Phone! Reconnecting and Regenerating the Extended Family in Africa*

GADO ALZOUMA, PHD

Introduction

If African populations have adopted mobile phones so rapidly (International Telecommunication Union, 2004; 2008), it is plausibly due to the compatibility between these devices and the social, economic, and cultural conditions in Africa. Thus, it is surprising that only a few authors (Alzouma, 2005; Donner, 2007; Hahn & Kibora, 2008) have examined the relationship between mobile phones and their users in countries like Niger, where the vast majority of the population is illiterate and, therefore, disabled in using information and communication technologies (ICTs). At the same time, socioeconomic status is a discriminating factor (Attewell, 2001; Zillien & Hargittai, 2009) because computing technology use requires significant financial, intellectual, physical, and cultural resources, which are unevenly distributed in Africa. As Winner (1986) notes, the design of communication technologies is "binding," because it requires user compliance with preconditions and social arrangements. Computers carry a "vision of the world," and this technology is well suited to societies with a literary tradition (Ong, 1982).

It is therefore important to understand how a device designed or invented in one cultural context is used in another, particularly the effects of that use on the relationships among members of an extended family, based on the differential meaning attached to the device and its use. Going beyond a purely instrumentalist or deterministic approach to technology, we can speculate a kind of codependence between technology and culture, in that technology may expand existing relationships between people, and at the same time people's use of technology may redefine the technology's meaning (in accordance with the principles that shape their own worldview). Thus, new mobile technologies not only expand preexisting forms of interaction between people (especially in societies characterized by oral traditions) but also regenerate ties between the members of communities particularly where, due to migration, distance between kin often leads to severed or loosened ties.

This chapter focuses on an understudied yet distinctive population of migrants who frequently use mobile phones to communicate with their families, while almost completely bypassing the Internet. The research is also unique in that it addresses African migrants who use their mobile phones as a way to forge, sustain, and renew the social relations that link them to their extended families, as they adapt to the changing context of work that forces them to globalize resulting in greater worker mobility. This changing context is particularly marked by technological innovations, economic integration, and changing labor markets. The globalization of communication technologies has hitherto been identified as an important factor in shaping current aspects of migration, and studying family networks illuminates the motivation to migrate, and new media links in migratory flows (Babalyan, 1997) Although African migrants have been widely studied, their relationship with mobile phones has not been accorded the same level of attention. Little attention has been paid to the reasons why African migrants choose to use one technology over another, although the importance of mobile phones in connecting migrants with their homelands and sending remittances to families has been noted (Owusu, 2003; Tall, 2004). More important, scarce attention has been paid to the social characteristics of the migrants themselves, particularly their education level in relation to their choice of technology use, their residential origin (urban versus rural), and their economic status. This is probably because most African migrants that have been surveyed come predominantly from English-speaking African countries, such as Ghana and Nigeria, where migrant populations are relatively well educated compared to some of the Francophone African countries, which are among the world's poorest and least formally educated. The transnational studies of African migrants regarding ICTs have focused on migrants located in European countries or America (e.g., Bastian, 1999; Burrell & Anderson, 2008; Kadende-Kaiser, 2000; Tynes, 2007; Wright, 1996).

Although the connections among transnational migrant families have been extensively studied (e.g., Bryceson & Vuorela, 2002; Christou, 2006; König& de Regt, 2010; Reynolds & Zontini, 2006), the implications of information and communication technology (ICT) use on diverse forms of families, particularly the extended family—one of the essential social units in Africa—have rarely been examined. Modernization (including ICT use) has had a profound effect on family solidarity and child fostering, which are said to be fading with the rise of individualism (Ocholla-Ayayo, 1997; Wusu & Isiugo-Abanihe, 2006). Several studies (Aldous, 1962; Weisner, Bradley, & Kilbride, 1997) have examined how African families have adapted and evolved in light of worker migrations, which are affecting family structures negatively (Okoth-Ogendo, 2009), by staying away from each other for long periods of time if not indefinitely. In fact, several authors (Bachen, 2001; Chen & Katz, 2009; Sooryamoorthy, Miller, & Shrum, 2008) have noted the special role that mobile phones play in maintaining social ties among family members and have pointed out how the mobile phone differs, in this respect, from computer-mediated communication. The implication is that mobile telephony and Internet technologies may have "different consequences for the globalization process" (Sooryamoorthy, Miller, & Shrum, 2008, p. 729). The next section covers the socio-demographic characteristics of Niger, the Nigerien population, and Nigerien migrants, with a focus on the reasons for worker migration.

Migration and reasons for migration in Niger

Migration in Sub-Saharan Africa, particularly in Niger, is best conceptualized as a "circulation of migrants" (Adepoju, 2006, p. 26). Seasonal migration has a long tradition and is largely regarded by migrants as a circulation within a cultural space that sometimes encompasses their traditional homeland (i.e., Hausaland), which was divided by the British and French colonial powers and within which people from Niger and Nigeria still circulate without much regard for the established colonial boundaries (Miles, 1994). As Adepoju (2008) notes, "Seasonal, short-term and frontier workers regard their own movements as simply an extension across national boundaries of internal movements, seeing them as rural-rural migration" (p. 15). The traditional lands that Nigerien people migrate to are primarily in Nigeria, but also Cote d'Ivoire, Ghana, Togo, and Benin, and only recently Europe and the United States (Bakewell, 2009; Gervais, 1995). The reasons for these migrations are similar to those to everywhere in Sub-Saharan Africa, and particularly in Niger:

> rapid population and labour force growth, poverty and—not least—environmental deterioration. The region's fragile ecosystems, desertification and diminishing arable land have rendered many agricultural workers landless: pastoralists have been compelled to migrate to coastal regions, towns and cities or even to neighbouring countries simply to survive. (Adepoju, 2008, p. 13)

Not only is Niger the poorest country in the world with the lowest human development index, at 0.340 (United Nations Development Program, 2009), it is also the country in which the fertility rate (births per women) and the annual rate of population growth are among the highest in the world (United Nations Development Program, 2009). Almost 50% of its people are under 15 years old. (Population Reference Bureau, 2010) These indicators—the rural origin and poverty of the migrant population along with a high illiteracy rate—help explain why Nigerien migrants in Sub-Saharan Africa primarily use mobile phones. The migrants in this study are peasants of various professions from rural Niger living in Yola, Nigeria, particularly small business owners who work on an occasional, seasonal, or definitive basis.

The poverty of this area is significant when it comes to accessing information technologies. For example, in 2006 there were only 29,925 subscriptions to landline phones among a population of more than 13,000,000 people. Niger also has the lowest ICT development indicator in the world (International Telecommunication Union, 2009): In 2005, per 100 inhabitants, only 0.07% owned computers and 0.17% owned landline phones, and in 2010 the Internet penetration was 0.7% (Internet World Stats, 2011). These low rates of computer ownership and use are largely due to the prohibitive costs and explain why collective use of computers at cybercafés and workplaces are the only options for most residents. Other contributing factors are illiteracy and the lack of foreign language proficiency. In Niger, more than 65% of people age 15 and over are illiterate (UNICEF, 2010). Although Web sites and software in local languages exist, people have still to learn to read in their own languages before they can achieve proficiency in a foreign language and in the use of innovative devices, such as voice-activated software.

Although the number of landline phones and computer owners is low, the number of Nigeriens who own a mobile phone is comparatively higher—in 2009, more than 30% owned mobile phones. However, disparities do exist between rural areas and urban areas, as only 13% of rural people in Niger use mobile phones whereas 52% of urban dwellers use them (Institut National de la Statistique [INS] & Programme des Nations Unies pour le Développement [PNUD], 2009). One of the main advantages of the mobile phone for rural illiterate people is that it does not require knowledge of reading and writing or the knowledge of a foreign language. In addition, its low cost and its relatively easy operation explain the phenomenal diffusion of this technology in Niger and in Africa.

Studying Nigerien migrants in Yola

Thirty male migrant workers were interviewed in Yola, Nigeria, for this study; the interviews were conducted in Hausa and Zarma, the two main languages of Niger. This sample cannot be said to be "representative," as in-depth conversations sought to discover subjective meanings of technology use and its associated feelings and the representations and behaviors associated with them. Although the interviewees belong to different ethnic groups (Hausa, Fulani, Zarma) with varying family structures and member roles, they all maintain extended family ties.

Some migrants in the study left their extended families as well as their immediate nuclear family (or family of procreation) behind them and are living in Yola alone. These are migrants who return home as frequently as they can afford to do so. Others have their immediate nuclear families with them in Yola, either because they married a local woman or because they brought their wives and children with them from Niger. These migrants are, therefore, in a situation where they could potentially lose contact with their extended family. They are especially interesting for this study because they allow us to gauge the impact of mobile phone technology in potentially helping migrants reconnect with their families and to regenerate a sense of community, as they are traditionally identified by assimilation theories as being in "a process of permanent resettlement that involved cutting all ties and shifting allegiances to the host country" (Burrell & Anderson, 2008). Of course, as discussed below, these homeland and family ties are almost never completely cut. However, the opportunities for contact and frequency of contacts are greatly diminished and impaired by various factors, such as the cost of transportation, poverty, and also over time, the diminishing interest of migrants in maintaining those ties.

It should also be noted that the interviewees mostly come from peasant families and rural areas, and most received a traditional Koranic (Islamic) education; only a few attended a modern Western-style school. In terms of occupation, most of the interviewees were traders, as well as two cobblers, one Islamic school teacher, and a watchman. They were all male, since I was not permitted to interview women (Muslim women are generally not permitted to interact with men who are not relatives). Still, it should be noted that female migration is not an unknown phenomenon in Niger, particularly among the Wodaabe and Tuareg. Also, Hausa women traders are numerous—they are generally married women who come to sell or buy products and who only stay for short periods of time before going back to their families.

The youngest of the interviewees was 35 years old, and the oldest was 65. They all came to Nigeria after 1970 in the wake of the oil-led expansion and construction boom, which was accompanied by growth in the infrastructure, education, and allied sectors, attracting skilled and unskilled migrants (Adepoju, 2006). All the interviewees came to the country more than 10 years before the conversations, though in many cases, it was impossible to obtain the exact migration dates since most were illiterate and only associated the date of their arrival with a main event that took place in that year. Time spent away from the family always has an emotional dimension that is filled with nostalgia, a feeling that is probably extenuated for these migrants partly because of the imprecision that surrounds the experience, and also because for many of them, "going back home" has always remained a goal.

Maintaining family ties before the advent of ICTs

Long before the advent of new communication technologies, most migrants tried to maintain as close contact as possible with their families and their homeland, usually by short-term regular "visits" either by the migrant or some member of his immediate family from Yola. For example, when asked if he goes back home from time to time, Ibrahim says:

> I go and come back as often as possible. Sometimes, I go twice in a year, depending on the circumstances. Sometimes a member of my family goes on my behalf. I have never made an interval of two years without going back home.

Although the regularity of these visits differs from one immigrant to the other ("the last time I went home was in 1999," according to Djibril), most identify face-to-face contact as the preferred mode of maintaining ties. Indeed, visits are rarely casual and mostly occur in relation to some social event (such as marriages, baptisms, mourning events, and visits to ailing parents) that requires a physical presence to demonstrate membership and loyalty to the community. As the 45-year-old Annabi explains the situation:

> It would be unpardonable when a close relative such as a father, a brother, or a sister, an uncle dies or is very sick and you don't show up. In these occasions, even if you have been absent twenty years, people expect you to pay them a visit. Otherwise you would be considered definitively lost for the family, not only because you are far away, but because everybody will turn their back on you and this would be a shame that people would go telling each other for years to come.

Even when "going back home" is not possible, migrants try to keep in touch through others, especially with people from their original village who come to Yola for seasonal trading and who may "drop in" to visit, sometimes unexpectedly. It is even more interesting that children from the extended family are sometimes sent to Yola, and the migrant becomes their guardian, ensures their education, and when they become adults, "finds them a spouse," often among his own children, through a process that not only helps maintain but constantly recreates the ties of the extended family. This ongoing circulation of the members of the extended family is a powerful vanguard against the family's dissolution and ranges from providing lodging and assistance, to hosting seasonal migrants from the same village, to cooperating and integrating with the host family's children—resulting in a transnational family that crosses borders in an attempt to maintain the traditional extended family in its "physical" and "human" reality.

Before (and after) the introduction of new communication technologies, face-to-face interactions were the preferred means by which family ties were maintained, as well as letters and communication delivered through intermediaries—the seasonal migrants who constantly circulated between the homeland and the host country. Older technologies, such as the radio, also played an important role in this process. For example, in Niger, weekly broadcast concerts of the Voix du Sahel, the national broadcasting company, were opportunities for the Nigerien diaspora to greet and show their affection to those who remained home through popular songs dedicated to them and broadcast on their behalf. Those who were in Niger often went directly to the radio station to orally transmit greetings and appreciation to distant family and village members.

These various types of exchanges and interactions should lead us to question technological deterministic theories that allege a "transformative" impact of mobile phones on societies, as there is no clear-cut before and after in the nature of these relationships, the character of the family ties, the people involved, and the general configuration of the family. As Kelly and Waldinger (quoted by Devriendt, 2008, translated by the author) indicate, migrants from the past until today "financially supported their parents back home, were communicating regularly with them, were involved in the politics of their countries of origin, and returned there either temporarily or permanently" (p. 24). However, the newer technologies probably expanded these relationships and regenerated ties that were weakened and even eliminated. Physical distance has always constituted an obstacle to the continuity of group membership and identity, which can only be sustained if the intensity and frequency of communication between members remains strong enough.

The mobile phone as a mark of status

We also asked migrants to identify their strongest motivation for acquiring a mobile phone. The motivations they reported include maintaining contact with family members, maintaining business relations, and the desire to "not be left behind" by others who were using mobile phones. As Ibrahim puts it:

> The first reason why I decided to get a mobile phone was to be informed about my relatives at home, but also to follow up with my business partners and other relations. It helps me to call my people and receive their calls without somebody's knowledge. Thus, I can avoid intermediaries and talk freely with my contacts. They know about me and I know about them. These are the things that

> motivated me the most. Also, you know, today, even in the remote areas people have mobile phones. It is a shame for one to not have a mobile phone.

The social pressure to conform, exerted by extended family members, peer groups, colleagues, and business partners, is one of the biggest motivations they reported for getting a mobile phone. In the context of Niger, this pressure takes on a particular meaning with regard to migrant status. In Nigerien villages, migrants are often associated with the "new" and particularly with new technologies. Thus, in every village, people are able to identify the man who brought the first flashlight, radio, television, bicycle, or motorcycle to the village—with attendant prestige. Often, being a migrant is associated with coming back home with "the gadget of the year," and a sense of failure is felt by those who cannot afford the coveted device. Thus, technology contributes to the definition of the identity of migrants and differentiates them from those who stayed at home. Early mobile phones were not only a "fashion statement" but also a status marker, a mark of distinction that had to do with "the socioeconomic aspects of its early adopters" (Katz & Sugiyama, 2005, p. 64). In particular, owning a mobile phone in Niger is strongly associated with the idea of modernity and also with the idea of the "outside world," and users are assumed to be more "developed" and "advanced" than Nigerien villagers. Thus, the use of mobile phones is reinforcing the traditional role of migrants in the family cluster, not only as "providers" but also as the embodiment of change and having a social and economic distinction that then extends to their families.

The transnational extended family and the mobile phone

Whereas the traditional extended family in Niger is a multigenerational group of kin usually living together in the same compound, the transnational extended family is de-territorialized. This change in residential pattern has many consequences for the family's general configuration, its internal structure, and the relationships between the members. The traditional family pattern not only consists of members living together in the village, but it also often includes several nuclear families that compose the extended family; these nuclear families sometimes have little autonomy and are embedded within an expanded household compound. This extended family household is headed by a senior member who has authority over others, and the ties of the extended family prevail over the individual or nuclear family ties. This primacy of the seniority system is possible primarily because the members live in the same residential complex.

Well before the transnational extended family gained attention, scholars noted that the monetization of the economy, the urbanization of society, the progressive nuclearization of families, and the influence of individualism and individual expressions had eroded extended family ties (Bascom, 1955;

Comhaire, 1956; Wirth, 1938). This trend has accelerated with globalization, with former extended family members settling far away from home and creating their own families free from the control exerted within the confines of the extended family compound. However, this emancipation has rarely amounted to cutting ties with the extended family (Aldous, 1962). As discussed above, the ties to extended family were maintained through various means, and the de-territorialized units now formed a network, the nodal center of which can be identified as being the original extended family at home, at least referentially in the social representations of its members. By abolishing the inaccessibility caused by the distance and physical constraints related to far-away travel, ICTs have eased the major obstacle to the sustainability of community groups. They have also extended the spatial boundaries of the group and its dimensions, which, through electronic networks, have become virtually infinite, opening the possibility of constantly integrating new members, which makes it possible to speak of globalization in terms of a network whose contours are the dimensions of the world.

Thus, the extended Nigerien family comprises not only migrants and those who stay at home but also members living in other countries—who are all connected through mobile communication. For example, most of our respondents underlined the fact that they stay in touch with relatives in other countries (primarily in Nigeria but also in Benin, Togo, Cote d'Ivoire, and beyond). This was a major change because in the past they were rarely able to maintain contact with cousins, nieces, nephews, and uncles living in countries other than Niger. For example, as Issoufou says:

> Yes! I have a brother in the USA, in New York. I do call him from time to time but he never visits me and I never visit him. As you know, it is difficult to get the visa. Also, he cannot come because his own visa has expired many years ago. So our main way of communicating is the mobile phone.

And Soumana:

> Yes. I have a cousin in Lagos. We call each other almost every month. We get news of our respective families and news of the big family at home .

These interviewees describe how the frequency of phone contact helps to maintain the ongoing exchanges between family members and how the intensity of those exchanges strengthens family relations. Some migrants also own mobile phones or use the head of the family's mobile phone whenever possible. As Musa describes his experience:

> In my family, my elder son, my wife, and I all have mobile phones. When there is a family event back home in Niger, for example when somebody dies, they all

> call to make heard their own voices because this is important in expressing their condolences and sorrow. Sometimes I expressly tell them to call because this or this happened in the family. I want them to know people, and I want people to know them.

Among all the reasons why members of an extended family call each other, one of the most important is the request for and circulation of monies. Moussa describes the interaction in his family as follows:

> I usually send money through intermediaries, people I know and who are going back home. However, most of the time, it is the members of my extended family back home who call me to ask for something. Nowadays, they use mobile phone to call me. I too use mobile phone to call them I sent that and that, who I gave the money to. After the money reaches them, they are usually the ones who call me to tell me that they received it.

In most countries, including many African countries, remittances are sometimes sent through agencies, such as Western Union, or through banking services, such as Safaricom (branded M-Pesa) and Zain in Kenya, which are mobile money transfer technologies (Kamotho Njenga, 2009). At the time of this study, no such service existed in Nigeria or Niger (with the exception of Western Union, which is barely known in these countries). Besides, for security concerns it is only possible to receive money in Nigeria through Western Union or MoneyGram but not to send money. Even traditional money transfer banking services were barely used among Nigerien migrants, and this in part has to do with illiteracy and the absence of banks in Nigerien villages. Intermediaries were and still are the main way of sending and receiving money for Nigerien migrants in Yola.

The effects of remittances on family ties are very important in Niger. According to some informants, yearly seasonal crops harvested in the Niger rural areas are no longer enough to provide for the livelihood of most families because the number of people in these families is growing, while productivity is low and the land is shrinking. Some people are even landless. That is why migrants have to supplement the income from family crops grown back home with money sent to buy food (Adepoju, cited in Findley et al., 1995). Most male migrants from the region of Boboye in the Western region of Niger, for example, leave their wives and children at home for one or more years to be able to send them money to buy food. Those who fail to send money are exposed to an eventuality of their wives' leaving them and their families dispersing. In Djibril's opinion:

> There is no true family without food. If you cannot send your wife and children money to buy food, then you have no longer a family because everybody will leave you.

As can be seen here, not only are family ties at stake, but also the identity and the status of the migrant, as providing money has to do with what it means to be the true family head, a father, and a husband. According to Issoufou, if he does not leave his country to provide for his family, "People would think that there is something wrong with him." Migration, therefore, becomes a kind of rite of passage for every man. This continuing societal and family pressure is associated with expectations of remittances. Thus, the impact of remittances on the family ties is not only a "physical" effect of keeping the family intact but also signifies the family's identity and status.

Mobile phones help ease the separation anxiety of dispersed family members that happens even while receiving remittances. Although the children of migrants left back home are often taken care of by the members of the extended family and the community in general, women and children left alone for many years certainly suffer the "care drain" caused by absence of husbands and father (Ehrenreich & Hochschild, 2003). Sociological and psychological studies of transnational families (Parreñas, 2005) have also noted the positive effects that information and communication technologies have on transnational psychological stability when these devices are used to stay in touch with the families left behind. Nigerien migrants in Yola are well aware of those positive effects on family life and stress the importance of mobile phone use in the daily exchange of emotional resources.

Beyond communication with the nuclear family, marriages, baptisms for newborns, and bereavement ceremonies for family members are all opportunities to express solidarity and participate in collective expressions of emotional concern through remittances and "well-wishing" calls. Here again, Dodo describes mobile phones as playing an important role in relaying emotional sentiments:

> For example, when a member of the family is getting old without being married and when we know that this is due to lack of money, it happens that somebody takes the initiative to call everybody everywhere and ask them to send whatever they can send in order to help this relative to get married. Also, when something bad happens to one of the members of the family, if we can solve the problem with money, we ask all those who can to contribute. You know, sometimes the problem is so important that it is the name of the family that comes at stake. It would be shameful for people to know that we had the money and we did nothing.

However, it would be misleading to think that altruistic behavior and family concerns are shared by all family members, particularly as the rise of individualism is frequently lamented by extended families members. Kinship obligations and communal ties are not always strong enough to result in what Finch and Mason (1991) call a "normative agreement . . . about 'the proper thing to do' for relatives" (p. 345), as sharing and giving are not always felt to be obligatory. As Djibril puts it:

> You know, nowadays many people are selfish. There are some among us who are very rich, but they never give anything. That is just the way they are. Not everybody behaves the same way. Some are concerned, and some are not.

One of the main disadvantages identified is that the mobile phone, in some way, places respondents back "under the eyes of the community," even if they are located thousands of kilometers away. Before the advent of newer communication technologies, kin separated by a long physical distance could easily fall into the category of "distant kin" (people whose kinship ties are very weak). But with the advent of the mobile phone, migrants may now reevaluate obligations vis-a-vis other members of the family, as they can be easily reached and may be embarrassed to say "no" to family requests. As Musa describes:

> Another disadvantage is that now whoever wants money just calls me, and I cannot say that I did not hear what they said. Sometimes people call to give news of the family, and they finish by asking me money and I cannot say that I did not hear them. It is very embarrassing even when you decide to ignore them.

On the other hand, the main advantage of the mobile phone identified by the respondents is its reliability. "You are directly speaking to the person you are communicating with and you can directly hear from them," says Issoufou. This statement has major implications for people who are illiterate and who, before the advent of the mobile phone, rarely used a telephone to communicate. Rather, they dictated letters in their original language, which were then translated by intermediary letter writers. This process posed several problems in terms of accuracy, reliability, and privacy. In the words of Issoufou:

> When I did not have a mobile phone, I used to communicate through intermediaries, but I always carefully chose them because you have to be confident in the person you are talking to. These are family matters, and nobody wants people outside the family to know what is going on in your own family. There were many people who could be intermediaries that I avoided. Sometimes I had to wait a long time before I found a man sufficiently reliable. Nowadays, we don't have that problem. With mobile phones, you are directly speaking to the person you are communicating with, and you can directly hear from them.

Thus, the mobile phone helped the migrants to circumvent these intermediaries and encouraged a higher level of authenticity and fidelity of the message.

Conclusion

Poverty and illiteracy largely explain the patterns of worker migration in Niger and also correlate with the use of new communication technology. Previous studies have focused on the use of the Internet among African migrants in Europe, but those migrants who are illiterate and who live in Sub-Saharan countries mostly prefer using mobile phones to communicate with the members of their extended families in their native countries or elsewhere. They do so not only because of the low cost of the device but also because the mobile phone allows them to maintain and expand ties that were maintained before the advent of these technologies. Sending and receiving remittances through intermediaries and the related concerns about reliability and trust were reshaped by confirmation calls made between family members. In the same way, the constraints made necessary by physical distances were diminished by the use of the mobile phone. The direct, immediate reach of kin has important consequences for reciprocal kinship obligations and family commitment, creating disadvantages for some within a context of orality. This change creates a communal atmosphere in illiterate communities, where people who do not know how to read and write feel more comfortable when it comes to family trust and reliability and kinship obligations. Last, it should be reiterated that, in the context of globalizing work and worker mobility, mobile phones are affording some people the opportunities to maintain and regenerate family ties.

References

Adepoju, A. (2006). Internal and international migration within Africa. In P. D. Kok, J. Gelderblom, J. Oucho, & J. van Zyl (Eds.), *Migration in South and Southern Africa: Dynamics and determinants.* Cape Town, South Africa: Human Sciences Research Council.

Adepoju, A. (2008). Migration in Sub-Saharan Africa: A background paper commissioned by the Nordic Africa Institute for the Swedish Government White Paper on Africa. Retrieved from http://www.sweden.gov.se/content/1/c6/08/88/66/730473a9.pdf

Aldous, J. (1962). Urbanization, the extended family, and kinship ties in West Africa. Social Forces, *41*(1), 6–12.

Alzouma, G. (2005). Myths of digital technology in Africa: Leapfrogging development? *Global Media and Communication, 1*(3), 339–356.

Attewell, P. (2001). The first and second digital divides. *Sociology of Education, 74,* 252–59.

Babalyan, A. M. H. (1997, September). *Globalization and migration.* Paper presented at the European Solidarity Conference on the Philippines: Responding to Globalization, Zurich, Switzerland. Retrieved from http://www.philsol.nl/solcon/Anny-Misa.htm

Bachen, C. (2001). The family in the networked society: A summary of research on the American family. *STS NEXUS, 1*(1), Retrieved from http://www.scu.edu/sts/nexus/winter2001/BachenArticle.cfm

Bakewell, O. (2009). *South-south migration and human development: Reflections on African experiences.* United Nations Development Program, Human Development Reports (Research Paper 2009/7).

Bascom, W. (1955). Urbanization among the Yoruba. *American Journal of Sociology, 60*, 446–454.

Bastian, M. L. (1999). Nationalism in a virtual space: Immigrant Nigerians on the Internet. *West Africa Review 1*(1). Retrieved from http://www.africaknowledgeproject.org/index.php/war/article/view/400/

Bryceson, D. F., & Vuorela, U. (2002).*The transnational family: New European frontiers and global networks.* Oxford, United Kingdom: Berg.

Burrell, J., & Anderson, K. (2008). 'I have great desires to look beyond my world': Trajectories of information andcommunication technology use among Ghanaians living abroad. *New Media & Society, 10*, 203–224.

Chen, Y. F., & Katz, J. E. (2009). Extending family to school life: College students' use of the mobile phone.*International Journal of Human-Computer Studies, 67*, 179–191.

Christou, A. (2006). Deciphering diaspora-translating transnationalism: Family dynamics, identity constructions and the legacy of 'home' in second-generation Greek-American return migration. *Ethnic and Racial Studies, 9*(6), 1040–1056.

Comhaire, J. L. (1956). Economic changes and the extended family. *Annals of the American Academy of Political and Social Science, 305*, 45–52.

Devriendt, A. (2008). Les Maliens de Montreuil, des "quêteurs de passerelles". Mémoire de Master 1 de Géographie, parcours "Pays émergents et en développement". Université Paris. Retrieved from http://www.africanti.org/IMG/memoires/memoire_devriendt.pdf

Donner, J. (2007). The rules of beeping: Exchanging messages via intentional ''missed calls'' on mobile phones. *Journal of Computer-Mediated Communication, 13*(1), 1–22.

Ehrenreich, B., & Hochschild, A. (2003). *Global woman: Nannies, maids and sex workers in the new economy.* London, United Kingdom: Granta Books.

Finch, J., & Mason, J. (1991). Obligations of kinship in contemporary Britain: Is there normative agreement? *British Journal of Sociology, 42*(3), 345–367.

Findley, S. E., Traore, S., Ouedraogo, D., & Diarra, S. (1995). Emigration from the Sahel. *International Migration, 33*, 469–520.

Gervais, M. (1995). Structural adjustment in Niger: Implementations, effects and determining political factors. *Review of African Political Economy, 22*(63): 27–42

Hahn, H. P., & Kibora, L. (2008). The domestication of the mobile phone: Oral society and new ICT in Burkina Faso. *Journal of Modern African Studies, 46*, 87–109.

Institut National de la Statistique (INS) & Programme des Nations Unies Pour le Développement (PNUD). (2009). *Impact des le téléphonie mobile sur les conditions de vie des utilisateurs et des intervenants du marché.* Retrieved from http://www.pnud.ne/rap_eval/RapportFinal_Etude_telephonie_pauvrete.pdf

International Telecommunication Union. (2004, April). *Africa: The world's fastest growing mobile market.* Retrieved from http://www.itu.int/newsarchive/press_releases/2004/ 04.html

International Telecommunication Union. (2008). *African telecommunication/ICT indicators 2008: At a cross-roads* (8th ed.). Geneva, Switzerland: Author.

International Telecommunication Union. (2009). *Measuring the information society: The ICT development index.* Geneva, Switzerland: Author.

Internet World Stats: Usage and population Statistics. (2011). *Internet Users in the World: Distribution by World Regions—2011.* Retrieved from http://www.internetworldstats.com/stats.htm

Kadende-Kaiser, R. M. (2000). Interpreting language and cultural discourse: Internet communication among Burundians in the diaspora. *Africa Today, 47*(2), 121–48.

Kamotho Njenga, A. D. (2009). *Mobile phone banking: Usage experiences in Kenya.* Paper presented at the W3C Workshop on the Africa Perspective on the Role of Mobile Technologies in Fostering Social Development, Maputo, Mozambique.

Katz, J. E., & Sugiyama, S. (2005). Mobile phones as fashion statements: The co-creation of mobile communication's public meaning. In R. Ling & P. Pedersen (Eds.), *Mobile Communications: Re-negotiation of the social sphere* (pp. 63–81). Surrey, United Kingdom: Springer.

König, R. S., & de Regt, M. (2010). Family dynamics in transnational African migration to Europe: An introduction. *African and Black Diaspora: An International Journal, 3*(1), 1–15.

Miles, W. F. S. (1994). *Hausaland divided: Colonialism and independence in Nigeria and Niger.* Ithaca, NY: Cornell University Press.

Ocholla-Ayayo, A .B. C. (1997). The African family between tradition and modernity. In A. Adepoju (Ed.), *Family, population and development in Africa* (pp. 60–67). London, United Kingdom: Zed Books.

Okoth-Ogendo, H. W. O. (2009). The effect of migration on family structures in Sub-Saharan Africa. *International Migration, 27*(2), 309–317.

Ong, W. J. (1982). *Orality and literacy: The technologizing of the world.* London, United Kingdom: Methuen.

Owusu, T. Y. (2003). Transnationalism among African immigrants in North America: The case of Ghanaians in Canada. *Journal of International Migration and Integration, 4*(3), 395–413.

Parreñas, R. S.(2005). *Children of global migration: Transnational families and gendered woes.* Stanford CA: Stanford University Press.

Population Reference Bureau. (2010). 2010 World population Data-Niger: Demographic and health highlights. Retrieved from http://www.prb.org/Countries/Niger.aspx

Reynolds, T., & Zontini, E. (2006). A comparative study of care and provision across Caribbean and Italian transnational families. *Families & Social Capital, ESRC Research Group Working Paper, 16*, London South Bank University.

Sooryamoorthy, R., Miller, P. B., & Shrum, W. (2008). Untangling the technology cluster: Mobile telephony, Internet use and the location of social ties. *New Media & Society, 10*(5), 729–749.

Tall, S. M. (2004). Senegalese émigrés: New information and communication technologies. *Review of African Political Economy, 31*(99), 31–49.

Tynes, R. (2007). Nation-building and the diaspora on Leonenet: A case study of Sierra-Leone in cyberspace. *New Media & Society, 9*(3), 497–518.

UNICEF. (2010). At a Glance: Niger. Retrieved from http://www.unicef.org/infobycountry/ niger_statistics.html

United Nations Development Program (2009). Human development report 2009. *Overcoming Barriers: Human mobility and development.* Retrieved from http://hdr.undp.org/en/reports/global/hdr2009/

Weisner, T., Bradley, C., & Kilbride, P. L. (1997). *African families and the crisis of social change.* Westport, CT, and London, United Kingdom: Bergin & Garvey.

Winner, L. (1986). *The whale and the reactor.* Chicago, IL: University of Chicago Press.

Wirth, L. (1938). Urbanism as a way of life. *American Journal of Sociology, 44*(1), 1–24.

Wright, H. K. (1996). Email in African studies. *Convergence: The Journal of Research into New Media Technologies, 2*(1), 19–29.

Wusu, O., & Isiugo-Abanihe, U. (2006). Interconnections among changing family structure, Childrearing and fertility behaviour among the Ogu, Southwestern Nigeria: A qualitative study. *Demographic Research, 14,* article 8, 139–156. Retrieved from http://www.demographic-research.org/Volumes/Vol14/8/

Zillien, N., & Hargittai, E. (2009). Digital distinction: Status-specific Internet uses. *Social Science Quarterly, 90*(2), 274–291.

13. When Indian Women Text Message: Culture, Identity, and Emerging Interpersonal Norms of New Media

ROBERT SHUTER, PHD

Introduction

Although text messaging has exploded in popularity worldwide, there is scant research on the social and interpersonal norms guiding its use, particularly when communicators send or read texts while conversing with others (Pew, 2010). In a recent cross-cultural study of text messaging that I conducted, I found that "textiquettes"—emerging interpersonal norms of text messaging—that differed between cultures were quickly developing in India and the United States and appeared to be linked to indigenous cultural values in each society (Shuter & Chattopadhyay, 2010). Although different textiquettes were identified for each country, several findings regarding the specific difference of the text messages of Indian women were particularly puzzling, which serves as the genesis of the study reported in this chapter: Why did women in India engage in text-messaging patterns that were significantly different than their female and male counterparts in both countries?

This author found that women in India sent and read texts more often when they were alone and also that they received more negative reactions from conversational partners when they sent texts in their presence. Both patterns were significantly different for Indian men who rarely sent or read texts when alone—preferring instead to text message around family members—and reported seldom receiving negative reactions from others when texting. Similarly, US women and men sent and read text messages predominantly around friends, much less frequently when alone, and rarely, if ever, reported receiving negative reactions for texting when conversing with others. Although it may not be surprising that Indian women preferred texting alone given the negative reactions they reported receiving from others, it was not clear from the results why they received such negative reactions from others, significantly more than did men. Why do Indian women prefer sending or reading text when alone and with friends? This chapter explores these compelling questions and offers a sociocultural explanation that attempts to link interpersonal norms of text messaging to identity, gender roles, and power/hierarchy within Indian society.

Text messaging/new media and gender in India

Mobile phones are more plentiful in India than any country on the globe, with more than 400 million cell phone users, which is about one third of India's population (Giridharadas, 2009). In India, mobile phones often replace computers since there are 65 times more cell phone connections than Internet links. Despite the penetration of mobile phones throughout India, women's ownership of phones may be significantly lower than men's, although there is no reliable data on the number of female subscribers in the country. However, data on mobile adoption in South Asia, including Pakistan, Bangladesh, and India, suggest that more than twice as many males tend to acquire phones than females (Johar, 2009). Moreover, 74% of Indian women who own mobile phones report that men control the budgets and decisions regarding mobile technology (Zainudeen, Iqbal, & Samarajiva, 2010). A recent Stanford University study conducted in India finds that mobile phones significantly decrease domestic violence and empower women by giving them immediate access to social services and family planning agencies (Lee, 2009). For example, more than 50% of calls to social service agencies from women originated from mobile phones, according to the Stanford study, which prompted the investigators to conclude that mobile phones may increase Indian women's autonomy and economic independence if they had more access to them.

In addition to mobile phones, there is ample evidence that women in developing countries like India have less access to a variety of information and communication technologies (ICTs)—which has been referred to as the *gender divide* in ICTs. (Gurumurthy, 2004; Primo, 2003; Zainudeen, Iqbal, & Samarajiva, 2010). For example, only 22% of Internet users in Asia were identified as female in a study conducted in 2000, as compared to 38% in Latin America and 6% in the Middle East (Hafkin & Taggart, 2001). In a 2010 investigation in India, 23% of Internet users were women and just 8% of married Indian women with children used the Internet (Johnson, 2010). Gender disparities in Internet access in developing countries have been linked to lower levels of education and literacy among women, which is further complicated by poverty and traditional cultural values that restrict the autonomy of women (Karl, 1995). In India, for instance, home computers were found to be used predominantly by men and children regardless of income, largely because a computer is often viewed in India as masculine technology that requires sophisticated technological skills perceived to be less compatible with women (Johnson, 2010).

In terms of text messaging, there is minimal research on its social functions in Indian society and even less on the digital divide. What we do know is that text messaging, referred to as SMS in India, is very popular and is used for a variety of purposes including interacting with family and friends, shopping, locating employment, and even communicating opinions during live television shows (Giridharadas, 2009). Small Indian businesses often use text messaging because it is inexpensive and the only electronic connection the enterprise has. It is interesting that text messaging is also sometimes used by males in India to meet women, which traditionally has been arranged by parents (Donner, 2009). SMS is employed in India to send monetary remittances to electronic payment facilities and, in certain parts of the country, utilized in elections for voting. Zainudeen, Iqbal, & Samarajiva (2010) find no significant differences in SMS use in India among poor men and women. A curious finding is that women in India are able to type their own text messages regardless of their literacy. Mobile access in India, according to these investigators, is essential to using SMS, and as long as there is a gender mobile divide, Indian women will have less access to SMS and its increasingly important social functions.

Because India is still a patriarchal society, it is not surprising that women are treated differently than men with respect to information and communication technology (ICT) access and ownership. Professional women in India express frustration at being employed in the digital economy in a patriarchal society—a sign of changing women's roles in India—and, yet, being responsible for taking care of home, children, spouses, aging parents, and extended families members (Kelkar, Shrestha, & Veena, 2005). A woman's identity in India has traditionally been tied to her family—a daughter to her parents, a wife to her husband, a mother to her son, a daughter-in-law to her husband's parents—and to executing the duties associated with the roles they play in these relationships (Kabeer, 2005). With the explosion of urbanization and globalization in India in the 20th and 21st centuries, this traditional view of Indian women has been challenged (Ng & Mitter, 2005). Now, Indian women in middle and upper income homes are encouraged to further their education, attain undergraduate and graduate degrees, and compete with men in the marketplace for professional careers. Bhasin (2000) argues that because of the patriarchal nature of Indian society, even professional and college-educated women must satisfy myriad social expectations of family life as they pursue career goals, as well as subordinate themselves to their husbands, who continue to exert control over their daily lives, which seems to include decision making about, access to, and ownership of ICTs.

Text messaging and college-educated Indian women: Studies 1 and 2

The investigation (Study 1) that identified the divergent text-messaging patterns of Indian women was a quantitative study, which secured its data through the text-messaging logs completed by 23 college-educated Indian females between the ages of 18 and 25 living in India. Although the logs contained a significant amount of data derived from an examinination of the participants' recorded responses to survey questions, which were answered after they sent and received a certain number of text messages, Study 1 was limited methodologically because it did not use qualitative approaches to generate initial data or follow-up findings. As a result, it was determined that Study 2, the investigation conducted for this chapter, would be completed qualitatively, using in-depth interviews to explore the results of selected text messages by Indian women identified in Study 1. Sixteen college-educated Indian women between the ages of 18 and 25 who were living in the US were interviewed, each for two-and-a-half hours. All interviewees had lived in the US for less than two years, had frequently engaged in text messaging while living in India, and were college undergraduate or graduate students attending the same Midwestern university. Thirteen interviewees were single and three were married. The interviews explored several topics concerning Indian women's attitudes and behaviors about text messaging, generated from the results of Study 1, in greater depth: the specific negative reactions reported by the Indian women and the source of these reactions, as well as why these women tend to send and read text messages when alone and with friends.

Interviews were conducted in two stages: First, college-educated Indian women were contacted with the assistance of Campus International at a Midwestern university. Utilizing Seidman's (1991) approach for conducting in-depth interviewing, each interviewee first filled out an open-ended survey that requested information about the person's life history, including social and family background, marital status, education, work history, personal and professional interests, duration of stay in the US, and the frequency and nature of her text-messaging behavior while living in India. Based on survey responses, open-ended questions were developed for each interviewee that explored sociocultural factors highlighted in the interviewee's background that may have influenced the person's text-messaging behavior while living in India. Two-and-a-half-hour interviews provided sufficient time to also probe, in a more general way, the interviewees' perspectives on why the Indian women in Study 1 reported receiving negative reactions from conversational partners when they sent and read text messages. Data were analyzed after each interview and preliminary themes isolated, and then the data were refined and modified again—a process outlined by Schatzmann & Strauss (1973). Four themes were identified regarding the text-messaging patterns of Indian women, each of which is discussed in relation to the following condensed questions explored in this investigation:

1. What types of negative reactions do Indian women receive from conversational partners when sending or reading text messages, and what precipitates these reactions?
2. Why do Indian women tend to send and read text messages when they are alone and with friends, unlike their Indian male counterparts who generally text message among family members and rarely alone?

Results

All interviewees agreed that Indian women, both married and single, were more restricted than Indian men in their text-messaging behavior. Interviewees were not surprised that Indian women in Study 1 reported receiving negative reactions from conversational partners when they sent and received text messages; in fact, all interviewees had experienced similar reactions at some time from conversational partners when text messaging while conversing with others. Interviewees also identified with subjects in Study 1 who reported that they often sent and read texts when alone and with friends rather than around family members. Data analysis identified four themes according to which negative reactions from conversational partners may have been triggered, causing Indian women to text message when alone: (a) family, gender, and text messaging; (b) marriage, gender, and text messaging; (c) "eve teasing," or public sexual harassment, gender, and text messaging; and (d) text messaging, social media, and gender.

Family, gender, and text messaging

Interviewees reported that the family restricted and monitored text messaging of Indian women significantly more than they did of Indian men. In Indian society, women generally live at home until marriage, a tradition that springs from the historical view of Indian women as requiring protection from the vicissitudes of society and the advances of unfamiliar males who are seen to threaten to compromise the purity of females. This "conservative" view of women, as interviewees termed it, is omnipresent in contemporary India and was prevalent in each of their households even though they were raised in families where one or both parents were college educated. Protection of women, interviewees agreed, was particularly keen for single females, with teenage girls and women in their early 20s receiving very close monitoring from parents, male siblings, and extended family members. Interviewees reported that they grew up in what one respondent called a "cocoon," in which parents ensured that their teenage daughters were constantly supervised and monitored, including being driven to and from school, escorted to events by male escorts who were usually family members, and—with the advent of new media—restricted and scrutinized when engaging in text messaging.

Indian parents were very suspicious when their daughters, rarely their sons, sent or read a text message in the home. The primary concern of parents, according to interviewees, was that their daughters would initiate text relationships with male strangers who the family did not know and had no hand in arranging. Most interviewees recalled that whenever they received a text message while at home, their mothers would instantly ask them the identity of the sender, and, if the response was not satisfactory, mothers would request to examine their phones, attempting to ascertain the gender of the sender. It is interesting that interviewees reported that they never declined when parents asked to examine their text messages and believed at the time, and at the time of the interview, that their parents had their best interests at heart.

To avoid conflicts with their parents, many interviewees reported that they engaged in a variety of text-messaging strategies to conceal text messages sent and received from their parents. For example, young single women living at home would often wait until late at night, when their parents were sleeping, to either send or read text messages. Messages from males that interviewees wanted to retain were often filed in their mobiles under girls' names and phone numbers, just in case parents examined their phones. For many interviewees, the safest route was to delete all text messages everyday from their phones, a common precaution of Indian females to ensure that messages from senders, regardless of gender, would not be found and misinterpreted by parents.

Some of the interviewees lived in joint family systems, which further complicated their text messaging while at home. In joint families, several generations live together, often in the same domicile, with grandparents, because of their age, exerting a good deal of control and making many decisions. Grandparents were often unable to use new technology and criticized it, blaming it for many social problems, including young women eloping with strangers, a major family issue in India. Since text messaging is not well understood or viewed favorably by grandparents, they policed its use in the home, often forbidding young women, and even boys, to engage in it. It is not surprising that Indian women avoided text messaging in front of grandparents and other family members, preferring to text when alone and with close friends.

It should be noted that young single Indian women living in dormitories at universities across India were not plagued by family surveillance of their text messages. On the contrary, these women reported that, once they left their homes for university life, parents were unable to monitor their text messages. These women were able to text message whomever they pleased, regardless of gender, and not worry about the watchful eyes of parents. They also were more apt to text message in public settings and around friends. On returning home for vacations or summer break, however, Indian women were, again, subjected to close scrutiny by parents and questioned about their text messaging, according to interviewees. It is estimated that less than 15% of Indian women attending Indian colleges live in dormitories (Norton, 2009).

Researchers have suggested that traditional collective societies other than India also pose unique challenges for women when utilizing mobile devices and other forms of new media/technology. Ito and Okabe (2005) suggest that young Japanese women use mobiles to differentiate themselves and, hence, create boundaries with older generations of Japanese, who may not be as technologically friendly or savvy. Dheepa and Barani (2010) detail how technology, knowledge, and transfer can empower women in particular; however, the values of traditional collectives societies are often obstacles to women's use of technology. Singh's (2001) research also demonstrates the empowering potential of new media/technology for women in traditional collective societies, suggesting that once women become skilled at using the Internet, for example, it becomes a "tool" for securing goods and services and increasing economic empowerment.

Marriage, gender, and text messaging

Interviewees reported that married Indian women often received negative reactions when they sent or read text messages in the home when husbands, children, or extended family members were nearby. Inextricably tied to family, Indian women, whether college educated or not, are viewed instrumentally—their identity, self-worth, and social value emerge from the quality of care they provide to family members and their dedication and loyalty to the extended family. Kakkar (1998) notes that Indian girls are socialized from childhood into believing that successful marriages require that they assume a subservient role and learn to be modest, unquestioning, and self-denying. Any activity or task that interferes with the primary responsibility of Indian women—caring for husband and family—is perceived unfavorably, and text messaging appears to fit in this category.

It is not surprising that Indian women reported that their husbands were often critical of them when they sent a text message in their presence. Husbands frequently asked, according to interviewees, who they were texting and, sometimes, admonished their wives for sending or reading a text. Indian men, however, considered it acceptable for themselves to text at home and assumed that they should not be questioned by their wives. In addition, when Indian women "date" their husbands-to-be in excursions generally arranged by both families, they were careful, according to interviewees, not to text message very much, if at all, concerned that their male partners would not like it. The gender-based disparity over text messages is common in Indian households and, according to interviewees, has caused married Indian women to text message in private, apart from husbands, children, and extended family members.

A newly married Indian woman is generally required by custom to leave her family home and live with her husband, his parents, and, in the case of a joint family, her husband's extended family members (Best & Maier, 2007). The mother-in-law is normally responsible for inducting her daughter-in-law into the culture of the family, which generally includes teaching her the duties and responsibilities associated with care of husband, home, extended family members, and children. Closely watched by their mother-in-laws, newly married Indian women are almost always cooperative and abide by the rules of the household, sometimes fearing their mothers-in-law. It is customary that mother-in-laws are not supportive of text messaging by their daughters-in-law, according to interviewees, because it is perceived as unrelated to household duties and may delay their completion. Moreover, since married Indian females are considered to be the primary family caregiver, any activity, including text messaging, that takes time away from child care is anathema to Indian culture. Mothers-in-law frequently let their daughters-in-law know when they have transgressed the rules, particularly when they sent or read a text message in the presence of husband or children. To avoid social censure for text messaging at home, Indian women, both married and single, frequently send and read text messages while alone or, sometimes, in public settings—street corners, coffee shops, and Internet cafes—which pose other challenges for them.

Eve teasing, gender and text messaging

Known as eve teasing in India, this form of communication refers to Indian men taunting, ridiculing, harassing, and sometimes accosting Indian women they do not know in public settings. A common practice in India, it is widespread, occurring in public transport, streets and alleyways, and university campuses—in fact, in any setting where, generally, two or more Indian males are gathered, and an Indian female, unescorted by a male, happens to walk by them. Typical eve teases range from whistling and ogling at a passing female to voicing loud suggestive comments and, sometimes, singing sexually explicit lyrics from Bollywood movies. Because women's groups in India have complained frequently about eve teasing, many municipal governments across the country have attempted to curtail it by providing women with separate railway cars, off bounds for men as well as "women-only" sections on public busses (Yardley, 2009).

Data from interviewees suggest that eve teasing is rooted in the patriarchal nature of Indian society and is directed especially at single Indian women on college campuses who challenge traditional female roles by pursuing university degrees and competing with men in the marketplace. In addition to Indian college women being verbally eve teased on campus, they are often subject to anonymous eve teases sent as text messages from male students attending their university. It is common for these text eve teases to be directed at freshman Indian women by senior Indian males who secure the women's phone numbers from freshmen men who are cajoled by upper classmen to turn over the information or risk ostracism. Text eve teases, according to interview data, are often flirtatious initially and directed at appearance and frequently then escalate to more sexually suggestive comments. Always anonymous, text eve teases vary in regularity, sometimes being just a single message from an unknown male but, occasionally, multiple messages being sent by the same person. Indian college women are often unsure how to respond to text eve teases; some, for example, simply ignore these messages, whereas others attempt to block senders from sending messages. As a last resort, women will change their phone numbers or even call the sender, which women are reluctant to do, fearing they will antagonize the person. Rarely, if ever, do college women report text eve teasers to university authorities or tell their parents.

Indian women are also subject to eve teasing when they engage in text messaging in public settings. Interviewees reported that when Indian men notice women texting in public, this activity becomes the subject and object of their eve teases, triggering random comments about who, what, or why they are text messaging. One interviewee reported that a male stranger who noticed her texting actually followed her and asked her repeatedly for her phone number. Another interviewee was dubbed "Miss SMS" by a group of males who shouted this out, among other taunts, whenever she passed them. It is not surprising that whereas some Indian women refrain from text messaging in public to avoid being eve teased, others simply ignore male comments and continue texting in public, and a few—according to interviewees—fight back, confronting the eve teaser and even slapping him. It is ironic that the Indian state of Kerala is advising women to utilize text messaging to report eve teasing by typing the word "Vanitha" in a text message and sending it to the women's commission. On receipt of the message, the commission contacts local police where the eve teasing is taking place, and they are supposed to help the woman in distress (Tharakan, 2010).

Text messaging, social media, and gender

Since text messaging poses unique challenges for Indian women, many prefer utilizing social network sites (SNSs) like Orkut and Facebook to send messages to friends and others. According to interviewees, women generally access social media sites at Internet cafes to ensure that family members are not aware of their activities. Orkut, a popular SNS in India, operates much like Facebook and provides individuals with virtual opportunities to "friend" people across the country and to chat with them on the site. This reduces the need to send text messages; hence, for Indian women living at home, SNSs eliminate potential conflicts with parents and other family members over text messaging.

It is interesting that interviewees reported that some women utilize home computers to access SNSs, particularly when parents and grandparents are not computer savvy, which is often the case in India (Maslak & Singhal, 2008). And because parents assume that their children must be able to operate computers to be successful at school, it is considered more acceptable for Indian college women to work at home on a computer than to engage in text messaging. Unlike text messaging, chatting on SNSs is private and, when completed, the text is removed from the screen once the person logs out of the site. As a result, Indian women are far less concerned that parents will locate and read past chats with friends. With respect to married women, it appears that even though operating a computer at home is more acceptable than text messaging, computers are still perceived to be the province of men and children, distracting females from their primary family responsibilities.

Discussion

Data strongly indicate that women and men in India are obliged to follow different interpersonal norms—referred to as textiquettes—when they send and read text messages. According to the results, single women living at home often send and read text messages when alone and with friends precisely because they expect negative reactions from parents and extended family members if they text in their presence. Married Indian women also avoid text messaging at home around husbands, children, or extended family member because they, too, are concerned about receiving negative reactions. It seems that Indian men can send and read text messages anywhere and anytime they choose, preferring to text message at home around family members and in public settings, contexts in which it is far more difficult for women to send and read texts (Shuter & Chattopadhyay, 2010).

Different textiquettes for women and men in India reflect a deeper *gender text-messaging divide* that appears to flow from the patriarchal nature of contemporary India.

The data suggests that when single and married women text message, it is viewed as a potential threat to Indian patriarchy: Families fear single women may become too independent when they text; husbands bristle at their wives texting rather than tending to family tasks; and males on street corners shout eve teases at the sight of Indian women operating mobile technology, apparently seeing them as a symbol of gender liberation. Text messaging, like all new media, has the potential to empower humankind if access is provided equally to women and men. However, as demonstrated in this chapter, equal access to mobile technology will not solve the gender text-messaging divide unless male and female textiquettes are also altered.

Conclusion

While this investigation explores text-messaging behaviors in India, it is apparent that women in traditional patriarchal societies also face similar obstacles when utilizing other types of new media/technology due to the indigenous beliefs and values of these societies. Faulkner and Arnold (1985) found early on that technology in developing societies is associated with maleness and that it is "incongruous," as they wrote, for traditional cultures to associate technology with femaleness. Singh (2001) suggests that women in traditional societies are initially uncomfortable utilizing new media/technology precisely because they view technology as the province of men. With encouragement, support, and training, women, according to Singh (2001), become adept at utilizing new media/technology and adapting them to their needs. Wajcman (1991) offers a more global view of new media/technology, suggesting that technology itself is "gendered"; that is, the language of science and technology is conflated with masculinity in traditional and postmodern societies, which poses unique challenges for women who aspire to education and careers in technology. Hence, it may not be surprising that the uses of mobile devices and text messaging, as demonstrated in this study, are influenced significantly by the sociocultural forces that determine who uses them, how they are utilized, and the interpersonal norms that guide their use.

Equalizing male and female norms for text messaging essentially requires that Indian women become free to send and read text messages in the same settings as men without being concerned about receiving negative reactions from others. Because textiquettes, like all interpersonal norms, are inextricably tied to sociocultural forces, they are resistant to change, particularly when they are linked to long-held beliefs about identity, gender, and patriarchy. With the identities and roles of women in Indian society changing due to globalization and modernization, the transformation occurring significantly more in urban areas throughout India, there should be subsequent changes in gender-based interpersonal norms guiding text messaging. Although

India has made strides in the recent past and provided information technology to more women across the country (Sharma, 2003), this study indicates that significant change is still required not only to equalize gender access to mobile technology but also to alter cultural attitudes and interpersonal norms regarding how men and women use mobile devices particularly when text messaging.

References

Best, M., & Maier, S. (2007). Gender, culture, and ICT use in rural south India. *Gender Technology and Development, 11*(2), 137/155.

Bhasin, K. (2000). *Understanding gender*. New Delhi, India: Kali for Women.

Dheepa, T., & Barani, G. (2010). Emancipation of women through empowerment. *SIES Journal of Management, 6*(2), 94–103.

Donner, J. (2009). Mobile media on low-cost handsets: The resiliency of text messaging among small enterprises in India (and beyond). In G. Goggin & L. Hjorth (Eds.), *Mobile technologies: From telecommunications to media* (pp. 93–104). New York, NY: Routledge.

Faulkner, W., & Arnold, E. (1985) Introductory insights. In W. Faulkner & E. Arnold (Eds.), *Smothered by invention: Technology in women's lives* (pp. 1–17). London, United Kingdom: Pluto Press.

Giridharadas, A. (2009, May 7). In cellphone, India reveals an essence. *The New York Times.* Retrieved from http://anand.ly/articles/in-cellphone-india-reveals-an-essence

Gurumurthy, A. (2004). Gender and ICTs: Overview report. *Institute of Development Studies.* Retrieved from http://www.bridge.ids.ac.uk/reports/cep-icts-or.pdf

Hafkin, N., & Taggart, N. (2001). Gender, information technology, and developing countries: An analytic study. *Academy for Educational Development.* Retrieved from http://icd.aed.org/infocenter/gender.htm

Ito, M., & Okabe, D. (2005). Youth culture and the shaping of Japanese mobile media: Personalization and the Ketai Internet as multimedia. In M. Ito, D. Okabe, & M. Matsuda (Eds.), *Personal, portable, pedestrian: Mobile phones in Japanese life.* Cambridge, MA: MIT Press.

Johar, G. (2009, December 7). Indian women benefiting from mobile revolution. *Digital Opportunity.* Retrieved from http://www.digitalopportunity.org/news/indian-women-benefiting-from-mobile-revolution/

Johnson, V. (2010). Women and the Internet: A micro study in Chennai, India. *Indian Journal of Gender Studies, 17*(1), 151–163.

Kabeer, N. (2005). Gender equality and women's empowerment: A critical analysis of the third millennium development goal. *Gender and Development, 13*(1), 13–24.

Kakkar, S. (1998). Feminine identity in India. In R. Ghadially (Ed.), *Women in Indian society: A leader* (pp. 44–68). New Delhi, India: Sage.

Karl, M. (1995). *Women and empowerment: Participation and decision making*. London, United Kingdom: Zed Books.

Kelkar, G., Shrestha, G., & Veena, N. (2005). Women's agency and the IT industry in India. In C. Ng & S. Mitter (Eds.), *Gender and the digital economy: Perspectives from the developing world* (pp. 110–131). Thousand Oaks, CA: Sage.

Lee, D. (2009). The impact of mobil phones on the status of women in India. Retrieved from mobile.active/org/files/file_uploads/MobilePhoesandWomeninIndia.pdf

Maslak, M.A., & Singhal, G. (2008). The identity of educated women in India: Confluence or divergence? *Gender and Education, 20*(5), 481–493.

Ng, C., & Mitter, S. (2005). Valuing women's voices: Call center workers in Malaysia and India. In C. Ng & S. Mitter (Eds.), *Gender and the digital economy: Perspectives from the developing world* (pp.132—158). Thousand Oaks, CA: Sage.

Norton, A. (2009, November 24). Co-ed dorms linked to more drinking, sex. *The New York Times*. Retrieved from http://www.reuters.com/article/2009/11/24/us-co-ed-dorms-idUSTRE5AN4DS20091124

Pew Internet & American Life Project (2010). Mobile access 2010. Retrieved from http://www.pewinternet.org/~/media//Files/Reports/2010/PIP_Mobile_Access_2010.pdf

Primo, N. (2003). Gender issues in the information society. *Publication for the World Summit on the Information Society*. Paris, France: UNESCO.

Schatzmann, L., & Strauss, A. L. (1973). *Field research: Strategies for a natural sociology*. Englewood Cliffs, NJ: Prentice-Hall.

Seidman, I. E. (1991). *Interviewing as qualitative research: A guide for researchers in education and the social science*. New York, NY: Teachers College Press.

Sharma, V. (2003). *Women empowerment through information technology*. New Delhi, India: Author's Press.

Shuter, R., & Chattopadhyay, S. (2010). Emerging interpersonal norms of text messaging in India and the United States. *Journal of Intercultural Communication Research, 39*(2), 121–145.

Singh, S. (2001). Gender and the use of the Internet at home. *New Media & Society, 3*(4), 395–416.

Tharakan, A. (2010, March 1). SMS policing to save women from eve teasers. *True/Slant*. Retrieved from http://www.trueslant.com/abytharakan/2010/03/01/sms-policing-to-save-women-from-eve-teasers/

Wajcman, J. (1991). *Feminism confronts technology*. North Sydney, New South Wales, Australia: Simon & Schuster.

Yardley, J. (2009, September 16). India women find new peace in rail commute. *The New York Times*, A1.

Zainudeen, A., Iqbal, T., & Samarajiva, R. (2010). Who's got the phone? Gender and the use of the telephone at the bottom of the pyramid. *New Media & Society, 12*(4), 549–566.

14. *To Browse or Not to Browse: Perceptions of the Danger of the Internet by Ultra-Orthodox Jewish Women*

Azi Lev-on, PhD
Rivka Neriya-Ben Shahar, PhD

Introduction

The ultra-Orthodox Jewish community is a minority in Israeli society and maintains a complicated relationship with the majority (Efron, 2003). Most ultra-Orthodox people have adopted a segregation model, distancing themselves from the surrounding society and zealously preserving their traditional ways of life (Orbe, 1998). Of particular concern to the community is the preservation of its unique characteristics, which are perceived to be threatened by external influences from modern secular society—including the influences of ICTs and new media. This chapter considers Jewish ultra-Orthodox women who use modern technologies for purposes that are illegitimate according to their community—browsing Internet forums or working in "technological hothouses" (technology-oriented factories/offices that are adapted to the needs and lifestyle of the ultra-Orthodox Haredi women, providing them protected environments). The chapter focuses on the encounter between orthodoxy, new media, and gender and analyzes Orthodox women's perceptions of the impact of their Internet use on themselves and others, demonstrating a *third-person effect* (TPE) with regard to the perceived dangers of the Internet. The correlations and possible implications of this TPE are discussed in this chapter.

The ultra-Orthodox community

The ultra-Orthodox, or Haredi, Jewish community constitutes about 6.7% of the adult (age 20 and up) Jewish population in Israel (Israel Central Bureau of Statistics, 2007). The Jewish religion and all of its principles, instructions, and limitations form the private and public life of community members. Its all-encompassing religiosity and obligation to study the Torah differentiate the ultra-Orthodox and ultra-Orthodox people from other religious sectors in the Jewish society (Friedman, 1991; Liebman, 1999). Haredi Jews consider themselves a singular and distinctive part of the Jewish world. They strictly adhere to the dictates of the halakha (Jewish law) that were developed over the centuries by Jewish rabbinical authorities (Friedman, 1991). This stands in sharp contrast to conservative and reform Jews, who recognize the conflicts between the traditional and modern worlds and make a conscious effort to adapt the Jewish lifestyle to the values and norms of the modern Western world (Heilman, 2000).

The ultra-Orthodox are a minority in Israeli society and maintain a complicated relationship with the majority (Efron, 2003). Scholars point to three central models that characterize minority–majority relationships over time: assimilation, integration, and segregation (Lee & Tse, 1994; Wilson & Gutierrez, 1995). Most ultra-Orthodox people choose the segregation model, zealously preserving their traditional ways of life and distancing themselves from the surrounding society (Orbe, 1998). This segregation from the majority society turns the ultra-Orthodox community into an *enclave culture* (Sivan, 1991), which is exemplified by its being a minority that concerns itself with preserving its unique characteristics and that is disinterested in nurturing relationships with the surrounding majority for fear of external influences (Berry, 1990). The ultra-Orthodox community use a few strategies to create a buffer between secular society and their enclave culture. These include geographical segregation (living in separate neighborhoods and areas), educational segregation (maintaining a separate educational system), judicial segregation (maintaining a separate system of courts, including for arbitration and judicial proceedings), and dietary segregation (consuming food that passes the strictest guidelines of Jewish law).

The ultra-Orthodox community preserves the key positions in society for males and leaves women in the periphery (El-Or, 1995, 1997; Friedman, 1998). The traditionalist worldview emphasizes, as an ideal, the place of the ultra-Orthodox woman in the private sphere. All household functions, such as child rearing and the constant upkeep of the house, are considered primarily to be the obligation of women. One of the outstanding characteristics of the ultra-Orthodox domestic sphere is the abundance of children: The average ultra-Orthodox woman has 7.7 children, as opposed to 2.6 children that Jewish women in Israel have on average (Gurovich & Cohen-Kastro, 2004).

And yet, ultra-Orthodox women are no longer found exclusively in the private sphere. The key reason is the transformation of the ultra-Orthodox community in Israel into a "society of learners," in which the majority of men do not work to provide for their families but exclusively study the Torah (Friedman, 1991; 1995). In order to provide for their families in place of their husbands, women can be found in growing numbers in the public domain, learning and working. According to Central Bureau of Statistics data from 2005, 55% of ultra-Orthodox women were employed outside their homes, compared to only 44% of ultra-Orthodox men (Israel Central Bureau of Statistics, 2006).

In recent years, new routes are developing to channel ultra-Orthodox women into "new" professions. Institutions of higher education offer technical training, in addition to educational training, in an Orthodox "kosher" environment, devoid of masculine or secular presence. In addition, Centers for the Development of Occupations for the Ultra-Orthodox have been open since 2006. These centers provide job placement and guidance for the ultra-Orthodox by the ultra-Orthodox. After the first few months of activity of the centers, it became apparent that 71% of the applicants were women, with some level of professional or post-high school training (Schwartz, 2008). In recent years, a few technological hothouses have been established, which function as protected habitats for ultra-Orthodox women; these are offices/factories staffed exclusively by ultra-Orthodox women that meet the wider needs of an ultra-Orthodox family. Some of the data gathered for this research was collected in such environments.

The ultra-Orthodox women who go out to work in these technological hothouses are situated in a delicate position: Although they are educated that their proper place is in the private sphere, their husband's participation in the society of learners forces them to work and function in the public sphere. Experiences in the workplace may alter women's conceptions, for example, of their role in the household and of the proper scope of the authority of rabbinical and community leaders, which may lead them to decide to develop a career—at the expanse of the traditional obligations to their spouses and families. They may also become better acquainted with the legal system, their rights, and the functioning of the institutions of the modern welfare state. Thus, women's employment in such technological environments may result in multiple challenges to family and community life.

The Internet and the ultra-Orthodox in Israel

The self-isolation of the ultra-Orthodox community manifests, among other things, in the mass media. The secular media are generally banned; it is forbidden to own a television set, a device which is called "the device of impurity" or "that device." The radio is also nicknamed "the device" and is considered illegitimate.

The admission of modern technologies into traditional communities is oftentimes accompanied by wariness and suspicion, and all the more so in regard to the Internet. For some, the Internet is perceived as a tool that allows new religious and spiritual experiences and provides for believers' religious and social needs (Campbell, 2005a, 2005b; Cobb, 1998; Ess, Kawabata, & Kurosaki, 2007). In the eyes of many among the ultra-Orthodox, the Internet is not viewed as means for entertainment or leisure but rather as an efficient way to access information and services. In addition, using the Internet, families can keep in touch and communicate with ultra-Orthodox groups in Israel and abroad. Likewise, using the Internet, one can access an enormous wealth of religious knowledge, online classes, and rabbinical questions and answers.

On the other hand, a few studies report opposition from religious and traditional communities to Internet usage (Campbell, 2004; Dawson, 2005; Marty & Appleby, 1991). Among other things, the leaders of these communities are concerned about novel challenges to their authority and to the hierarchal communal system. They see the Internet, which enables access to infinite worlds of content as well as anonymous communication, as easily being able to expose the ultra-Orthodox to works of heresy, sexual or violent content, and other materials that desecrate the name of God and threaten traditional values (Livio & Tenenboim-Weinblatt, 2007).

The objection to Internet access and usage by the Orthodox led to the establishment of a special rabbinical committee regarding "the issue of breaches in computers," which on January 7, 2000, publicized a Torah opinion stating that "every man of Israel should know, that the connection to Internet or television places is, God forbid, the continuation of the generations of in grave danger, and it is a terrible breach in the holiness of Israel." As opposed to television, the Internet is presented as a device whose danger "is a thousand times more severe, and is liable to bring destruction, God forbid, to the nation of Israel" (Zarfati & Blais, 2002, p. 50). As a result of this concern, all Internet usage was forbidden, even for assistance in providing a livelihood.

However, the prohibition was one that the public could not uphold. For example, a number of groups in the ultra-Orthodox community, such as the Chasidic sects of Breslov and Rachlin (Zarfati & Blais, 2002), continued to use the Internet. It appears that although the voices calling for the total rejection of this new technology are heard, the measure to which they are actually applied is questionable. For example, a survey by the Shiluv Research Center (2007) shows that 60% of the ultra-Orthodox use computers, and among them 57% (approximately one third of the ultra-Orthodox public) use the Internet.

Faced with this, and also taking into consideration the growth of the ultra-Orthodox business sector and with it the growing pressure for computerized communications, another Rabbinic Committee for Matters of Communications was established in 2006. In December 2007, the committee permitted, for the first time in the ultra-Orthodox sector, the use of the Internet, but the permission was only granted for supervised access to a small number of Web sites for business purposes. At time of writing of this chapter, the kosher Internet project is still in its infancy.

A popular activity among ultra-Orthodox Internet users is surfing on dedicated forums (Rose, 2007; Tydor Baumel-Schwartz, 2009). The leading ultra-Orthodox Israeli Web portal for forums is BeChadrei Charedim, receiving some 250,000 unique visitors every month. Rose (2007) claims that ultra-Orthodox forums are the only means of communication that succeed in "peeking, without being harmed, into the cloud of hidden secrets that surrounds the controversies within the ultra-Orthodox circles." In addition to the "mixed" forums (open to both men and women), special forums were created that are designated for women only and are only open to women. Even though there are fewer forums for women than for men, they arouse great interest, because they constitute a unique platform for ultra-Orthodox women to express themselves and discuss the topics that interest them alongside other ultra-Orthodox women (Tydor Baumel-Schwartz, 2009).

Elsewhere, we conducted two related studies that analyze ultra-Orthodox women's perception of the Internet, as well as their Internet usage patterns. The first study targeted women who are members of closed online forums, and the second study targeted women working in technological hothouses. We found these two sites to be fascinating arenas on which to study the intersection of gender, orthodoxy, and new media (see Lev-On & Neriya Ben-Shahar, 2009; Neriya Ben-Shahar & Lev-On, forthcoming). Some interesting observations we made are that in technological hothouses ultra-Orthodox women are employed in the public sphere in technological occupations, even though their education teaches them that their natural place is in the private sphere; ultra-Orthodox women belong to a closed and conservative community, whose perception of the new technology that they are exposed to is complex; and the hothouses are owned by secular employers and yet are adjusted to the needs of the ultra-Orthodox population.

The 'third-person' phenomenon

In this chapter, we focus on the existence of a TPE (Davison, 1983) among the study populations. A third-party effect occurs when people perceive media messages to be significantly more influential on others than on themselves.

Third-person studies have examined, among other things, perceptions of the influence of mass media messages regarding pornography and violence and show that individuals perceive themselves as more resilient to the negative influences of the media than others (for a meta-analysis of third-person research, see Sun, Pan, & Shen, 2008).

Studies have demonstrated that the TPE exists on the Internet, in a variety of domains and contexts. It exists, for example, regarding the perceived effect of exposure to pornographic content online (Lee & Tamborini, 2005; Lo & Wei, 2002; Zhao & Cai, 2008; Wu & Koo, 2001), in the views of the potential harm resulting from exchange in Internet auction sites (Yang, 2005), and in the impact of playing online games (Zhong, 2009). Li (2008) finds a TPE regarding perceived online threats, which consists of a cluster composed of six items: virus attacks, hacker attacks, identification theft, credit card theft, privacy intrusion, and online insult.

Last, two additional studies demonstrate that people also perceive a greater influence of online media stories on others than on themselves (Banning & Sweetser, 2007) and, in a similar sense, perceive a greater impact of the interactions that take place on online social networks on others than on themselves (Zhang & Daugherty, 2009). All of these findings add up to a robust picture of the existence of third-party effects in a variety of spheres on the Internet.

Another significant and consistent finding is that in a variety of contexts, the TPE has behavioral implications. It turns out that the more one is concerned for the adverse consequences of media content on third parties, the more one would support limiting access to this technology—less so for the fear of his or her own and mainly for shielding vulnerable others (see Xu & Gonzenbach, 2008). This effect was found online as well (see Yang, 2005).

To utilize the above-mentioned studies, we ask if the tendency of ultra-Orthodox women, the subjects in our studies, to attribute greater vulnerability to third parties to the "negative" impact of the Internet can be traced. It would be interesting to learn if the subjects, exposed to new technology in their workplaces or while browsing online forums, perceive themselves to be more or less resilient to the impact of the Internet than their reference group: other members of the ultra-Orthodox community.

Methodology

To study the TPE among the study population, we used a questionnaire composed of two main parts: (a) a list of statements in which subjects were asked to rate a number of items on a scale of five options, ranging between "strongly disagree" and "strongly agree," and (b) a few socioeconomic and demographic questions.

The first section of the questionnaire was broken into four subsections addressing (a) conceptions of the place of the Internet in the ultra-Orthodox community, (b) conceptions of the place of the Internet in the lives of ultra-Orthodox women, (c) connections formed online by ultra-Orthodox and nonultra-Orthodox women and men, and (d) the extent to which one shares information on his or her online pursuits with his or her spouse and friends.

The second part of the questionnaire included questions addressing (a) employment outside of the home (of both the woman and her spouse), (b) occupation, (c) availability of a computer at home, (d) availability of an Internet connection at home, (e) amount of browsing women's forums, (f) level of education (number of formal years of study and academic degree), (g) personal status, (h) number of children, (i) age, (j) birth country of the woman and of her father, (k) level of religious observance, (l) political stance, and (m) income level.

Research procedure

As explained, the study targeted ultra-Orthodox women at two sites: women who browse closed online forums and women working in technological hothouses.

Regarding women using the first type of study site, closed forums online dedicated to ultra-Orthodox women, we started by mapping the forums and choosing the relevant ones. After filtering out forums for religious but nonultra-Orthodox women, as well as open forums for the ultra-Orthodox population, we were left with four closed forums for ultra-Orthodox women. The relevant population is ultra-Orthodox women that browse in closed forums online. We cannot know how many women belong to this population, and what their characteristics are, and hence a probabilistic sample is irrelevant. Instead, we decided to perform a nonprobabilistic volunteer sampling along the lines described above. The online survey on the SurveyMonkey Web site was filled out by 53 women ($N = 53$).

Regarding women working at the second type of study site, technological hothouses, we started by identifying and mapping sites from which data could be collected, contacted the management to receive permission to distribute the questionnaires, and then handed out questionnaires at five such hothouses and collected the data. The survey was filled by 156 women ($N = 156$).

Despite the differences between the two settings and the two populations, we found it appropriate to look at both of them in a single chapter, in light of their common significance as unique settings that allow the exposure of ultra-Orthodox women to technologically oriented public spaces. Still, due to the differences between the populations, we present the relevant results from both studies side by side rather than in aggregate.

In the following questionnaire, "agreement" with a certain statement is calculated by the percentage of the answers that indicate "very much agree," "agree," and "somewhat agree" (answers 3–5), and the "agreement level" is the average number indicated in answers (between 1 and 5). It was decided that labeling one of the choices as "undetermined" should not occur, in order to place subjects in a position of agreement or disagreement. The mean level of agreement with each statement is also presented.

Findings

The Internet as harmful and dangerous to the ultra-Orthodox community

The term "dangerous" was chosen as a potential description of the Internet on the questionnaire, as it is a part of the ultra-Orthodox view toward modern technology, which threatens the borders of its "enclave culture" (Sivan, 1991). In Population 1 (forums), the majority of women view the Internet as dangerous to the ultra-Orthodox community. Indeed, 74% (39 women) of the women agreed with the statement "I think the Internet is a danger to the ultra-Orthodox lifestyle" (mean = 3.5). Furthermore, 64% (34 women) agree with the statement "I think the Internet is as dangerous as the television since it enables hearing and seeing forbidden content" (mean = 3.9), and 62% (33 women) agree with the statement "In my opinion, the Internet is dangerous like the cell phone since it enables contacting other people" (mean = 3.1). Let us recall that "agreement" involves checking one of the options "very much agree," "agree," or "somewhat agree" in relation to the corresponding statement.

Among Population 2 (hothouses), 90% (137 women) agree with the statement "I think that the Internet is a danger to the ultra-Orthodox lifestyle," 95.5% (149 women) of the women agreed with the statement "I see the Internet as dangerous just like television because it enables hearing and seeing forbidden content," and 75% (116 women) agree with the statement "In my opinion, the Internet is dangerous like the cellular phone since it enables contacting other people."

Third-person effect

Among Population 1 (forums), 88% (46 women) agree with the statement "I think that the Internet can weaken *people* in terms of religion" (an average agreement level of 4.2), but only 48% (26 women) agreed with the statement "I feel that the Internet weakens *me* in terms of religion" (an average agreement level of 2.6).

Among Population 2 (hothouses), 92% (141 women) agreed with the statement "I think that the Internet can weaken *people* in terms of religion" (average level of agreement: 4.4). In contrast, 56% (86 women) agree with the statement "I feel that the Internet weakens *me* in terms of religion" (average level of agreement: 3.02).

Next, the existence and character of a TPE among individual subjects were analyzed. The TPE is defined as the perceived danger of the Internet to others minus the perceived danger to self and computed in the following way: For each subject i, TPE (i) is the perceived danger of the Internet to others according to person i, minus the perceived danger to self (i). Figure 1 shows the distribution of TPEs among the two study populations. (Note that a small number of subjects attribute more vulnerability to themselves than to others, in which case their TPE value has been designated as negative.) The mean TPE for Population 1 (Forums) is 1.58 (1.63 when negative values are removed). The mean TPE for Population 2 (Hothouses) is 1.36 (1.46 when negative values are removed).

T-tests demonstrate no significant differences between the means of Populations 1 and 2. We also used T-tests to look for differences between various groups of the means of individual TPEs. No significant differences were found in either population between women whose husbands are employed and those whose husband study, between women who have a computer at home and those who do not, between women who have Internet at home and those who do not, between those who were born in Israel and those who were not, and between those who have an academic degree from a university

Figure 1. Distribution of individual third-person effect (TPE) values

or a college and those who do not. We also tested for differences across a host of other variables but found no significant differences. Thus, it can be confidently argued that the TPE that was observed was not unique to certain clusters of the study population.

Last, we used Pearson correlation tests to check for correlations between TPE and various statements. Among Population 1, negative correlations were found between TPE and the statements "I use the Internet as a tool to meet new friends" ($r = -.33$, $P < .05$) and "My husband knows about some of the sites that I browse" ($r = -.32$, $P < .05$). Among the second study population, composed of women employed in computerized environments, a positive correlation was found between TPE and the statements "I think that Internet access is allowed in workplaces only" ($r = .2$, $P < .05$), and "Rabbis in my circle allow use of the Internet for work-related purposes only" ($r = .2$, $P < .05$). Among this population, negative correlations were found between TPE and the statements "Rabbis in my circle do not allow to use the Internet for just any purpose" ($r = -.16$, $P < .05$) and "The Internet enables me rest and relaxation" ($r = -.24$, $P < .01$). Note that there was no overlap between the correlations found in the two populations.

Discussion and conclusions

This chapter focuses on the encounter between orthodoxy, new media, and gender, through the study of ultra-Orthodox women that browse Internet forums or work in technological hothouses. Special emphasis was given to the existence of a TPE.

The Haredim, like other fundamentalist communities, adopt and use technological innovations but at the same time remain suspicious and reserved about the modern and scientific processes involved in their creation and relentlessly try to mitigate and control the possible adverse effects of new technologies on their communities. Our findings indeed demonstrate that participants perceive the Internet as harmful and dangerous to the ultra-Orthodox lifestyle. The women point out that the Internet can weaken the ultra-Orthodox community and themselves in terms of their religiosity, among other things, through exposure (either willing or unintentional) to dangerous content.

The study also suggests the importance of the TPE in understanding the slow processes of penetration of new technologies into conservative communities. Whereas the Internet is perceived by community members as a danger for oneself, it is perceived as an even greater danger for other community members. According to the findings, 74% of subjects perceived other community members to be more vulnerable than themselves to the negative impact of the Internet; the mean TPE—the perceived danger of the Internet to others minus the perceived danger to self—was around 1.5 in both study populations, on a scale of 0 to 5. The differences between the perceived impact of the Internet on oneself and others are not mediated by any of the demographic variables that were measured.

The findings are interesting because participants in the current study are placed in an advantageous position relative to other community members, from which they can use, observe, and reflect upon the possible impact of the Internet for their community. Unlike the majority of the ultra-Orthodox community members that still do not use the Internet, the participants in our study are positioned in the crossroads between modernity and Orthodoxy. But in spite of their positive personal experiences with the new media (Lev-On & Neriya Ben-Shahar, 2009; Neriya Ben-Shahar and Lev-On, forthcoming), they are well aware of its use as well as potential abuse.

The TPE is manifest by the ultra-Orthodox individual's perception of herself and her community. The communal fears about the hazardous potential of the Internet are reflected in the answers provided by participants, but their personal acquaintance with new technology also generates perceptions of themselves as less vulnerable than others. It should be kept in mind that the women who participated in the study live in a demanding environment. Unlike men, who by and large take part in the society of learners and spend many hours each day in the "divine city," women appropriate the "earthly city," face the complexities of daily life firsthand, and may recognize the advantages of the technologies that are portrayed as the "ultimate evil" by rabbinical and community leaders who do not use them. Women may be more aware of the tensions between ideological obligations and pragmatic demands and the need to accommodate them. Thus, as they use the Internet, they continue to quote the rabbinical authorities that denounce it. The TPE may be a result of the rationalization of the dissonance between rabbinical decrees and technologies opportunities. This suggests that the TPE may be a route for ultra-Orthodox women to accommodate the technological and communicative arena of the 21st century with its porous and open borders—the arena having enclaves that continue to fence themselves from it and nonetheless containing individuals who can see "behind the fences."

Conclusion

Ours is the first study to uncover the TPE with regard to new media in a small group of ultra-Orthodox women, whose views of technology are especially complex. Although our findings are telling, they are by no means comprehensive. Future studies will sample the ultra-Orthodox community at large to see if our findings can be extrapolated to the entire community, particularly to those parts of the community less familiar with new media, whose judgments of its potentials and perils may not be based on firsthand experiences. Other studies have found that third-person perceptions may be affected by the perceived likelihood of exposure to the relevant content (Sun, Pan, & Shen, 2008), so it is interesting to study the perception of the Internet and the presence and magnitude of the TPE among parts of the community less familiar with new technology.

Future studies might also explore the possible behavioral implications of TPEs in conservative communities to learn if the members of closed communities who are exposed to new media still justify limitations and censorship, due to the perceived fragility of third parties.

References

Banning, S. A., & Sweetser, K. D. (2007). How much do they think it affects them and whom do they believe? Comparing the third-person effect and credibility of blogs and traditional media. *Communication Quarterly*, *55*, 451–466.

Berry, J. W. (1990). Psychology of acculturation: Understanding individuals moving between cultures. In R.W. Brislin (Ed.), *Applied cross-cultural psychology: Cross-cultural research and methodology series*, (pp. 232–253). Newbery Park, CA: Sage.

Campbell, H. (2004). Challenges created by online religious networks. *Journal of Media and Religion*, *3*(2), 81–99.

Campbell, H. (2005a). Considering spiritual dimensions within computer-mediated communication studies. *New Media & Society*, *7*(1), 110–134.

Campbell, H. (2005b). Making space for religion in Internet studies. *The Information Society*, *21*(4), 309–15.

Cobb, J. (1998). *Cybergrace: The search for god in the digital world.* New York, NY: Crown.

Davison, W. P. (1983). The third-person effect in communication. *Public Opinion Quarterly*, *47*(1), 1–15.

Dawson, L. L. (2005). The mediation of religious experience in cyberspace: A preliminary analysis. In M. Hojsgaard & M. Warburg (Eds.), *Religion in cyberspace.* London, United Kingdom: Routledge.

Efron, J. E. (2003). *Real Jews: Secular versus Ultra-Orthodox and the struggle for Jewish identity in Israel.* New York, NY: Basic Books.

El-Or, T. (1995). Ultra-Orthodox Jewish women. In S. Deshen (Ed.), *Israeli Judaism: The sociology of religion in Israel* (pp. 149–169). New Brunswick, NJ: Transaction.

El-Or, T. (1997). Visibility and possibilities: Ultra-Orthodox Jewish women between the domestic and public spheres. *Women's Studies International Forum*, *20*(5), 149–169.

Ess, C., Kawabata, K., & Kurosaki, H. (2007). Cross-cultural perspectives on religion and computer-mediated communication. *Journal of Computer-Mediated Communication*, *12*(3), 939–955.

Friedman, M. (1991). *Haredi (ultra-Orthodox) society: Sources, trends and processes.* Jerusalem, Israel: The Jerusalem Institute for Israel Studies.

Friedman M. (1995). The ultra-Orthodox woman. In Y. Atzmon (Ed.), *A view into the lives of women in Jewish societies* (pp. 273–290). Jerusalem, Israel: The Zalman Shazar Center for Jewish History.

Friedman, M. (1998). The King's daughter's glory is outside: The Haredi woman. In D. Ariel, M. Laibowitz, & Y. Mazor (Eds.), *Am I blessed to be a woman?* (pp. 189–205). Tel Aviv, Israel: Yediot Ahronot, Hemed Books.

Gurovich, N., & Cohen-Kastro, E. (2004). Ultra-Orthodox Jews: Geographic distribution and demographic, social and economic characteristics of the ultra-Orthodox Jewish population in Israel 1996–2001. *Working paper series, 5.* Jerusalem, Israel: The Central Bureau of Statistics.

Heilman, S. C. (2000). *Defenders of the faith: Inside ultra Orthodox Jewry.* Berkeley: University of California Press.

Israel Central Bureau of Statistics. (2006). *Statistical abstract of Israel.* Retrieved from http://www.cbs.gov.il

Israel Central Bureau of Statistics. (2007). *Statistical abstract of Israel.* Retrieved from http://www.cbs.gov.il

Lee, B., & Tamborini, R. (2005) Third-person effect and Internet pornography: The influence of collectivism and Internet self-efficacy. *Journal of Communication*, *55*(2), 292–310.

Lee, W., & Tse, D. (1994). Changing media consumption in a new home: Acculturation patterns among Hong Kong immigrants to Canada. *Journal of Advertising*, *23*(1). 57–70.

Lev-On, A., & Neriya Ben-Shahar, R. (2009). A forum of their own: Views about the Internet among ultra-Orthodox Jewish women who browse designated closed forums. *Media Frames, 4,* 67–106.

Li, X. (2008). Third-person effect, optimistic bias and sufficiency resource in Internet use. *Journal of Communication*, *58*, 568–587.

Liebman, C. S. (1999). Tradition, Judaism, and the Jewish religion in contemporary Israeli society. In J. Wertheimer (Ed.), *The uses of tradition: Jewish continuity in the modern era* (pp. 411–429). New York, NY & Jerusalem, Israel: The Jewish Theological Seminary of America.

Livio, O., & Tenenboim-Weinblatt, K. (2007). Discursive legitimation of a controversial technology: Ultra-Orthodox Jewish women in Israel and the Internet. *The Communication Review*, *10*(1), 29–56.

Lo, V., & Wei, R. (2002). Third-person effect, gender, and pornography on the Internet. *Journal of Broadcasting & Electronic Media*, *46*, 13–33.

Marty, M. E., & Appleby, S. R. (Eds.) (1991). *Fundamentalism observed* (Vol. 1). Chicago, IL: University of Chicago Press.

Neriya Ben-Shahar, R., & Lev-On, A. (forthcoming). Open spaces? Perceptions of the Internet among ultra-Orthodox women working in computerized environments. *International Journal of Communication.*

Orbe, M. (1998). *Constructing co-culture theory: An explanation of culture, power and communication.* Thousand Oaks, CA: Sage.

Rose, A. (2007). The Haredim and the Internet—Enemies, a love story? *Eretz Acheret, 41.* Retrieved from http://acheret.co.il/?cmd=articles.124&act=read&id=356

Schwartz, E. (2008). *Encouraging ultra-Orthodox women employment.* Jerusalem, Israel: Israeli Knesset Center for Information and Research.

Shiluv Research Center. (2007, July). Paper presented at The Potential for Success in the Haredi Sector, Israel.

Sivan, E. (1991). Enclave Culture. *Alpayim, 4,* 45–98.

Sun Y., Pan, Z., & Shen, L. (2008). Understanding the third-person perception: Evidence from a meta-analysis. *Journal of Communication, 58,* 280–300.

Tydor Baumel-Schwartz, J. (2009). Frum surfing: Orthodox Jewish women's Internet forums as a historical and cultural phenomenon. *Journal of Jewish Identities, 2*(1). 1–30.

Wilson, C., & Gutierrez, F. (1995). *Race, multiculturalism and the media: From mass to class communication.* Thousand Oaks, CA: Sage.

Wu, W., & Koo, S. H. (2001). Perceived effects of sexually explicit Internet content: The third-person effect in Singapore. *Journalism & Mass Communication Quarterly, 78,* 260–274.

Xu , J., & Gonzenbach, W. J. (2008). Does a perceptual discrepancy lead to action? A meta-analysis of the behavioral component of the third-person effect. *International Journal of Public Opinion Research, 20,* 375–385.

Yang, K. C. C. (2005). Consumers' attitudes toward regulation of Internet auction sites: A third-person effect perspective. *Internet Research, 15,* 359–377.

Zarfati, A., & Blais, D. (2002). Between "cultural enclave" and "virtual enclave": Haredi society and the digital media. *Kesher, 32,* 47–55.

Zhang, J., & Daugherty, T. (2009). Third-person effect and social networking: Implications for online marketing and word-of-mouth communication. *American Journal of Business, 24,* 53–63.

Zhao, X. Q., & Cai, X. M. (2008). From self-enhancement to supporting censorship: The third-person effect process in the case of Internet pornography. *Mass Communication & Society, 11,* 437–462.

Zhong, Z. J. (2009). Third-person perceptions and online games: A comparison of perceived antisocial and prosocial game effects. *Journal of Computer-Mediated Communication, 14,* 286–306.

15. From the Coffee Table Album to the Mobile Phone: A Portuguese Case Study

Carla Ganito, PhD
Cátia Ferreira, PhD

Introduction

Family photos remain a primary means of personal memories, of telling a story about ourselves. They are part of what Foucault (1988) characterizes as *technologies of the self*: "[technologies] which permit individuals to effect by their own means or with the help of others a certain number of operations on their own bodies and souls, thoughts, conduct, and way of being" (p. 18). Throughout history, we have developed different technologies to record our own stories. From the first writings to traditional photo albums, the way we store our individual and collective memories have undergone profound changes. Nevertheless, the main goal of keeping memories remains the same: to preserve our own narratives. In an increasingly digital context, where print is giving way to digital storage, what is happening to the family story itself? What is happening to the storyteller? In many societies, women have been the traditional guardians of family memories (Garde-Hansen, Hoskins, & Reading, 2009; Reading, 2008). Will the gendered domestication of mobile camera phones inhibit or change the role of women in their families as guardians of visual memory?

Background

This chapter explores the domestication of mobile camera phones by adult Portuguese women and its effect on the storage and sharing of family memories. Our empirical findings come from an exploratory case study of the Portuguese mobile phone context using a mixed methodology. The qualitative approach is dominant, however, as previous research had made clear that quantifying the overall trends of mobile phone usage gives little insight into the mobile phone biography and trajectory in women's lives and the social, cultural, institutional, and economic factors that lead to such a trajectory.

One of the first motivations for studying the gendering of the mobile camera phone was the empirical observation that Portuguese women feel very comfortable using mobile phones. By the time we conceived the study, most women carried a mobile phone and treated it as a mundane object. Second, statistical studies demonstrate that, in Portugal, women have adopted mobile phones (unlike other technologies) at the same rate as men. Further comparative studies on gender and mobile phone use had uncovered no meaningful differences between mobile phone use by men and women and offered no explanation for this surprising phenomenon. This chapter aims to fill this gap, and thus the research is not centered on differences between men's and women's mobile phone use but rather on women's specific experiences, leaving space for its contradictory effects and meanings for different groups of women.

Portugal is a particularly interesting context in which to study the use of mobile phones given that its high penetration rate (145% by mid-2010) is one of the highest in Europe (whose average penetration rate is 123%; ANACOM, 2010). Moreover, although Portugal is a small market, with just over 10 million inhabitants, the Portuguese have been particularly innovative in adopting new technologies and applications. Many companies have selected Portugal as a test bed for new products and services (see, for example, Vodafone, 2011). From the gender perspective, Portugal is also considered unique in Europe because though it does not yet provide satisfactory social services, such as schools and maternity benefits, it is nevertheless the European country in which the greatest proportion of women work full-time. In 2007, 62% of women aged 15 to 64 were employed, which exceeds the European average of 58% (European Communities, 2008). The convergence of these characteristics makes Portugal an interesting sociotechnical system for studying the gendering of the mobile camera phone.

Photography and memory

The role of the family album

Gye (2004) argues that family histories have always been fragile and momentary and that "the fixity of the photo album and the diary or chronicle, both print-based memory machines, have obscured . . ." (that reality. We might argue that *new* forms of digital photography, the latest being the mobile camera phones, make that fragility visible.

Photography plays an active role in supporting a range of social functions related to the sharing and storing of historical memory. According to Chalfen , there are five essential social uses of photography: (a) shared, (b) household, (c) work, (d) wallet, and (e) tourist photography. Sharing photographs is a common practice among relatives and friends; we tend to share photographs to share experiences and stories. Household photography demonstrates a different and more private dimension of the social use of photography in which we make use of photographs to embellish our living spaces with our memories. Work photography has three subcategories: pictures taken during work-time, pictures taken with colleagues outside the workplace, and private or organizational pictures displayed in the workplace. Wallet photography is also a traditional social use of photography. Chalfen notes that it is common practice to carry a few snapshots in one's wallet, particularly of close relatives. Last, tourist photography consists of photographs taken of popular tourist sites—"souvenir photographs"—and photographs taken by people during their trips. These social uses of photography are imprinted with cultural traditions, and the different uses are appropriated in different ways and have different meanings for different cultures. Photography clearly exists, therefore, as a social artifact for the construction and maintenance of individual and collective memory and shapes belonging, exclusivity, social order, and community (Zelizer, 1995).

Photography is one of the most important elements for private memorialization, and family photo albums have been traditional sites of "cross-generational exchange and intrafamilial communication". In many societies, collections of photographs and organized albums play a central role in the perpetuation of family stories. Families allow pictures to register key moments in their shared history, and through albums, families organize their own stories. According to Van House, Davis, Takhteyev, Ames, & Finn (2004), "photos reflect social relationships but they also help to construct and maintain them". Family albums can therefore be understood as tools for the maintenance of collective memory. Due to its narrative essence, the act of sharing a family album may be seen as an act of transfer—"[acts] that make remembering in common possible".

If we consider family albums as means of transferring family memory, we are seeing them as part of family narratives and therefore of family cultural identity. Sharing a family album is an act of family storytelling. Cross-generational exchange through family albums is the most common. It is traditional that grandparents share family photographs with children to present to them the family history (Favart-Jardon, 2002; Van House, Davis, Ames, Finn, & Viswanathan, 2005).

The social uses of photography are also gendered. De Lauretis argues that gender is a product of various social media, including television, mobile phones, radio, and newspapers. Her concept of *technology of gender* is rooted in Foucault's (1978) *technology of sex*, which she extends, proposing that technologies of gender concern not only "how the representation of gender is constructed by the given technology, but also how it becomes absorbed subjectively by each individual whom that technology addresses". With the advent of photography, women quickly came to play the role of guardians of family pictures. Compiling and sharing a photo album quickly became a feminine activity, a tradition passed on from mother to daughter (Garde-Hansen, Hoskins, & Reading, 2009; Reading, 2008). This gendered approach to technology is the result of cultural expectations: Women are considered to be the natural guardians of "family life"—those who bear and raise children and who organize and store family memories (Leite, 2000; Lins de Barros, 1989). Thus, argues Langford (2001), "The photographic album and the mother are often linked".

Digital photographs are now replacing traditional printed photographic images. Even though digital photographs can be easily printed, they are most commonly shared in their digital format (Gye, 2007). Systems of storing photographs are also changing, and it is important to understand how this can lead to new modes of family storytelling. In addition to photography, cinema, and television, mobile phones are a new technology that has played a significant role as a technology of memory. Through this new medium, people are exploring new ways of recording their memories. Photographs captured on mobile phones are replacing not only traditional print photos but also those taken with digital cameras (Gye, 2007; Hjorth, 2006; Van House, Davis, Ames, Finn, & Viswanathan, 2005).

In order to understand the impact of mobile phone photographs on family storytelling, it is important to first understand the artifact of the mobile phone and how we are appropriating it in our daily lives. Does the ubiquity of mobile phones contribute to the growth of their use as a photographic device? Who takes and keeps these photographs: women or men? How might the gendered use of mobile phones contribute to a change in family storytelling?

The mobile camera phone and impulse photography

Mobile phones now play a significant role in our daily lives, and without them we tend to feel lost, disconnected, and anxious—indicating a strong emotional connection (Katz & Aakhus, 2002; Lasen, 2005; Ling, 2004). They provide a multiplicity of conveniences that range from control over time and space to emotional reassurance and privacy. In a recent survey, 83% of women respondents agreed that they felt calmer when they had their mobile phone with them, and 53% agreed that the mobile phone was only useful if it was always turned on. But the mobile phone is also a source of paradox, with 63% of women stating it offered new ways for others to control them, and 72% reporting that they frequently turn off their mobile phone so that calls do not interfere with their personal relationships (Cardoso, Gomes, & Espanha, 2006).

Mobile phones are also taking on a "Swiss Army Knife character," as they incorporate an increasingly wide range of tools and media:

> The mobile phone is an artifact that is not only changing but is also multiform, multifunctional. It has become an agenda, pager, calculator, and Internet terminal, video game, watch, alarm clock, radio, camera and more. In this sense, it shares with other technical objects the destiny, so well understood by Baudrillard many years ago, of becoming more complex than individuals' behavior relating to those devices. The mobile phone's multifunctionality, as well as that of other ICTs testifies to the desire for reunification that modern society expresses in the face of its opposing tendency to divide, fragment and pulverize. Arnold, 2003, p. 153

Mobile camera phones have been hugely successful, always available to take pictures, a dynamic method of carrying and accessing our personal histories. Like an old friend, they are always with us, ready to capture new memories and preserving treasured memories in images. Integrated cameras are nonintrusive and allow impulse photography. The mobile phone surely owes its extremely fast adoption rate to its untethered nature, satisfying one of human's most obvious needs—to communicate on the move. Its incorporation of the camera seems only natural and consistent with the mobile nature of photography itself. Media archaeologist Errki Huhtamo (2004) has even claimed that the very first "mobile medium" was amateur photography.

Compared to other photographic devices, mobile camera phones provide a new convenience and tend "to be used more frequently as a kind of archive of a personal trajectory or viewpoint on the world, a collection of fragments of everyday life" Okabe & Ito, 2006, p. 99). Moreover, as Gye (2007) argues, "Changes to the ways in which we capture, store and disseminate personal photographs through the use of devices like camera phones will have important repercussions for how we understand who we are and how we remember our past" (p. 279). The meaning of the mobile phone is not just utilitarian or instrumental but also emotional. We can regard the mobile phone as an affective technology, an object of mediation, demonstration, and communication of feelings and emotions. And pictures play an important role in that they function as objects of personalization that transform the device into a conveyor of identity and self-expression and a source of reassurance. These pictures become "a key to our emotional understanding of ourselves and the world" (Gibbons, 2007, p. 4).

Changing the frame: The decline of the family photograph album

The proliferation of mobile camera phones has brought with it enhanced fragmentation: It multiplies the options for capturing pictures but also fragments traditional archiving practices. As Gye (2004) notes, "Where once an individual or family's memories were restricted to limited numbers of photographs and documents now we can acquire vast repositories of mediated memories to send into the future" (p. 4).

Although the mobile phone has the potential to allow sharing and public display of photographs, some theorists have argued that in fact "the story" stops being shared, public, and on display, and instead becomes fragmented, individual, and even more private than before. According to Bourdieu, the viewing of the album is a tribal ritual, a magical ceremony. Sontag (1977) also defines the taking of pictures as a "rite of family life". Mobile phones are changing the practice of those rituals and arguably removing some of the magic, as flipping through the pages of a photo album is not quite the same as clicking the mouse to view pictures on a computer screen or flipping through picture galleries on someone else's mobile phone. Even the physical act is harder, being that most people regard their mobile phone as a highly private device, which makes it hard to pass around for picture sharing (Gye, 2007).

To some degree, the sharing of photographs taken on mobile phones has also been constrained by cost. After taking a picture, the user is faced with two options: to download the pictures to the computer or to spend some money on messaging services or Internet usage to either upload it to a social networking site (SNS) or send it through an e-mail account. On the other hand, the increase in Internet-enabled mobile phones is accelerating the practice of shared photographs, as most mobile phones allow us to immediately contact our social network of close friends and relatives and facilitate connection to our online networks through social network services such as Facebook or Twitter.

The outcome is a reduction in *household photography*, as most photographs taken on a mobile phone will not be printed and placed in frames or albums. As one of our interviewees, Carla, a 30-year-old single woman who lives alone with no children, states:

> I never printed pictures that I have taken with the mobile phone, only the ones I take with the digital camera. I never got to it. I have downloaded some from the mobile phone to the computer but then I never print them. The pictures that I do not print out are left in the computer and each time I want to find one I lose a lot of time searching for it and every time that happens I think I should make a digital album but I never do it.

The traditional family album is a frame that stresses "chronology, continuity, and repetition within and across generations . . . predictable framings and messages" (Hirsch, 1997, p. 214). This frame has been disturbed by the introduction of digital photography, which increased opportunities for photographing the mundane and now by mobile phones that can record the unexpected, the unpredictable, the unplanned—the true "Kodak moment" becomes a "Nokia moment." Mobile phone photographs are set in a different frame—one of impulse, fragmentation, and fragility.

Pictures of Portugal: A case study of mobile camera phone use by Portuguese women

Methodology

We used a methodology that combined elements of both quantitative and qualitative approaches. Quantitative data was taken from the results of a survey "Network Society in Portugal 2006" (Cardoso, Gomes, & Espanha, 2006). This is a questionnaire-based survey, distributed to a representative sample of the Portuguese population age eight and above, living in mainland Portugal. The market research company MetrisGfK conducted the fieldwork during the first semester of 2006. From the survey, we used the module referring to mobile phones.

We drew qualitative data from 37 in-depth interviews with Portuguese women. Using semistructured interviews allowed us to navigate the questions as appropriate to the flow of the conversation and to follow up on interesting comments with additional questions/probes, as we were "interested in thoughts and feelings that are often not articulated as stable opinions or preferences" and which cannot be captured by more direct methods.

We aggregated these women into the following seven groups corresponding to a life course approach that uses a modified version of the market research study of Portuguese consumers: single dependent, young independent, nesting, mother, single mothers, mature independent, and empty nests. Women in the "single dependent" life stage are above 18 years old but depend on their families financially and live with them. "Young independent" women live with their families but have control over their purchasing decisions. The "nesting" life stage is determined by the beginning of a cohabitation relationship that might or not be formally constituted as marriage. The "motherhood" category (contrary to the Marktest study) aggregates women with children of different ages, although we acknowledge some differences in the use of the mobile phone depending on their children's age. "Single mothers" are either widowed, divorced, or separated. "Mature independent women" are over 35 years old, have no children and have never been involved in a cohabitation relationship or are divorced or separated. Last, "empty nesters" are women whose children have left home and who have retired or are in a career plateau. To select women for the seven groups, we constructed a convenience sample, as it was the one that best fit the "ideal types" strategy of analysis. Some interviewees were selected by the "snowball" method of sampling from a pool of urban heterosexual women.

As an analytic strategy, we used "ideal types"—a term developed by Max Weber in association with the construction of pure cases to illustrate a conceptual category. In Weber's work, ideal types are fictions, but in our research, following Turkle's (1984) methodology in her study of computer cultures, we have isolated real cases that serve the same function—to highlight particular aspects of the gendering of the mobile camera phone. The ideal type analysis followed the methodology used by Soulet.

Quantitative results

From the quantitative survey, we found that Portuguese women own less sophisticated mobile phones than men and therefore that slightly fewer women than men own a mobile camera phone. Nonetheless, a camera was one of the most important add-on features for women: Picture taking ranked third in the list of most used phone features, surpassed only by text messaging (SMS) and voice calls. It is more common for younger women to own mobile camera phones because they have more sophisticated devices. Women with higher incomes also own more camera phones, as well as women with higher education levels.

In terms of usage patterns, our research results show that mobile camera phones are mainly used by women to show that a person was actually present at a certain event or to keep a souvenir. In fact, we found that "keeping memories" is the only purpose for which women use mobile camera phones more than men. This is more common in young, active women, whereas it is most common for retired women to use their phones to show images of objects or people to friends, family members, and others. Retired women have less access to social networking sites, so mobile phones likely function as a "mobile album" for them. These women still characterized "keeping memories" or "showing presence" as a more formal picture-taking activity, which might even be performed for them by other family members.

Qualitative results

Our qualitative study confirms that after voice and text, women find cameras to be the most interesting feature to have on mobile phones, but the reasons for this differ across life stages. In early life stages, women value the camera as a tool for keeping up with their social life. Mobile phone cameras were useful for their last-minute social plans.

When women become mothers, camera phones become important for taking family photos, and this is extended to empty nesters as they take care of grandchildren. As Carla D., a "mother," tells:

> Now that I am a mother I use the MMS a lot. I also used it while I was pregnant to show my growing belly. The mobile phone is only when I want to send the pictures. I use the digital camera to record more planned moments but even then I take a picture with the mobile phone to send and then I will use the digital camera. But I do not organize the pictures in the mobile phone.

Even when men are the photographers, we found that women remain responsible for telling the story, for keeping the record, and for organizing a coherent, chronological account of family history. Based on Carla D.'s account, the mobile phone made her a guardian of family memories and the author of the family story. Another interviewee, Sandra, a "nester," indicates that men were no longer responsible for the camera, at least not for all cameras, because the mobile phones are seen as individual gadgets of "personal property":

> I take a lot of pictures. In the past I used to make albums, before I had a digital camera. Now I only do it for special occasions—a Christmas gift or anniversaries. When I lived with my parents I was the one that organized the albums. Before me, my mother took the pictures and organized them but only when I was little. Sergio [husband] loves to take pictures but I never saw him printing them or organizing them. All the picture frames we have—it was me who chose them.

Discussing printed family photographic albums, Langford (2001) argues that "the showing and telling of an album is a performance . . . Viewing the album in company must be considered the normal spectatorial experience". We found, however, that as photographs are increasingly being archived in computers, thumb drives, and hard disks, many women have lost their social role as photo organizers and as storytellers—as performers of this oral family history. As Ana C., a "mother," describes:

> I keep the pictures in the mobile phone and in the computer. I usually see them in the computer but it is not usual to show them to other people.

Some women also describe their use of mobile phone cameras as a last resort when no better solution is available, such as for unplanned events or mundane, ordinary situations—acting as a personal diary of everyday life. Carla A., a "mature independent," describes her experience:

> I use the mobile phone for spontaneous situations, when I do not have the camera with me. I usually prefer the camera but when I do not have it I resort to the mobile phone.

Because the price of text messages is perceived as high, and most of the women had low Internet usage, their photo sharing was restricted to rare occasions—much like one would share a wallet photo, as a way of remembrance. Based on our interviewees' responses, mobile photos appear to be transversally contingent and temporary, with many women reporting that they either did not know how to download their pictures or had no intention of doing so. They rarely or ever printed them out, reinforcing that digital photography and now mobile phone photography are bringing an end to the traditional practice of compiling family photo albums. Fátima, an "empty nester" shares:

> In the previous phone I had lovely pictures of my grandson, one in particularly of him crying on his first day of school. I kept asking my children to copy it to the computer but they never did it and one day the mobile phone fell and a car ran over it and I lost all my photos.

Conclusion

The central role of family albums is changing. With digital technologies increasingly becoming part of our lives, new types of photographs and albums are emerging. The dematerialization of family albums and the increasing tendency to capture photographs using mobile camera phones is contributing to a change in the storage of family memories. Although we have argued that the mobile camera phone was another step toward the end of the traditional family album and its ritual of archiving and presenting an individual and collective memory, we are not suggesting that there is an erosion of tradition or of a cohesive ritual. Instead, we argue that the mobile phone offers another site to "capture, share and archive the digital representations of experiences . . . new modes of engagement with cultural traditional and ritual pertaining to the act of remembrance"(Keep, 2009, p. 61). The promises of unlimited sharing have yet to be realized, however. Contrary to popular belief, remembrance is still bound by constraints of time, place, and gender. The business models and the technical requirements constraining mobile communications currently prevent wider sharing of personal and collective memories.

The immaterial nature of mobile phone photographs changes the ritual of the family album in the capturing, archiving, and the sharing of it. Pictures are now located in a highly personalized device. Unlike digital cameras that are usually used by several family members, the mobile phone is the most private technology and usually used only by those who own it. Mobile phones are contributing to the individualization of a traditionally collective ritual in which family storytelling happened through family photographs. Nevertheless, new platforms that might help control this change have appeared, such as digital archives available online on Flickr or Facebook. These relatively new digital archives might serve as helpful tools to allow us to build networks on which we can share our individual and collective memories.

Interviews with Portuguese women have allowed us to understand the impact of mobile phones in our lives and how they are contributing to social and personal changes. Traditional family archives, family photo albums, are being replaced by dematerialized albums, created through new media devices that allow us to capture and store our memories in digital format. Women remain the guardians of family memories, but now they may also be the producers, as well as the organizers and protectors, of those memories. Our study of Portuguese women verifies trends observed in studies conducted in other parts of the world: in South Korea (Hjorth, 2007; Lee, 2005) and in Great Britain (Reading, 2008). New technologies, like mobile phones, are contributing to a shift in the storage of personal memories. Family photo albums are becoming more and more fragmented and fragile without order or sequence; nevertheless, women remain the guardians of family memories.

References

ANACOM. (2010). *Statistics: Mobile Networks and Services.* Retrieved from http://www.anacom.pt/render.jsp?contentId=1042267

Araújo, V., Cardoso, G., & Espanha, R. (2009). *A sociedade em rede em Portugal 2008. Apropriação do telemóvel na sociedade em rede.* Lisboa, Portugal: Obercom.

Arnold, M. (2003). On the phenomenology of technology: The "Janus-faces" of mobile phones. *Information and Organization, 13*, 231–256.

Bourdieu, P. (1991). *Photography: A middle-brow art.* Stanford, CA: Stanford University Press.

Cardoso, G., Espanha, R., & Gomes, C. (2006). *Inquérito a sociedade em rede em Portugal 2006* [Survey of the network society in Portugal 2006]. Lisbon, Portugal: CIES.

Cardoso, G., Gomes, M. D. C., & Espanha, R. (2006). *The Network Society in Portugal 2006 Survey.* Retrieved from http://obercom.pt/en/client/?newsId=68&fileName=fr2_uk.pdf

Chalfen, R. (1991). *Turning leaves: The photograph collections of two Japanese American families.* Albuquerque: University of New Mexico Press.

Chalfen, R. (1997). Japanese home media as popular culture. *Proceedings of the Japanese Popular Culture Conference 1997, Victoria, BC, Canada.* Retrieved from http://www.richardchalfen.com/wip-jhm.html

Connerton, P. (1989). *How societies remember.* Cambridge, United Kingdom: Cambridge University Press.

European Communities. (2008). *The life of women and men in Europe. A statistical portrait.* Luxembourg: European Communities.

Favart-Jardon, E. (2002). Women's 'family speech': A trigenerational study of family memory. *Current Sociology, 50*, 309–319.

Fortunati, L. (2009). Gender and the mobile phone. In G. Goggin & L. Hjorth (Eds.), *Mobile technologies: From telecommunications to media* (pp. 23–34). New York, NY: Routledge.

Foucault, M. (1978). *The history of sexuality. An introduction* (R. Hurley, Trans.; Vol. I). London, United Kingdom: Alan Lane.

Foucault, M. (1988). Technologies of the self. In L. H. Martin, H. Gutman, & P. H. Hutton (Eds.), *Technologies of the self: A seminar with Michel Foucault.* London, United Kingdom: Tavistock, 16–49.

Ganito, C. (2007). As mulheres e os telemóveis: Uma relação por explorar. *Comunicação & Cultura, 3*, 41–58.

Ganito, C. (2008). *Moving acts: Transforming gender*. Paper presented at the Sixth International Workshop on Phenomenology, Organisation and Technology, Oxford, United Kingdom.

Garde-Hansen, J., Hoskins, A., & Reading, A. (Eds.) (2009). *Save as . . . digital memories*. New York, NY: Palgrave Macmillan.

Gibbons, J. (2007). *Contemporary art and memory: Images of recollection and remembrance*. London, United Kingdom: I. B. Tauris.

Gye, L. (2004). *Moments of coalescence: Family history and new technologies of remembrance*. Paper presented at Text and Sound 2004: The Yet Unseen: Rendering Stories, RMIT School of Creative Media, Melbourne, Australia. Retrieved from http://search.informit.com.au/documentSummary;dn=874767194596354;res=E-LIBRARY

Gye, L. (2007). Picture this: Mobile camera phones on personal photographic practices. *Continuum: Journal of Media and Cultural Studies, 21*(2), 279–288.

Hirsch, M. (1997). *Family frames: Photography narrative and postmemory*. Cambridge, MA, & London, United Kingdom: Harvard University Press.

Hjorth, L. (2006). *Snapshots of almost contact: Gendered camera phone practices and a case study in Seoul, Korea*. Paper presented at Cultural Space and Public Sphere in Asia, Seoul, Korea.

Hjorth, L. (2007). Snapshots of almost contact: The rise of camera phone practices and a case study in Seoul, Korea. *Continuum: Journal of Media and Cultural Studies, 21*(2), 227–238.

Huhtamo, E. (2004). An archaeology of mobile media. *Proceedings of the 12th International Symposium on Electronic Art*, Helsinki, Finland.

Katz, J., & Aakhus, M. (2002). *Perpetual contact: Mobile communication, private talk, public performance*. Cambridge, United Kingdom: Cambridge University Press.

Keep, D. (2009). The portable shrine. Remembrance, memorial, and the mobile phone. *Australian Journal of Communication, 36*(1), 61–72.

Langford, M. (2001). *Suspended conversations: The afterlife of memory in photographic albums*. Montreal, Canada: McGill-Queen's University Press.

Lasen, A. (2004). Affective technologies. Emotions and mobile phone. *Receiver, 11*(3). Retrieved from http://www.receiver.vodafone.com/archive/index.html

Lasen, A. (2005). 'History repeating?' A comparison of the launch and uses of fixed and mobile phones. In L. Hamil & A. Lasen (Eds.), *Mobile world: Past, present and future*. Surrey, United Kingdom: Springer.

Lauretis, T. de. (1987). *Technologies of gender: Essays on theory, film and fiction*. Bloomington: Indiana University Press.

Lee, D. (2005). Women's creation of camera phone culture. *The Fibreculture Journal, 6*. Retrieved from http://journal.fibreculture.org/issue6/issue6_donghoo.html

Leite, M. M. (2000). *Retratos de família* (2nd ed.). Brazil: Editora da Universidade de São Paulo.

Ling, R. (2004). *The mobile connection: The cell phone's impact on society*. San Francisco, CA: Morgan Kaufmann.

Lins de Barros, M. M. (1989). Memória e família: Estudos históricos. *Vértice, 3*(2), 29–42.

Marktest. (2006). *O perfil do consumidor Português* [The Profile of the Portuguese Consumer]. Lisbon, Portugal: Marktest.

Mortimer, J. T., & Shanahan, M. J. (Eds.). (2004). *Handbook of the life course*. New York, NY: Springer.

Okabe, D., & Ito, M. (2006). Everyday contacts of camera phone use: Steps toward technosocial ethnographic frameworks. In J. R. Hoflich & M. Hartmann (Eds.), *Mobile communication in everyday life: Ethnographic views, observations and reflections*. Berlin, Germany: Frank & Timme.

Reading, A. (2008). The mobile family gallery? Gender, memory and the cameraphone. *Trames, 12*(3), 355–365.

Sontag, S. (1977). *On photography*. London, United Kingdom: Penguin.

Soulet, M.-H. (2002). *Gérer sa consommation. Drogues dures et enjeu de conventionnalité*. Fribourg, Switzerland: Éditions Universitaires Fribourg Suisse.

Tashakkori, A., & Teddlie, C. (1998). *Mixed methodology: Combining qualitative and quantitative approaches* (Vol. 46). Thousand Oaks, CA; London, United Kingdom; & New Delhi, India: Sage.

Torres, A. C., Silva, F. V. D., Monteiro, T. L., & Cabrita, M. (Eds.). (2004). *Homens e mulheres entre família e trabalho* [Men and women between family and work]. Lisbon, Portugal: DEPP—Departamento de Estudos, Estatística.

Turkle, S. (1984). *The second self: Computers and the human spirit*. London, United Kingdom: Granada.

Van House, N., Davis, M., Takhteyev, Y., Ames, M., & Finn, M. (2004). *The social uses of personal photography: Methods for projecting future imaging applications*. Working Papers. Retrieved from http://people.ischool.berkeley.edu/~vanhouse/van%20house_et_al_2004b%20.pdf

Van House, N., Davis, M., Ames, M., Finn, M., & Viswanathan, V. (2005). The uses of personal networked digital imaging: An empirical study of cameraphone photos and sharing. *Proceedings of the Conference on Human Factors in Computing Systems (CHI 2005), Portland, Oregon*. Retrieved from http://people.ischool.berkeley.edu/~vanhouse/van_house_chi_short.pdf

Vodafone. (2011). Vodafone casa. Retrieved from http://www.vodafone.pt/main/Particulares/vodafonecasa/

Zelizer, B. (1995). Reading the past against the grain: The shape of memory studies. *Critical Studies in Mass Communication, 12*(2), 214–239.

Section Four: Engaging Politics

For many, the most pressing questions regarding the global and intercultural adoption of new media are those concerning sociopolitical impact: Are new media permitting the development of the utopian "global village" that McLuhan envisaged and fulfilling the "social transformation" projections of early utopian theorists? Is access to the Internet offering greater access to education or permitting new forms of political engagement? Are Internet and communication technologies (ICTs) being deployed to strengthen the public sphere? Are new media bringing the peripheralized population closer to the center?

It is perhaps unsurprising that the investigation of the global implications of new media on the cultures of political engagement in this "third age" of the Internet (Wellman, 2009) provide evidence of the same dialectical tensions found elsewhere in this collection. Although figures show that the digital divide is slowly closing, simple access to the Internet is not necessarily transforming privileged elites into social justice-oriented actors or global citizens; at the same time that new media are permitting new modes of cultural engagement and allowing new voices to be heard, engagement with Web 2.0 may also be facilitating mass integration of commercial interests into the public sphere and disrupting tried and tested modes of political engagement. The chapters in this section offer case studies of these continuing tensions.

In the first chapter, Konrad Ng offers a heartening example of transformation at the center: the role of new visual and social media in reinscribing popular notions of "who is a citizen" in the United States. Ng points out that Asian bodies and cultural forms have long played a role in US life but explains that the recognition of Asian American contributions and the circulation of multidimensional representations of the Asian American experience have been obscured by enduring stereotypes. How, he asks, are new media platforms and programs transforming the political and cultural presence exercised on behalf of Asian America? To explore this question, he examines three contemporary instances of Asian American new media practice: the rise of independent Asian film production coupled with widespread film support and promotion mediated through online social networks; the new media initiatives (software development, social networks) at the Center for Asian American Media; and the recent meteoric rise to stardom of the Asian American boy band Legaci, supported to a large extent by their effective self-promotion strategies on YouTube. Asian Americans, he concludes, are embracing new media to engage in novel forms of intercultural communication: community organizing, creating alternative cultural spaces, and repopulating the American imaginary with alternative representations of Asian American identity.

In the second chapter of this section, by contrast, Nickesia Gordon and Kristin Sorenson report on studies of new media's impact on societies of "the periphery": societies in the so-called developing world. These authors take as their starting point proclamations by international bodies such as the United Nations Development Program (2010) that the Internet is a powerful tool in the promotion of "peace and security, development and human rights," (3rd para.) and offer two contrasting studies of new media uptake and impact in the developing world: Internet use in Jamaica and Web 2.0 and multimedia use in Chile. The authors ask: Who benefits? Whose voices are heard? They report that although Internet penetration rates in Jamaica have risen in the past decade, users are overwhelmingly the urban and educated elite, and there is no evidence of increased access to education (especially tertiary education); even more tellingly, a survey of Jamaican Internet users uncovers that usage is overwhelmingly recreational (not political). In Jamaica, at least, the economic, educational, and developmental opportunities offered by the Internet are not yet being harnessed by or for the poor majority. In Chile, on the other hand, new media and Web 2.0 appear to be permitting new forms of public engagement in the public sphere in a postdictatorship state still coming to terms with its recent political history. The authors compare the tightly controlled mainstream reporting of former dictator Augusto Pinochet's arrest and later death with the proliferation of alternative perspectives expressed in online forums and through the online sharing of journalist/filmmaker's "handheld camera movie," *El día en que murió Pinochet* (*The Day Pinochet Died*). In this country, new media have expanded access to the public sphere, they argue, not least by giving voice to the poor, the homeless, the mentally ill, the blind, women, children, the elderly, and members of the lesbian, gay, bisexual, transgender, and queer (LGBTQ) community—all groups of people whose representations have either been highly contained or erased altogether in the mainstream media.

The third and fourth chapters in this section stand as an interesting dyad, illustrating a point emphasized powerfully by a range of authors in this collection: that the uptake and impact of new media is critically influenced by local sociopolitical and historic context, as well as by culture, ownership, regulation, and socioeconomic trends. In the third chapter, Irina Privalova presents

Web 2.0 and user-generated content as a social and political watershed for a population whose deep mistrust of mainstream news reporting has been fuelled by tight control of the media in the Soviet era and by the continuing media control by the Russian government and their corporate allies up to the present day. Privalova offers an axiological analysis of the Russian uptake of new media, connecting style and participation to deeply held values and communicative practices. She also highlights cases in which "citizen journalism" has upset the political status quo: revelations of systemic police corruption made by a police officer in a YouTube video and a citywide public health panic triggered by a malicious blog post. Whereas she suggests that the latter situation illustrates a challenge of critical media literacy for an unaccustomed population, the former situation offers a novel form of political engagement in the current Russian sociopolitical context.

By contrast, Herbert Hrachovec describes, in the fourth chapter of the section, the impressive implementation of new media as a political and communications platform in a different sociopolitical milieu: the world of Austrian higher education and student protests. Hrachovec describes how the rapid establishment of a website and social networks transformed the face of a 2009 student protest, catalyzing unprecedented reporting and "participation." Upon closer observation, however, this author cautions against forming simplistic conclusions. He highlights the irony of student activists embracing a suite of commercialized new media platforms and embracing the very corporatization of university governance that they claim to be resisting. With more worrisome implications, Hrachovec characterizes the group communication process of this uprising as diffusing the protest's political force. Where, at one time, mass protests relied on leadership and centralized communications, this Web 2.0-driven protest self-consciously avoided identifying leaders—resulting in a presentation of unfocused, collaboratively drafted "demands," leaving no one for the "opposition" to address or mediate with. Is Web 2.0, he wonders, allowing corporations to enter politics by (another) backdoor, while many users speak but few listen?

16. Asian American New Media Communication as Cultural Engagement: e-mail, Vlog/Blogs, Mobile Applications, Social Networks, and YouTube

KONRAD NG, PHD

Introduction

How are new media platforms and programs transforming the political and cultural presence exercised on behalf of Asian America? Asian bodies and cultural forms have long played a role in American life, yet the recognition of Asian American contributions and the circulation of multidimensional representations of the Asian American experience remain obscured by the endurance of particular stereotypes. Ono and Pham (2009) contend that the dominant media representations of Asians and Asian Americans are as "others that include exoticized women, asexual men, a yellow peril threat to the United States, a forever foreigner, and/or model minority" (p. 173). Cultural notions that view Asian Americans as the model minority or intrinsically exotic and foreign means that Asian American voices can be lost, misrepresented, or exploited in US political and cultural discourse (Lee, 1999; Lowe, 1996; Palumbo-Liu, 1999). Aoki and Takeda (2008) note how the politics of recognition and representation have become part of the Asian American political agenda. They write, "Control over one's identity is an important prerequisite for political inclusion and opportunity. Stereotypes have deep roots. . . . [and] the struggle to eliminate them is an important part of Asian American battle for equality and acceptance" (p. 154). During the 2008 presidential election in America, the poor inclusion of Asian Americans in political and popular culture prompted Democratic congressional leaders to write letters to the top executives of CNN and MSNBC criticizing each network's poor coverage of Asian American and Pacific Islander voters and issues in relation to other constituencies. The letters affirmed the importance of Asian and Pacific Islander American participation in the political process and the media's responsibility in "recognizing or ignoring these voices" (Honda et al., 2008, 2008a).

The quality of popular representation as an index of civic inclusion has long been the concern of Asian American cinema. Since the formation of independent media arts collectives by Asian American activists, filmmaker, and scholars in the late 1960s and 1970s, cinema has been a key mode of communication for empowering Asian American communities through representation. Influenced by the de-colonial and anti-imperial politics and aesthetics of Third Cinema (Mimura, 2009) and the ethnic consciousness activism of the Asian American movement (Wei, 1993), film became used as "a method for social change, instead of as a showcase of the artist [and] . . . produced a cinema . . . concerned with . . . identity politics, historical injustice, and contemporary racism" (Xing, 1998, p. 41). Asian American activists joined movements initiated by Latino and African Americans to use "cinema as a mirror and provocateur" (Tajima, 1991, p. 11) for community advocacy. They produced works that circulate narratives and imagery more attuned to the Asian American experience and question the stereotypes and modes of production that limited opportunities for the group's representation. Feng (2002) argues that the political spirit of Asian American film and video is not purely oppositional but rather expresses an ambivalent dis-identification and critique of the conventions of American cinema. That is, Asian American filmmakers are conscious of the ways that Hollywood cinema, from its representation to its production, exhibition, and distribution, have marginalized Asian American voices, yet they continue to use the medium and industry as a form of political engagement. Asian American film and video uses "cinema to critique cinema, using a mode of communication to convey messages that subvert that mode" (Feng, 2002, p. 15).

This chapter explores how Asian Americans use contemporary information and communication technologies to continue the cultural politics of Asian American representation inaugurated by Asian American cinema. New media modes of communications such as e-mail, Web logs (vlogs and blogs), social network sites (SNSs), YouTube, and applications for mobile devices have enabled grassroots approaches to repopulating the cultural and political imaginary. These self-representations strive to be more congruent with lived experience and provide provocative exercises that question prevailing stereotypes. The new media landscape has enabled forms of communication that have greater impact and reach than the traditional medium of cinema. Indeed, the possibility that grassroots digital media production can change cultural values is a feature of what Jenkins (2006) calls our current age of "convergence culture"—a media landscape in which the roles of cultural producer, consumer, and citizen are merging and media content flows across a range of platforms and devices. Innovations in digital and online technologies have altered the practices of the media industry and initiated new forms of participation and collaboration.

The advent of new information and communication technologies means that Asian Americans can create innovative forms of grassroots activism and cultural production. As Ono and Pham (2009) argue, the new media landscape has "proven fruitful for Asian Americans, since access to traditional media venues often prove to be too difficult, too costly, and too restrictive for their voice to be heard" (p. 155). Asian American new media communication, I contend, is about the flow of content across media platforms and the formation of independent cultural industries and mission-driven digital networks that encourage community participation in shaping and amplifying the representation of the Asian American experience. To consider this moment in intercultural communication history, I want to present three cases of Asian American new media practice that are redefining the political and cultural engagement of minority communities: First, I look at the grassroots new media promotional campaign of filmmaker Justin Lin to reveal how filmmakers engage in online community organizing that ties their work to cultural activism. Second, I explore the Center for Asian American Media's investment in social media and gaming initiatives as part of its mission to produce rich and diverse representations of Asian American life for public media. Last, I look at Asian American cultural production on YouTube as a way of producing new Asian American representation. The scholarship on new media cultures and how the new media landscape has motivated novel practices in the formation of Asian American identity remains relatively new. The rapid pace of technological change has meant that studies of the relationship between race and new media will remain slightly behind the phenomena, but there is also a persistent bias in framing Asian Americans in a privileged relationship with the Internet. The popular focus on how racial minorities can be victims of the digital divide through lack of access to the Internet collides with the stereotype of Asian Americans as a model minority in such a way that Asian Americans are often precluded as subjects of studies of the Internet; Asian Americans are seen as having a privileged relationship with contemporary information and communication technologies.

Nakamura (2007) extends the justification for an Asian American cyberanalytics by reminding us that Asian Americans "inhabit a range of positions in relation to the Internet. . . . [and] live on both ends of the networked information economy, both as low- and high-skilled workers and as consumers" (pp. 171–172). The task, then, is to examine how present-day intersections of communication, technology, and culture are redefining racial identity and cultural discourse in the US and the role that Asian Americans play in producing their identity and affecting the cultural economies of their representation. I share the optimism of Nakamura, who believes that

> the examination of the deployment of the Internet as a racial and gendered visual cultural practice might help us to take a closer look at the ways that cultural resistance to normative gender, racial, and national narratives might be enabled in new digitally interactive spaces. (p. 34)

I join Ono and Pham (2009) in examining Asian American independent and vernacular media—that is, media practices that offer an "alternative to dominant media and posit[s] complex images and visions of Asian Americans that may be oppositional to dominant media. . . . [and are] Asian American friendly, specific, and political on some level, and operate as an ego-function to inspire" (p. 123).

Case studies of Asian American new media practice

Asian American independent media and Justin Lin

The New York Times film critic Manohla Dargis (2010) offers an insightful reflection on the large-scale changes in American independent cinema in the digital age. After a period of commercial investment in independent film by Hollywood studios, during which filmmakers could go "from Sundance to Scorsese" (para. 2), Dargis notes how the aesthetics, stories and economic fortunes of independent film are now being shaped by new media forms like social-networking programs, handheld devices, and the "do-it-yourself" (DIY) culture that has been enabled by new information and communication technologies. Programs such as Facebook and technologies such as mobile devices have created a "virtual infrastructure" of self-distribution and promotion that cultivates an audience that is interactive and participatory and supports the film as a choice in lifestyle. Dargis's description of independent cinema in the digital age echoes the social changes of convergence culture whereby new media technologies are redefining the role of cultural consumers by enabling greater participation in media consumption and production. Her article prompts consideration of the transformations in Asian American cinema, a tradition in American film that is synonymous with independence because of its reliance on an independent mode of production, its themes, its aesthetic form, and the politics of its imperative. Here, I want to consider filmmaker Justin Lin. Lin's young film career is typical of independent cinema's arc "from Sundance to Scorsese." Lin made personal and financial sacrifices to make his debut feature film, *Better Luck Tomorrow* (2002), a story about a group of suburban Asian American high school students who use their status as overachievers to enable their illicit activities. The film premiered at the 2002 Sundance Film Festival to critical acclaim, which led to a distribution deal with MTV Films (a subsidiary of Paramount Pictures) and, eventually, to steady work for Lin and some of the film's actors at Hollywood studios like Universal Pictures and Touchstone Pictures. After Lin made commercial films like *Annapolis* (2006) and *The Fast and the Furious: Tokyo Drift* (2006), he returned to the world of independent cinema to make *Finishing the Game* (2007), a mockumentary about the search for an actor to replace Bruce Lee after his unexpected death during the filming of *Game of Death* (1978). My interest in Lin lies not in his merits as a filmmaker. Rather, I am interested in Lin's adoption of digital age DIY tactics and strategies that went beyond simply "breaking out," as most independent filmmakers aspire to do. Lin used his films as platforms for forms of Asian American community organizing and engagement that sought to impact the representation of Asian America in American film history and popular culture.

Although *Better Luck Tomorrow* received studio distribution, the film's distribution budget, a crucial ingredient in a film's box office success, was small. Without studio support, the film needed support from the community to demonstrate the commercial viability of Asian American cinema and the market impact of Asian Americans. The cast and crew of the film and Asian American media organizations started a grassroots e-mail campaign to supplement the promotion of the film by the film's Web site. Their participation went beyond self-promotion; they linked support of the film with investing in future opportunities for Asian American actors, crew, and film projects in an unfriendly industry. In an article published in the *Los Angeles Times*, lead actor Parry Shen stated that more than box office success was at stake: "If we fall short, it's going to be a shame. . . . It's going to be a long, long time before we get a chance like this again" (Yoshino, 2003). David Magdael, co-director of the Los Angeles Asian Pacific Film Festival, used the Asian Pacific American (APA) First Weekend Film Club (www.asianamericanfilm.com/firstweekend) to encourage Asian American support of the film. The APA First Weekend Film Club was an e-mail distribution list modeled on underground marketing campaigns for African American films in the late 1990s. The film club's Web site equipped e-mail recipients with media tools to further promote and support the film. The *Better Luck Tomorrow* Web site sold merchandise, listed screening dates, press clippings, and other media items for use by supporters. The promotional campaign also celebrated "superfans" as models of robust support for the film. On its opening weekend, *Better Luck Tomorrow* "bested MTV's historical per screen benchmark [and] . . . led the weekend box office on a per screen basis" (MTV Films, 2003). The grassroots mass e-mail campaign and online tactics behind *Better Luck Tomorrow* reflect the kind of savvy marketing techniques that are available in the digital age, but it is the imperative to organize the community that is unique; the Asian American community was asked to demonstrate its box office power and organizational capacity to improve Asian American opportunities in entertainment as well as the community's social standing.

Innovations in information and communication technology enabled a more expansive campaign for Lin's next Asian American film, *Finishing the Game*. The strategy and digital media tactics used to promote *Better Luck Tomorrow*, a Web site and e-mail campaign, were enhanced by Web logs (vlogs/blogs), social media sites, and YouTube videos. Here, the use of vlogs/blogs is noteworthy for the possibilities it suggests for Asian American cultural engagement. The Web logs stood out in the ways they presented personal experiences in opposition to stereotypical constructions of Asian American identity. Likewise, vlogs/blogs are interactive; readers can post comments, circulate entries to a wider audience, or repurpose the material. The original Web site for *Finishing the Game*, YouOffendMeYouOffendMyFamily (www.youoffendmeyouoffendmyfamily.com), now functions as a vlog/blog that promotes Asian American cinematic empowerment. Earlier versions of the film's original Web site remain on the Asian media arts site, AliveNotDead (www.alivenotdead.com/ftg/index.php) and on Myspace (www.myspace.com/finishingthegame). Most of the vlog/blog postings by the film's cast and crew were irreverent, but an important narrative thread emerges: a grassroots critical engagement of Asian America politics, culture, and community, and the use of personal experience to organize the community. For example, the Finishing the Game Myspace page states that the "Asian American community's national grassroots movement [turned] BETTER LUCK TOMORROW [into]. . . . a symbol of political empowerment through fair and self-media representation." The site also has an open letter from Lin asking fans of *Better Luck Tomorrow* to support *Finishing the Game* because "[i]n an industry governed by box office receipts, there is strength in numbers [and we need]. . . . a clear message that we demand to see ourselves on screen as multi-dimensional characters" ("A Letter From Justin Lin," para 2.). On the AliveNotDead Web site, a post describes how the "real" success of *Better Luck Tomorrow* was the empowerment of Asian American community and its forms of grassroots social media community organizing. The letter, signed by members of the cast and crew, states that *Better Luck Tomorrow*

> was the first time Hollywood was able to see that Asian Americans (and African, Latino, & Caucasian Americans) wanted to see and support a film made by Asian Americans and starring Asian American actors. Hollywood was scratching their heads wondering how the heck it was possible that BLT had the highest per screen average of any film in the country their opening weekend (+$30,000) with virtually no marketing budget. . . . Asian Americans and Asian American cinema is here to stay! (http://www.kellyhu.com/ftgmovie/details.html, para. 4)

Figure 1. Blog post for *Finishing the Game*

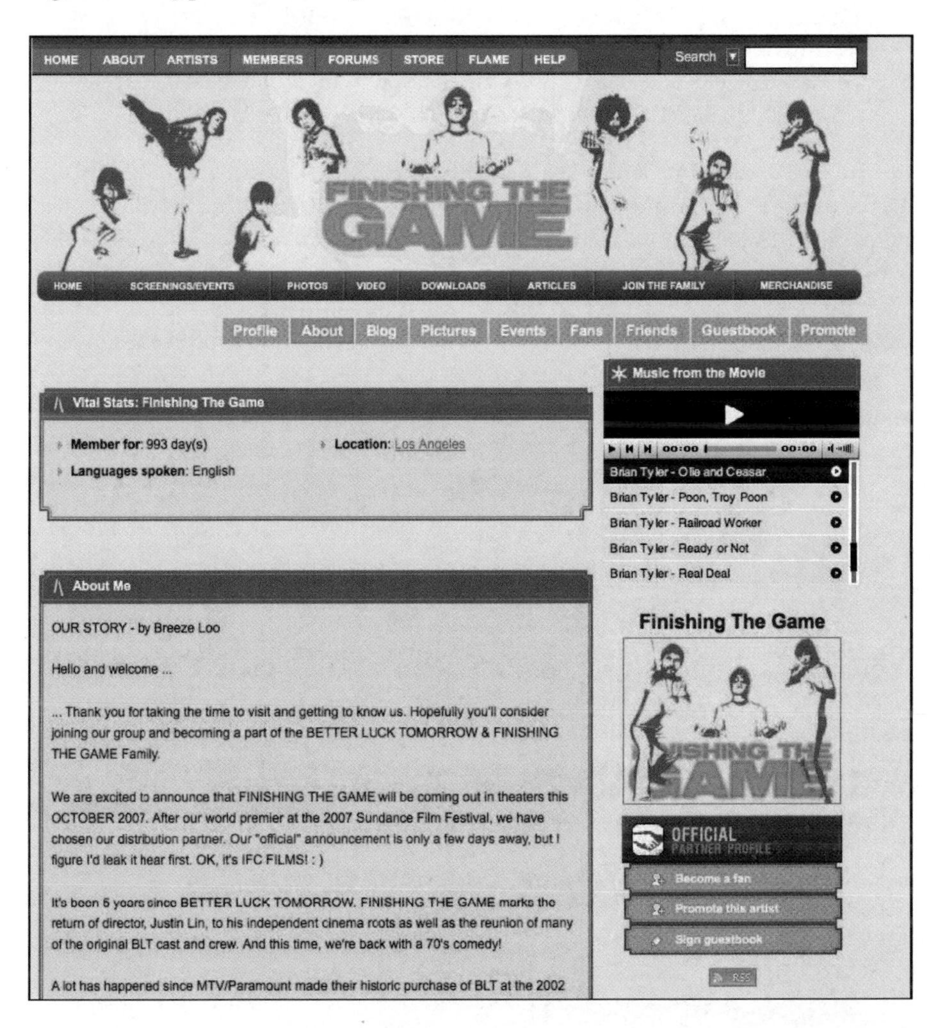

The blog contains YouTube videos of the film's grassroots promotional tour to key Asian American community organizations, academic programs, and cities with notable Asian American populations. Postings such as "WE <3 (love) NYC- FINISHING THE GAME premiere wrap up!" are typical for how the blog showcases how the community rallied around a single imperative. Another posting reads:

> WE as an Asian American community have the economic power to push for our own media dissemination. In regular English, it means: WE HAVE THE

> NUMBERS and THE MONEY to TELL HOLLYWOOD that. . . . ASIAN AMERICANS WANT TO SEE OUR OWN REPRESENTATION ON SCREEN. . . . We demand three-dimensional characters that truly represent how we are as a regular community. (www.alivenotdead.com/ftgmovie/WE-3-NYC-FINISHING-THE-GAME-premiere-wrap-up--profile-38424.html)

In a blog post by Lin on the YouOffendMeYouOffendMyFamily Web site, he recounts a conversation with a veteran Asian American film producer who called filmmakers who make Asian American films "retarded" on the basis that it didn't make good business or professional sense. Lin's reply outlines the ethical imperatives that have informed his films and is rooted in his experience in the film industry: He suggests that his aim is to introduce a range of

Figure 2. Blog post for *Finishing the Game*

multidimensional Asian American characters and show how Asian America is a source of American cinema and a commercially influential audience. As he puts it:

> [V]ery rarely do Asian American films get picked up for traditional distribution . . . . which is still the standard of measure for any film. And because of that, there hasn't been enough Asian American films deemed profitable to establish a pattern for a business plan. . . . I continue to have stories, some with Asian American characters, issues and themes I want to explore. And in order for me to do so I have to approach it on an independent level. Is it good business? No. Am I "retarded"? I'm not sure. I do know one thing though, I don't need to go to Vegas to lose money, I'm just going to keep making Asian American films. (www.youoffendmeyouoffendmy-family.com/am-i-retarded-for-making-asian-american-films/ para. 2, 4)

The online campaigns to rally around *Better Luck Tomorrow* and *Finishing the Game* present an ethic of community organizing that achieves objectives beyond promoting the film itself. Lin and his supporters used new media to challenge prevailing representations of Asian Americans and the constraints of a media industry that have marginalized the Asian American experience.

New media initiatives by the Center for Asian American Media

The Center for Asian American Media (CAAM) is one of the major media organizations to emerge from the Asian American political and cultural activism of the late 1960s. Formerly known as the National Asian American Telecommunications Association, CAAM was founded after a conference involving Asian American activists, filmmakers, and producers who were concerned about the state of Asian American representation and access to public broadcasting. Created in 1981, the nascent organization tapped into funds from the Corporation for Public Broadcasting to support the development of ethnic public media and dedicated itself to being an advocate for the inclusion of the Asian American experience in public media. Since then, CAAM has produced, exhibited, and distributed Asian American works in public media and is part of a small minority public broadcasting consortium designated by the Corporation for Public Broadcasting to provide programming for the Public Broadcasting Service. In 2008, CAAM began launching digital media initiatives to complement its mission and address the changing demographics and means of media consumption by the public. CAAM Executive Director, Stephen Gong, observed that the election of President Barack Obama revealed a growing awareness of the racial/ethnic diversity of the American population, the relevance of young people, and the prominent role of social media and cultural activism. Gong states that CAAM's (2009) "commitment to embrace diversity as a core principle of our work requires that we engage

Figure 3. CAAM's iPhone application "Filipino or not?"

more deeply with its complexity and provide America's younger and more ethnically diverse audiences with rich and relevant content" (p. 2).

To this end, CAAM embarked on two innovative digital initiatives: First, the center developed an application for the Apple mobile device, the iPhone. The free game "Filipino or Not?" (Figure 3) portrays images of public figures who may or may not be of Filipino descent. After being given clues, players are asked to determine if the person is Filipino. The game offers a playful way to raise questions about race and representation in the media. The larger objective is to highlight the role Filipinos have played "in the media-making world. Often Filipinos in the media are assumed to be ambiguous or of another ethnicity; with this game we want to make visible the contributions of Filipinos in a playful way" (itunes.apple.com/us/app/filipino-or-not/id356945848?mt=8).

CAAM also created a SNS called Hapas.us (Figure 4) for people with a *hapa* identity—that is, a multiracial Asian identity. The aim of the SNS is to create an open forum to engage the multiracial Asian experience and start an archive for its changing representations.

People who join the site can use photos, videos, and text to present their story and express their opinions. Hapas.us does not attempt to limit the scope

Figure 4. CAAM's social network site Hapas.us

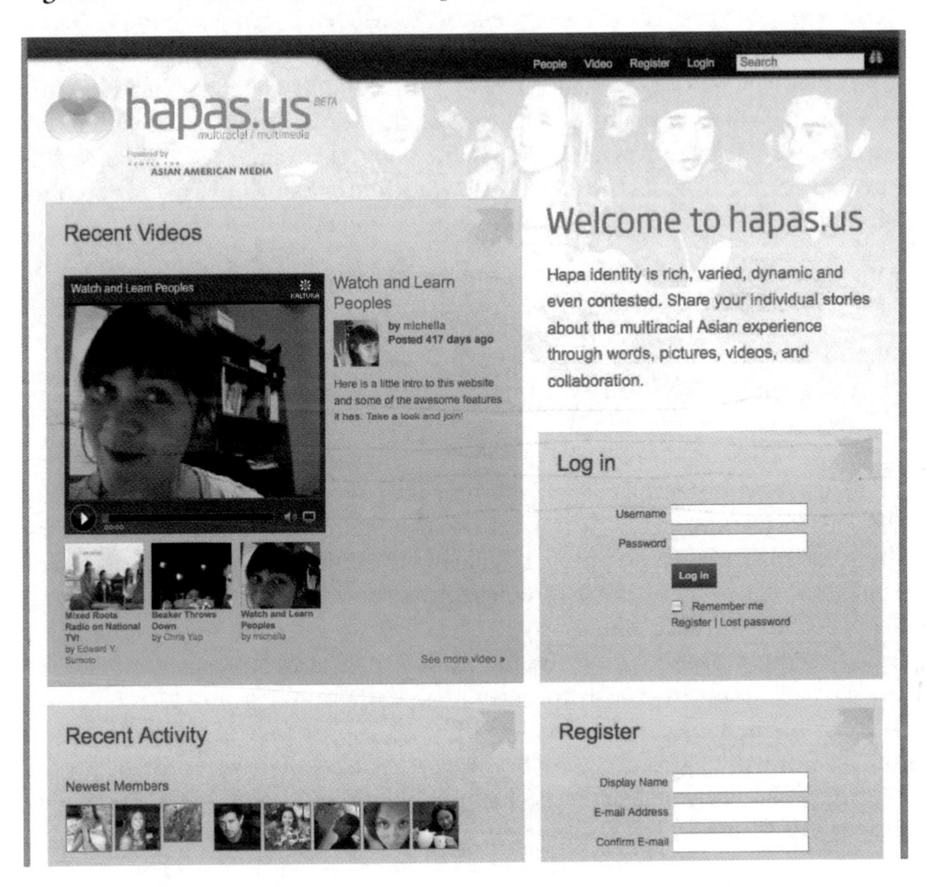

of hapa identity; rather, its solicitation of multiracial DIY cultural production seeks to represent the broadness of hapa identity and the real-life experiences of members. Although Hapas.us remains in a beta phase, it is representative of a larger movement of SNSs that provide alternative cultural spaces for Asian and Pacific Islander Americans.

The first SNS to enable this trajectory of organizing new media communities was the AsianAvenue Web site in 1997. AsianAvenue offered members an opportunity "to network, share a social lifestyle, find romance, entertainment, and even jobs." Similar to the e-mail campaigns undertaken by Justin Lin, AsianAvenue developed into a political movement when it appealed to members to be vigilant about the popular representation of Asian Americans. During the 1998 Winter Olympics, members of the site launched a campaign against MSNBC for reporting the gold medal win by figure skater Tara Lipinski, a White American woman and the silver medal win by Michelle Kwan, an Asian American woman, with the headline, "American Beats Out Kwan" (Ramirez, 1998). In 1999, members of AsianAvenue successfully protested ads for Skyy Vodka that employed stereotypical tropes of Asian women as submissive "china dolls" (Kane, 1999). Other significant SNSs include Iamkoreanamerican, whose aim is to showcase the "diversity and many intersecting stories of Korean Americans" and the native Hawaiian SNS Maoliworld. The emergence of SNSs and applications for mobile devices expands the terrain for Asian American political and cultural engagement.

YouTube apparatus and the rise of Legaci

One of the most influential platforms of the new media landscape is YouTube. Launched in February 2005, YouTube is an online repository where members of the public can upload, share, promote, and discuss short video clips. The platform has radically transformed how society consumes, produces, circulates, and integrates media content into our daily lives. In spite of its participatory ethos, collectivity, and reach, Jenkins (2006) expresses doubts about the platform's capacity for sustained civic engagement, contending that YouTube offers "an alternative perspective on what democracy might look like. . . . [but] we have a long way to go before we can achieve anything like a revitalized public sphere in the online world" (p. 273). However, the platform *has* allowed Asian Americans to become savvy cultural players. Here, the rise of R&B band, Legaci, is instructive. This group of four Filipino American singers from the San Francisco Bay area surpassed their perennial status of aspiring performers and achieved mainstream success as the back-up singers for popular teen pop-R&B singer, Justin Bieber. The ultimate place of Bieber and Legaci in music history remains to be seen, but after a decade working the traditional college and club circuit, producing self-released albums, and auditioning for television talent competitions, Legaci is now enjoying steady, higher profile work. What facilitated their ascent was the decision to self-promote on YouTube. They created a YouTube channel and posted videos of their cover performances. Legaci's cover performance of Justin Bieber's song "Baby" went viral and caught the attention of Bieber's manager who then hired them. *The New York Times* reporter, Josh Kun (2010), writes that Legaci "just might be the most visible yet invisible pop figures in the world" (para. 3) and contends that YouTube allows Asian American performers to pursue alternate routes to public exhibition and by so doing, challenge the stereotypes and conventions that have handicapped Asian Americans in the entertainment industry. Legaci's professional rise illustrates how YouTube can provide successful exposure and cultural critique. However, their ascent is also interesting because it reveals ways in which the *apparatus* of YouTube is also able to communicate exposure and critique. Kun reports that Legaci's style resonated with the "amateur aesthetic" of YouTube, "singing somebody else's songs in your bedroom or living room had long been part of Legaci's own upbringing. Its members all grew up singing pop and R&B hits on the karaoke machines of their Philippines-born parents." (para. 24). There is an additional meaning to the notion of YouTube's amateur aesthetic: It describes the genre of posting cover performances as opposed to original material, and the aesthetic properties of YouTube, that is, its "apparatus."

The treatment of YouTube as an apparatus draws on apparatus theory, a trajectory of film scholarship that emerged during the 1970s. Its premise is that the mechanics of cinematic representation—cinematography, editing, sound, and mise-en-scene—is ideological in how it composes an objective spectatorial position. Baudry (2004) writes that

> the ideological mechanism at work in the cinema seems thus to be concentrated in the relationship between the camera and the subject. The question is whether the former will permit the latter to constitute and seize itself in a particular mode of specular reflection. (p. 364)

Whereas Baudry argues that the apparatus of cinema is absolute, I take the contribution of apparatus theory to be the recognition that cinematic aesthetics have ideological effects; put differently, the relationship between the YouTube and the viewer has ideological dimensions. The amateur aesthetic of YouTube includes the mechanics of its DIY form of representation, the interactive features of the platform (spectators may comment, tag, rate the video), its ability to be repurposed, and the broad reach of its distribution and audience. Legaci has uploaded more than 40 videos onto YouTube, which all share common aesthetic properties. Most videos are shot in one long continuous take; use key lighting or location lighting; and are staged in everyday settings like a living room, restaurant, locker room, or lobby that is set up with microphones and/or musical instruments. The sound is clean and effectively captures the essence of the performance; the performers wear their regular clothes, not a costume or clothing that reflects a concerted effort to brand; and almost all of the videos are cover performances of popular R&B and pop songs. These aesthetic characteristics can be considered amateur relative to the professional production values of music videos broadcast on music video stations like MTV. Legaci chose aesthetic techniques that capture realistic self-representations and allow the viewer to appreciate the naturalness of their talent. The choice of posting cover performances of popular songs sung by non-Asian American artists allows Legaci to reappropriate the music as their own. Film theorist Andre Bazin (2004) suggests that the long take, when employed in conjunction with "deep focus"—where the entirety of a frame remains in focus—allows the viewer to direct his or her focus across the space of the scene and reflect on the action that unfolds in its totality. As a consequence, the meaning of the scene becomes open to interpretation as the viewer makes his or her own sense of the on-screen action and the meaning of the sequence.

The combination of aesthetic forces at work in Legaci's amateur YouTube videos encourages the viewer to consider how these empowered representations of Asian American identity depart from the stereotypes that disable the representation of Asian American in the industries of music and performance. The long take of amateur Asian American performers singing popular music enables the viewer to believe in the imagery and that there is a way to change the cultural economy of Asian American representation. In this sense, the information and communication technology of YouTube offers a new mode of cultural engagement for Asian Americans.

Conclusion

The convergence culture tactics and strategies exercised by and on behalf of Asian America could be read as examples of what Oren (2005) calls "bridging texts": "Asian American cultural production that gives voice to. . . . [racial] grievances [and abjection] *within* mainstream cultural texts" (emphasis in original; p. 340). For Oren, bridging texts reveal the registers of racial value in popular culture and, by so doing, assist in the genesis of political action. It is clear that new media has expanded the range of bridging text activities in pursuit of affecting the cultural currency of Asian American representations. What is notable is how information and communication technologies are affecting the political and cultural engagement exercised on behalf of Asian America. Through programs and platforms such as e-mail, vlogs/blogs, mobile applications, social networking, and YouTube, Asian Americans are engaging in novel forms of intercultural communication: community organizing, creating alternate cultural spaces, and repopulating the American imaginary with alternative representations of Asian American identity. I return to Nakamura's (2007) argument to study the formation of racial identity in relation to the discourse of cyberspace. She contends that popular analysis of the digital world tends to treat race as indexical of access to the Internet—one aspect of the so-called digital divide. This approach, Nakamura suggests, views racial identity in terms of the capacity to consume digital culture and, by doing so, neglects how people of color produce racial identities through Internet technologies and interfaces. She writes that "this is important information in the context of Internet users and their lived realities. Manifestations of expressive cultures on the Internet may thus provide an online oasis or refuge for users of color" (p. 183). The idea should not be how racial minorities may overcome barriers to access so that they may join "mainstream" America as consumers of the

Internet but how users of information and communication technologies can produce more richly layered and varied racial identities. In her study of Asian American Internet tactics, Nakamura contends that

> cyberspace functions as a vector for resistant cultural practices that allow Asian Americans to both use and produce cyberspace. Indeed, new media's potential when it comes to Asian Americans has much to do with the powerful ways in which it deploys interactivity to destabilize the distinctions between users and producers, as well as questioning a rigid and essentialized notion of Asian American "authenticity." (p. 185)

I want to frame the implications of this approach in relation to protests over the 2010 United States Census and the 2010 United States Health Reform Act.

In June 2009, Minnesota Representative Michele Bachmann participated in a series of interviews outlining her opposition to the 2010 US Census (Dinan, 2009; Media Matters for America, 2009; Kleefeld, 2009). In spite of the constitutionality of the census, assurances about privacy protections, and nonpartisan public service announcements and community-based video testimonials explaining the importance of the census in determining representation and funding, Bachmann and other public officials, such as Texas Representative Ron Paul (2010a) and former Georgia Representative Bob Barr (2009), rejected the census because of concerns over cost, the role of government, the scope of information submitted, and how such information would be used. One of the most aggressive statements against the census came from CNN political commentator and the Web site RedState's editor Erick Erickson. In response to the American Community Survey (ACS), the ongoing national survey conducted by the U.S. Census Bureau, Erickson threatens to "[p]ull out my wife's shotgun and see how that little ACS twerp likes being scared at the door" (Media Matters for America, 2010). Bachmann (2010), Paul (2010b), and Erickson (2010) expressed similar opposition to the 2010 US Health Reform Act and its proponents, which, like the census, they criticized because of cost, constitutionality, and the overreach of government power. *The New York Times* columnist Frank Rich (2010) offers a different take, stating that angry protests against the legislation were rooted in a "national existential reordering" (para. 9) and the "inexorable and immutable change in the very identity of America, not just its governance" (para. 7).

He writes that there were

> fears of disenfranchisement among a dwindling and threatened minority in the country no matter what policies were in play. . . . Demographics are avatars of a change bigger than any bill contemplated by Obama or Congress. The week before the health care vote, The Times reported that births to Asian, black and

> Hispanic women accounted for 48 percent of all births in America in the 12 months ending in July 2008. By 2012, the next presidential election year, non-Hispanic white births will be in the minority. The Tea Party movement is virtually all white. The Republicans haven't had a single African-American in the Senate or the House since 2003 and have had only three in total since 1935. Their anxieties about a rapidly changing America are well-grounded. (para. 12)

Was resistance to the legislation truly about cost, government, and politicians, or was it prompted by the inevitability of a profound demographic shift in which America is becoming more Asian, black, and Hispanic in composition (United States Census Bureau, 2008; Roberts, 2010)?

One way to address Rich's provocative hypothesis is to inflect the trajectory of its inquiry using precedent. What is lacking in the American imaginary, I suggest, is a robust visual heritage with which to understand the conception of the America that it is projected to be. *The New York Times* film critics, Manohla Dargis and A.O. Scott, contend that the positive images and narratives of *African* Americans offered in film and television helped America imagine an African American president before it occurred. These modes of intercultural communication wrote a "prehistory" for the election of President Barack Obama by offering "intriguing premonitions, quick-sketch pictures and sometimes richly realized portraits . . . grappling with issues of identity and the possibilities of power" (Dargis & Scott, 2010). As such, I want to suggest that Asian American new media performs a similarly important service: Such cultural production envisions the prehistory of the new America. In this sense, minority communities are using contemporary information and communication technologies to create cultural heritage for this age. The use of new media as the vehicle for minority self-representation may not be enough to address the anxieties of demographic change, and it may be idealistic of me to think so, but these creative media expressions do begin the process of envisioning a complex portrait that might become widespread, so that the inevitability of demographic transformation may be easier to accept. I cannot absolutely confirm the veracity of Frank Rich's hypothesis, but my hope is that if the fear of demographic change breeds anger, then these grassroots modes of expression about identity have value in helping us more fully embrace a different and more diverse portrait of America.

References

Aoki, A., & Takeda O. (2008). *Asian American politics.* Malden, MA: Polity Press.

Bachmann, M. (2010). Health care. Retrieved from http://bachmann.house.gov/Healthcare

Barr, B. (2009, September 8). Census goes too far with children. *The Atlanta Journal-Constitution.* Retrieved from http://blogs.ajc.com/bob-barr-blog/2009/09/08/census-goes-too-far-with-children/

Bazin, A. (2004). The evolution of the language of cinema. In L. Braudy & M. Coehn (Eds.), *Film theory and criticism* (6th ed.; pp. 41–53). New York, NY: Oxford University Press.

Baudry, J. (2004). Ideological effects of the basic cinematographic apparatus. In L. Braudy & M. Coehn (Eds.), *Film theory and criticism* (6th ed.; pp. 355–365). New York, NY: Oxford University Press.

Center for Asian American Media. (2009). *Annual report.* San Francisco, CA: Center for Asian American Media.

Dargis, M. (2010, January 29). Talking about a revolution (for a digital age). *The New York Times.* Retrieved from http://www.nytimes.com/2010/01/31/movies/31dargis.html

Dargis, M., & Scott, A. O. (2009, January 16). How the movies made a president. *The New York Times.* Retrieved from http://www.nytimes.com/2009/01/18/movies/18darg.html

Dinan, S. (2009, June 17). Exclusive: Minn. lawmaker vows not to complete census. *The Washington Times.* Retrieved from http://www.washingtontimes.com/news/2009/jun/17/exclusive-minn-lawmaker-fears-census-abuse

Erickson, E. (2010, March 21). Happy warriors still. *RedState.* Retrieved from http://www.redstate.com/erick/2010/03/21/happy-warriors-still

Feng, P. X. (2002). *Identities in motion: Asian American film and video.* Durham, NC: Duke University Press.

Honda, M., Scott, R., Faleomavaega, E., Abercrombie, N., Bordallo, M., Green, A., Becerra, X., Matsui, D. O., Hirono, M. K., & Wu, D. (2008, February 15). *Members of congress write to Jonathan Klein, CNN President.* Retrieved from http://www.aadt.us/AAPIdemocrat.html

Honda, M., Scott, R., Faleomavaega, E., Abercrombie, N., Bordallo, M., Green, A., Becerra, X., Matsui, D. O., Hirono, M. K., & Wu, D. (2008a, February 15). *Members of congress write to Phil Griffin, executive-in-charge, MSNBC.* Retrieved from http://www.democrats.org/a/2008/02/members_of_cong_2.php

Jenkins, H. (2006). *Convergence culture: When old and new media collide.* London, United Kingdom: New York University Press.

Kane, C. (1999, September 2). The media business: Advertising—Addenda; Vodka ad is stopped after racism protest. *The New York Times.* Retrieved from http://www.nytimes.com/1998/04/12/nyregion/neighborhood-report-new-york-on-line-one-site-14-ethnic-groups.html

Kleefeld, E. (2009, July 1). GOP Congressmen condemn Bachmann's call for defying the census. *Talking Points Memo.* Retrieved from http://tpmdc.talkingpointsmemo.com/2009/07/gop-congressmen-condemn-bachmanns-call-for-defying-the-census.php

Kun, J. (2010, June 18). Unexpected harmony. *The New York Times.* Retrieved from http://www.nytimes.com/2010/06/20/arts/music/20legaci.html

Lee, R. G. (1999). *Orientals: Asian Americans in popular culture.* Philadelphia, PA: Temple University Press.

Lowe, L. (1996). *Immigrant acts: On Asian American cultural politics.* Durham, NC: Duke University Press.

Media Matters for America. (2009, June 25). Beck asks Bachmann "the odds" the gov't will fine those who don't complete census, says he's "considered" not filling it out. Retrieved from http://mediamatters.org/mmtv/200906250039

Media Matters for America. (2010, April 1). CNN's Erickson: I'll "[p]ull out my wife's shotgun" if they try to arrest me for not filling out the American Community Survey. Retrieved from http://mediamatters.org/mmtv/201004010050

Mimura, G. M. (2009). *Ghostlife of third cinema: Asian American film and video.* Minneapolis: University of Minnesota Press.

MTV Films. (2003, April 18). MTV Films' 'Better Luck Tomorrow' expands to more than 100 markets on April 25 after opening with an amazing $27,752 per screen average [Message posted to PR News Wire new release list]. Retrieved from http://www.prnewswire.com/news-releases/mtv-films-better-luck-tomorrow-expands-to-more-than-100-markets-on-april-25-after-opening-with-an-amazing-27752-per-screen-average-70879077.html

Nakamura, L. (2007). *Digitizing race: Visual cultures of the Internet.* Minneapolis: University of Minnesota Press.

Ono, K. A., & Pham, V. (2009). *Asian Americans and the media.* Malden, MA: Polity Press.

Oren, T. G. (2005). Secret Asian man: Angry Asians and the politics of cultural visibility. In S. Davé, L. Nishime, & T. G. Oren (Eds.), *East Main Street: Asian American popular culture* (pp. 337–360). New York: NYU Press.

Palumbo-Liu, D. (1999). *Asian/American: Historical crossings of a racial frontier.* Palo Alto, CA: Stanford University Press.

Paul, R. (2010a, March 8). Census: A little too personal. Retrieved from http://www.freerepublic.com/focus/f-rlc/2466949/posts

Paul, R. (2010b, March 22). Ron Paul: We need free-market healthcare, NOT Obamacare. Retrieved from http://www.ronpaul.com/2010-03-22/we-need-free-market-healthcare-not-obamacare

Ramirez, A. (1998, April 12). Neighborhood report: New York on line; One site, 14 ethnic groups. *The New York Times.* Retrieved from http://www.nytimes.com/1998/04/12/nyregion/neighborhood-report-new-york-on-line-one-site-14-ethnic-groups.html

Rich, F. (2010, March 27). The rage is not about health care. *The New York Times.* Retrieved from http://www.nytimes.com/2010/03/28/opinion/28rich.html

Roberts, S. (2010, March 11). Births to minorities are approaching majority in U.S. *The New York Times.* Retrieved from http://www.nytimes.com/2010/03/12/us/12census.html

Tajima, R. (1991). Moving the image: Asian American independent filmmaking 1970–1990. In R. Leung (Ed.), *Moving the image: Independent Asian Pacific American media arts* (pp. 10–33). Los Angeles: UCLA Asian American Studies Center.

U.S. Census Bureau. (2008, August 14). An older and more diverse nation by midcentury. Retrieved from http://www.census.gov/newsroom/releases/archives/population/cb08-123.html

Wei, W. (1993). *The Asian American movement.* Philadelphia, PA: Temple University Press.

Xing, J. (1998). *Asian America through the lens: History, representation & identity.* Walnut Creek, CA: Altamira.

Yoshino, K. (2003, April 11). An e-mail push for 'Better Luck': Film about Asian American teens is one of a growing number of ethnic-focused and niche projects using grass-roots marketing efforts. *Los Angeles Times.* Retrieved from http://articles.latimes.com/2003/apr/11/business/fi-grassroots11

17. Jamaica and Chile Online: Accessing and Using the Internet in a Developing World Context

NICKESIA GORDON, PHD
KRISTIN SORENSEN, PHD

Introduction

Growing numbers of people in developing countries are going online. This growth has in part been attributed to developments in information infrastructure through telecommunications, as well as to the efforts of national governments and NGOs. Increased Internet usage in developing countries is often associated with socioeconomic development; there is an assumption that wherever people are logging on to the Internet, social and economic advancement is happening or is at least possible (Galperin & Bar, 2007; Sardar, 2007). These arguments are reminiscent of the "diffusion of innovations" model made popular by scholar Everett Rogers (2003), which sees individuals as targets of communication and is top down in its approach to creation and delivery of communication messages. In this model, communication is viewed as strictly a noun, that is, as something that *is* and not something that is *performed* or *acted on*. This limited view parallels traditional theories about national development.

Over the past two decades, however, some scholars (Mody, 2000; Melkote & Steeves, 2001) have challenged this view of communication in the developing world context, arguing for a more culturally appropriate, participatory, and thus contextualized approach. Such a view of development communication offers a useful frame for analysis of the spread of information and communication technologies (ICTs) in developing countries. Not only are local socioeconomic, political, and cultural factors important variables in how ICTs are being used, but they are often overlooked when considering how information technologies are designed and implemented. Such oversight can lead to the simplistic equation of information and communication technology (ICT) presence with socioeconomic and political progress in developing countries.

United Nations Under-Secretary-General for Economic and Social Affairs Sha Zukang (2009) illustrated the common practice of equating ICT adoption with economic progress at the 2009 World Telecommunication Policy Forum, when he declared, "For the United Nations, the Internet has become a powerful tool in our mission to promote peace and security, development and human rights. ICTs are also a factor of economic growth and development" (3rd para.).

Although the presence of and access to the Internet may provide opportunities for democratic expression and economic ventures, individuals who log on to the Internet do not necessarily do so for such reasons. Like people everywhere, individuals in developing countries use the Internet for varying reasons, and the cultural and sociopolitical contexts of the environment in which they live often color their motivations for usage. A contextualized perspective presents a critical caveat that tempers the assumption that mere usage of the Internet and ICTs is a sufficient indicator of development or increased political engagement.

This chapter offers a nuanced exploration of the association between increased Internet usage and socioeconomic and political development in the context of two developing states. It offers a comparative analysis between Internet usage in Jamaica and Chile, exploring how individuals living in each of these countries use the Internet. From the Jamaican perspective, this chapter compares apparent trends in Internet usage among Jamaicans with the development expectations of Internet use projected by international bodies such as the United Nations Development Program (UNDP) and the World Bank, as well as by the Jamaican government. From the Chilean point of view, this chapter examines how the Internet is playing a central role in the production, distribution, and reception of alternative political discourses.

The politics of "development"

The term "national development" is often used interchangeably with the word "development." It is a concept steeped in much controversy, given its emergence during the postwar and postcolonial era, in which many European countries were forced to grant independence to their colonial territories. With it evolved a particular discourse that associates Euro-America with progress and former colonies with stasis. The World Bank first posited the notion of development after World War II (Biccum, 2002) and, in doing so, established a dichotomous relationship between the developed and the underdeveloped;

that is, it construed development as inevitable for everyone yet only necessary to undertake in Third World contexts (Biccum, 2002). Such traditional perspectives on national development often associated it with modernization. Although contested for much of the latter half of the twentieth century (Barriteau, 1996; Mody, 2000; Melkote & Steeves, 2001), the dominant model dictating *how* national development was to be achieved entailed the application of strict economic measures wherein the degree of progress achieved by a nation-state was determined by changes in gross domestic product (GDP).

The belated realization, however, that sustained global development has not been achieved via economic measures has informed more recent efforts to redefine development theory and practice. At the same time, the role of ICTs in the development process has been equally interrogated. Heavily technocratic and GDP-centric practices have failed to enhance the development project and have actually worked to increase the communication divide rather than to bridge it (Dervin & Huesca, 1997; Rodriguez, 2000; Escobar, 2000). One of the most significant criticisms that scholars have leveled at the economic imperative guiding development and development communication has been the weak conceptualization of power and power structures that contextualize the environment in which development occurs and in which such discourse unfolds (Escobar, 1995, 2000; Fair & Shah, 1997; Wilkins, 2000; Hafez, 2008). Referring to outmoded conceptualizations of development in general, and specifically to communication for development, Gumucio Dagron (2003) argues:

> Over fifty years of failed attempts to promote development in Third World countries, particularly Africa and Latin America, have demonstrated that the paradigms of development could not be dictated by the North, and that the development agendas of bilateral and multilateral organizations had not taken into consideration social, political and cultural factors that determine social change and development. (4th para.)

The dynamics of power and how they affect development have largely been ignored or have been taken for granted as natural and immutable. New perspectives demand that development and development communication approaches "look for evidence of power dynamics in the political-economic structures that govern the distribution of resources across and within nations . . . or even in the assertion of particular ideologies that legitimize or resist dominant interests" (Wilkins, 2000, p. 2). In response, international financial institutions such as the World Bank have more recently embraced the notion of *human development* as one that more broadly addresses the meeting of basic human needs: access to health, education, shelter, and nutritious food. The United Nations Development Program (UNDP) has created the Human Development Index (HDI), according to which nation states are now ranked based on health conditions, literacy, and access to goods and services (Isbister, 2001).

In developing countries (as elsewhere), the Internet is seen as a powerful tool for information dissemination and the shaping of public opinion. It is viewed as having a role to play in helping national governments achieve their goals through public education about national development issues and by offering citizens avenues through which they may hold their governments accountable.

ICTs *can* play a critical developmental role in building and sustaining democracies, as well as in providing citizens with important information about health, the environment, rural development, and governance. Selected examples include the mobile remittance service M-Pesa, offered by Vodafone in Kenya (Rice, 2007), which allows its users—who are often excluded from the formal banking system—to transfer money to others, and OVI-Life, a mobile service in India that provides rural farmers with agricultural information, such as current market prices, enabling them to make more informed decisions about where and at what price to sell their produce (O'Brien, 2010). ICTs at least appear, then, to have the potential to facilitate the kind of development envisioned by the UNDP (2010): " . . . expanding the choices people have to lead lives that they value. [Development] is thus about much more than economic growth, which is only a means—if a very important one—of enlarging people's choices."

The questions remain: *Are* citizens of developing nations using ICTs to learn about national development and participate in new forms of political engagement? Are ICTs being deployed to strengthen the public sphere? Are the peripheralized moving closer to the center?

Logging on to progress? Jamaica

The Jamaican context

Popularized by former Jamaican Prime Minister P. J. Patterson during a reelection campaign, the notion of "logging on to progress" embodies the aspiration of developing societies such as Jamaica to reap the supposed economic and political benefits of ICTs. Patterson argues that the country's economic and social development is inextricably linked to the use and mastery of Internet technologies. In ambitious terms, Patterson (2000) paints a picture of Jamaica as a brave new Internet world:

> We envisage a cumulative process which, within a decade or less, will position Jamaica in the New Economy with strong advantages in computer and Internet services, and the ability to adapt quickly to, and absorb, new waves of

technological development. We project Jamaica to become renowned as the Silicon Beach. (p. 173)

Patterson asked Jamaicans to reimagine themselves as citizens of this new digital world, as if access to it was just a mouse click away. In this digital future, development is equated with connectivity, and progress is the inevitable byproduct of this coupling. The rapid growth of the Internet and ICTs in developing countries seems to confirm this perspective. Between 2000 and 2007, Africa and the Middle East had the fastest growing number of Internet users, increasing 875% and 920%, respectively, within that period (Graves, 2007). In 2010, Jamaica had a 55.5% Internet penetration rate with 1,581,100 users reported as of June of that year, increasing from 100,000 users (3.1% of the population) in 2000 (Deane, 2008).

As Hafez (2008) has pointed out, however, such quantitative data says very little about *who* uses the Internet, and for what purpose. Are users logging on simply to check their e-mail or to Tweet and update their Facebook statuses? These questions are important, as they help determine the relationship between usage of the Internet and clear development goals or criteria established by bodies such as the UNDP. Reflecting these criteria, are Jamaicans also using ICTs to expand their social and economic choices, to learn about important health issues, for educational purposes, to be more culturally aware, or to amplify their voices in the public sphere and hold their government accountable?

Investigating Jamaican internet usage

To explore Internet usage in the Jamaican context, an online survey was distributed using purposive and snowball sampling techniques, with the goal of consciously selecting subjects who met the study's needs (Baker, 1999, p. 138). These approaches provide the researcher with the capacity to actively select participants relevant to the research design and are useful in generating "information-rich cases" from which a researcher can discover, understand, and gain more insight into issues crucial to the study. These "networking" techniques have attracted increased attention in recent years and are commonly used by qualitative researchers to generate samples across fields such as sociology, psychology, and anthropology. Usually the researcher makes initial contact with a small group of people who are deemed relevant to the research topic; these respondents are then invited to designate other potential informants . . . and so on (Patton, 1990; Sarantakos, 1993). The Jamaica survey was originally distributed using a mailing list derived from a small group of contacts; it was also posted on a social networking site (SNS) for relevant participants to complete. Thirty-three respondents filled out the questionnaire over the course of two weeks. Given the small number of respondents, this study is exploratory. Results should be read as an *indication* of how Jamaicans use the Internet and cannot be generalized to the entire population. Continuing research will pursue these preliminary findings.

Internet access is an urban affair

Our survey reveals that Internet connectivity in Jamaica may be a predominantly urban phenomenon: 85% of Internet users who responded to the survey report that they live in an urban center, whereas only 52% of the total Jamaican population live in areas designated as urban (CIA, 2011). These findings reflect data from many other developing countries: Rural populations usually lag behind because large private operators are less likely to service low-income customers, particularly those living in rural areas where the backbone infrastructure may be more expensive to establish (Galperin & Bar, 2007). It can be inferred, then, that although Jamaica has seen a dramatic increase in Internet usage in the past decade, this should not be accepted as an indicator of widespread connectivity or economic progress.

Access increases according to social class and economic resources

Jamaican Internet usage also appears to be affected by socioeconomic factors. The majority of Internet users who responded reported that they lived in well-to-do areas of urban Jamaica and characterized themselves as business owners or professionals in very good middle- to senior-level jobs in the private and public sectors, earning above-average salaries. This unevenness in the distribution of Internet access reflects the unevenness of the socioeconomic structure of the society itself, demonstrating that Internet access is an indicator of the power dynamics in the political and economic structures that govern the distribution of resources within Jamaica.

There is a clear issue of Internet affordability for average Jamaicans. The average Jamaican household consumption (expenditure) in 2008/2009 was reported as J$205,000 (Planning Institute of Jamaica/Statistical Institute of Jamaica, 2010). A review of services offered by contemporary Internet service providers in Jamaica (see, for example, the LIME Web site, www.time-4lime.com) meanwhile indicates that monthly fees for home Internet access begin at J$1,700—equivalent to the average Jamaican per capita expenditure on housing or transportation (Planning Institute of Jamaica/Statistical Institute of Jamaica, 2010). It is clear that the cost of Internet connectivity puts it beyond the means of average Jamaicans, limiting the ability of ICTs to meaningfully change the lives of and create opportunities for the local citizenry.

Level of education often determines access

The educational level further contributes to the digital divide. Of survey respondents, 55% report having attained a bachelor's degree and 33% report having a master's degree, whereas 14% had only a high school diploma. Citizens with a tertiary-level education represent only a small fraction of Jamaican society. In 2008, only 24% of the population of "tertiary age" were enrolled in tertiary education, and this proportion had not increased significantly over the preceding 20 years (UNESCO Institute for Statistics, 2010). Because education and training are two of the most important elements for the development of human capital and permanent poverty reduction, this divide based on educational levels confirms that the mere presence of ICTs in a state is insufficient to guarantee effective (re)distribution of educational resources in the service of individual and national development.

Women and access

The majority of survey respondents (81%) were women. Although the sampling technique used may have predisposed the survey to such a skew, it is worth considering the implications of how women are using the Internet and which group of women actually have access to it. In 2001, women were reported to constitute 22% of all Internet users in Asia, 38% of Internet users in Latin America, and 6% of Middle Eastern users (Hafkin & Taggart, 2001). Global surveys of Internet users over the past 15 years have indicated that, globally, and in selected regions, the gender gap is closing (International Telecommunications Union, 2010), though women still lag behind in Internet use.

Although Internet access may be skewed by gender across Jamaica as a whole, what is significant in our survey findings is that Internet use is dominated by a small, urban educated elite. In other countries with similar ICT adoption patterns (Brazil, Mexico), urban women are also well represented (Hafkin & Taggart, 2001; Dholakia, R. R., Dholakhia, N., & Kshetri, N., 2004). Available statistics (OECD GID-DB, 2009), however, indicate that in Jamaica as a whole, on average women earn only 57% of the average male salary, and this in a country where 15–20% of the population are estimated to be living below the poverty line, depending on the measures used (CIA, 2011). In other words, our findings indicate that poor and rural women have little access to the potential development benefits of ICTs.

What are jamaicans doing online?

Overall, despite differences in who has access, it appears that Jamaicans predominantly use the Internet for recreational purposes, such as checking e-mail, chatting, or spending time on SNSs. The majority of survey respondents

(96%) report that they mostly use the Internet for checking e-mail. Only 37% indicate that they used the Internet for business purposes, and no one reports using it for political or advocacy purposes. Jamaicans who do have Internet access may be logging on frequently (96% reported logging on at least once a day), but their activities are not directed toward the kinds of development goals outlined by organizations such as the UNDP. There is little indication, for example, that individuals are logging on to find health information, to engage in civil society, or to lobby for any sort of social or political change.

In Jamaica, the envisaged Silicon Beach appears to have evolved into a digital ghetto occupied by an educated, relatively wealthy urban elite. Outside the ghetto, the rural and poor citizenry remain the "information have-nots"—those who are unable to harness the potential educational, social, and economic benefits of ICTs. This example illustrates why it is problematic to equate the increased presence or adoption of information technology in developing countries with social and economic progress.

Truth, justice, and new technologies in Chile

In Chile, meanwhile, increased access to and use of ICTs appears to be facilitating the democratization of this postrepressive nation. In postdictatorship Chile, the general rule for discussing sensitive themes in the national media, such as human rights violations committed during the Pinochet regime (from 1973 to1990), has been to avoid them. The reasons for this avoidance are multiple: continued governmental repression; a high concentration of ownership of mainstream media, largely by conservative owners; and nonexistent job security for writers and reporters. In addition, addressing divisive, painful memories is challenging and potentially volatile in postconflict societies. Despite continuing tight control of mainstream media outlets, however, avenues do now exist for alternative discourses, and most of these depend on the Internet. These have had some governmental backing. The center-left Concertación government, which held the presidency from 1990 until March 2010, supported expanding Internet access to all Chileans in an effort to create a more robust public sphere. In a report issued by Chile's Presidential Commission on New Information and Communication Technologies in 1999 (quoted in Tanner Hawkins, 2005), the authors state,

> The goal of advancing toward a developed, integrated information society cannot be reduced to its mere technological dimension. It implies, before everything, the renewing of will to drive a modernization with growing social integration; the aspiration to build an open and culturally advanced society; the desire to live in a participatory society that allows opportunity for all; as well as the intention to achieve an economic system capable of innovating and competing in a sustainable fashion, simultaneously elevating the social welfare of all Chileans. (p. 359)

With these ideals in mind, the government, with support from the private sector, attempted to bridge the digital divide during the early years of the Internet revolution. Community computer centers were created in poor neighborhoods. The telecommunications company Telefónica-CTC donated Internet connections to schools (Tanner Hawkins, 2005). In 2009, a new law was introduced to protect the rights of users "to navigate freely without barriers or discrimination of content." A spokesperson explains, "What is important is that along with defending this right of non-discrimination we also protect the creators of content, the innovators" (Gobierno dará urgencia, 2009). The Internet now offers new opportunities for some groups to come together across geographic barriers. Almost all Chilean media is now available online, and some of it is exclusively online. Chileans living abroad can now access their national media as quickly as Chileans in Chile.

A crucial moment occurred during General Pinochet's detention in the United Kingdom (from 1998 to 2000), in which the significant role of the Internet in opening up access to mainstream media and the public sphere became apparent. *La Tercera*, one of Chile's main commercial newspapers, hosted an ongoing forum on its Web site, in which users took the opportunity "to argue over the meanings of justice, reconciliation, forgiveness, truth, democracy, liberty, sovereignty, and human rights" (Tanner, 2001, p. 383). Tanner Hawkins performed a textual analysis of 1,670 letters sent to the forum between October 1998 and January 1999. She believes (2001, p. 388) that approximately one quarter of the letters she analyzed had been sent from inside of Chile, whereas that the other three quarters appeared to have been sent from outside the country. The majority of writers, regardless of location, identified as Chilean. Tanner Hawkins (2005) concludes:

> The use of the online forum by these Chileans suggests the online public spaces may do much more than create public opinions, as is suggested by public sphere theories. In this case, the forum was a place for forming popular or collective memories. (p. 402)

Moving beyond the mainstream: Chileans embrace alternative media

The unequal distribution of coverage, voices, and time given to various perspectives in the *mainstream* media reinforces the unequal distribution of agency and attention granted to divergent sectors of Chilean society. Remedi (2004) argues,

> For people to be empowered—to access the level of power—and to challenge *and* replace the current cultural hegemony, we need to be able to engage in other kinds of cultural production, that is, other than merely choosing from, or responding to mainstream cultural flows. (p. 514)

In addition to participating in online forums hosted by mainstream media, then, many Chileans, alienated from and by mainstream media, have taken it upon themselves to create alternative discourses that do confront Chile's recent past and other subjects that are treated as taboo. These discourses can be found in a variety of venues—periodicals, radio networks, and the streets—through demonstrations such as those performed by the street activist group Funa, whose members know that their actions will not be recorded by the mainstream media, so they usually come equipped with their own video cameras in order to document the events and then post them online.

The day Pinochet was arrested/the day Pinochet died

On October 16, 1998, General Pinochet was recovering from surgery at a London clinic when his recovery room was stormed by English police officers declaring him under arrest. Spanish judge Baltasar Garzón requested Pinochet's extradition to Spain to stand trial for human rights violations. The General's international detention encouraged some Chileans to persevere in their quest for more freedom of expression in their media, and most of these endeavors happened online.

Media coverage of Pinochet's death, six years after his return to Chile, did not begin on the afternoon of his passing on December 10, 2006. News crews were stationed outside of the hospital where he stayed for several days before he died, recording the candlelight vigils, rallies, and prayers performed by Pinochet's fans. Alongside this coverage, some of the most notorious moments from the demonstrations were captured by the Spanish news crews, whose own reporters were verbally and physically attacked by the demonstrators, who hated the Spaniards for their government's attempt to have Pinochet extradited. This footage went viral online.

On the afternoon of Pinochet's death, blogger and filmmaker Nicole Senerman ran into the streets of her Santiago neighborhood with a borrowed digital video camera and recorded reactions to the news, creating "*El día en que murió Pinochet* (*The Day Pinochet Died*)." which she then proceeded to post on the Internet as soon as editing was complete. Senerman captures the images and voices of people who are marginalized or entirely erased from mainstream media representations.

In scenes from her film, thick, black smoke hovers over Alameda, Santiago's main avenue. Senerman drops the camera to her side while it is still recording and runs away from the antiriot police. We see dozens of the frantic feet of protesters run away as the camera shakes wildly. We witness people trying to escape from tear gas. Suddenly, Senerman's camera is inside in a café, observing the news broadcasting from a TV set mounted to the wall. The news anchor is blaming the upheaval on anarchists and juvenile delinquents who had allegedly planned the riots. The news story offers a striking counterpoint to what we are able to witness through Senerman's point of view, and we are reminded that the official, mainstream version of the day's events offers a perspective that is tightly contained and excludes many of the social actors found on the streets.

Senerman poses a question to passersby: "What should the newspaper headlines be tomorrow?" As we hear their responses, we see the actual headlines, along with photos of a distinguished-looking Pinochet (alive), as well as a multitude of supporters in line to pay their final respects to him at his coffin. It is not surprising that the headlines offer different, tamer, more ambiguous frames than those suggested by the recommendations of people on the street. Whereas television footage is only available to a general public for a limited moment in time, thanks to the characteristics of the Internet, Senerman's film is available forever. Parts of *The Day Pinochet Died* are available on YouTube, with viewer comments included. Clips and links can also be found on many bloggers' pages. *The Clinic*, a satirical newspaper born during Pinochet's international detention, has an online version in addition to its paper version available at Chilean kiosks. Online, readers from across the globe can upload uncensored comments in response to any posted article. The film closes with a shot of Senerman's own arm, with the words "recording history" scribbled across it. It has become a crucial memory text in many Chileans' interpretations of this significant date in Chile.

In her study of 1998 Internet communications, Tanner Hawkins (2005) observes that whereas the Internet was potentially a tremendous tool for expanding the public sphere in Chile, opportunities for participation remained constrained, especially by economic barriers. During the time of her study (2005), only 1–2% of the population in Chile had Internet access (p. 388), increasing to 23.8% in 2002 (p. 352). Most of the contributors to Chilean online forums were men, which appears to confirm the historical data showing that globally men have outnumbered women in Internet use. Senerman's short film offers a startling example of the ways in which by 2006, new ICTs had continued to expand access to the public sphere. The poor, the homeless, the mentally ill, the blind, women, children, the elderly, and members of the lesbian, gay, bisexual, transgender, and queer (LGBTQ) community, all groups of people whose representations are either highly contained or erased altogether in the mainstream media, are clearly seen and heard in Senerman's film. These vulnerable populations also happen to be the groups of people who experienced the consequences of Pinochet's policies most severely. Senerman's film is a celebration of and by people without power. Even if many of these individuals do not use the Internet themselves, she has ensured that they will be seen and heard by others online.

Senerman's inclusion of television sound bites and newspaper headlines makes explicit the striking contrast between her version of the day's events and the versions offered by the right-leaning mainstream press.

Conclusion

It is perhaps unsurprising that Internet adoption and usage patterns in Jamaica and Chile—both "developing nations"—differ significantly as a result of differences in their political histories and climates, culture, ownership, regulation, and socioeconomic trends. Jamaica has experienced a reasonably stable political climate since its independence from Britain in 1962; it has never experienced a political dictatorship and traditional media has always offered an avenue for political expression (at least by the educated elite). It was not unexpected for us to find little political activism associated with Internet use by the educated urban elite in Jamaica. Chile enjoyed a relatively free press until 1973 but was highly censored during the 1973–1990 military dictatorship. Since its return to democracy in 1990, the country has struggled to revive a robust and diverse free press. The Internet has therefore offered Chileans a critical tool for political engagement, especially during periods of recent oppressive governance.

The differential impacts of ICT adoption in the case studies discussed in this chapter should also act as a reminder that claims that new technologies offer solutions to the socioeconomic woes faced by developing countries need to be better contextualized. There is no clear or natural correlation between Internet usage and national development or political engagement, only an indication of the *potential* of the former to significantly impact the latter. Gumucio Dagron (2003; 2004) highlights past failures by international bodies, such as the UN, to achieve development goals simply through the introduction of new technologies. Quantitative measures of "national development," such as GDP, mask system hierarchies of poverty and inequity within a population. It is widely documented, for example, that despite major gains made in the last decade, ICT networks and services are not effectively reaching the poor, particularly those living in rural areas (Galperin & Bar, 2007). As Sardar (2007) notes:

> New technology . . . always comes wrapped in hyperbole; but it seldom delivers what it says on the packaging. This is particularly true when sold to the world's poor. From pesticides to dams, genetic engineering to computers, almost every technology has been sold to people in the underprivileged world on the promise that it will make them rich. (p. 24)

Wisely deployed and developed, ICTs can be powerful tools for advancing human development (Burch, 1997), which can and should be monitored using criteria that go far beyond GDP. Moreover, we must clearly differentiate the potentialities (informative, educational, cultural, political, economic, etc.) offered by ICTs from their manifestations and actual impact on various contexts and social groups (Bonder, 2002).

References

Baker, T. L. (1999). *Doing social research.* New York, NY: McGraw-Hill.

Barriteau, V. E. (1996). Structural adjustment policies in the Caribbean: A feminist perspective. *National Women's Studies Association Journal, 8*(1), 142–156.

Biccum, A. (2002). Interrupting the discourse of development: On a collision course with postcolonial theory. *Culture, Theory and Critique, 43*(1), 33–50.

Bonder, G. (2002). *From access to appropriation: Women and ICT policies in Latin America and the Caribbean.* Paper presented at the United Nations Division for the Advancement of Women Expert Group Meeting on "Information and Communication Technologies and Their Impact on and Use as an Instrument for the Advancement and Empowerment of Women," Seoul, Republic of Korea, November 11–14, 2002. Retrieved from http://www.mujeresenred.net/zonaTIC/IMG/pdf/GBonder.pdf

Burch, S. (1997). *Latin American women take on the Internet.* Retrieved from http://www .connected.org/women/sally.html

CIA. (2011). *CIA World Factbook: Central America and Caribbean: Jamaica.* Retrieved from https://www.cia.gov/library/publications/the-world-factbook/geos/jm.html

Deane, R. (2008, June 29). Internet usage shoots to 55%—1.5 million Jamaicans surf the Web; Mullings aims for full penetration. *Jamaica Gleaner.* Retrieved from http://jamaica-gleaner.com/gleaner/20080629/business/business1.html

Dervin, B., & Huesca, R. (1997). Reaching for the communicating in participatory communication: A meta-theoretical analysis. *Journal of International Communication, 4* (2), 46–74.

Dholakia, R. R., Dholakhia, N., & Kshetri, N. (2004). Gender and Internet usage. In H. Bidgoli (Ed.), *The Internet Encyclopedia, Volume 1* (pp. 12–22). Hoboken, NJ: John Wiley & Sons.

Escobar, A. (1995). *Encountering development: The making and unmaking of the third world.* Trenton, NJ: Princeton University Press.

Escobar, A. (2000). Place, power and networks in globalization and postdevelopment. In K. G. Wilkins (Ed.), *Redeveloping communication for social change: Practice theory and power* (pp. 163–174). New York, NY: Rowman and Littlefield.

Fair, J., & Shah, H. (1997). Continuities and discontinuities in communication and development research since 1958. *Journal of International Communication, 4*(2), 3–23.

Galperin, H., & Bar, F. (2007). Diversifying network development through microtelcos. *Information Technologies and International Development 3*(2), 73–86.

Gobierno dará urgencia a proyecto de ley de Internet y neutralidad de la red. (2009, September 22). *La Tercera.* Retrieved from http://latercera.com/contenido/654_185201_9.shtml

Graves, P. (2007, December 10). *Independent media's vital role in development.* A report to the Center for International Media Assistance, Washington, DC. Retrieved from the Global Forum for Media Development Web site: http://gfmd.info/images/uploads/CIMA-Medias_Vital_Role_in_Development-Report.pdf

Gumucio Dagron, A. (2003). Fighting rural poverty. The role of ICTs. Keynote address given at *The World Summit on the Information Society, Geneva, Switzerland.* Retrieved from http://www.ifad.org/events/wsis/phase1/presentations/alfonso.htm

Gumucio Dagron, A. (2004). The long and winding road to alternative media. In A. Downing, D. McQuail, P. Schlesinger, & E. Wartella (Eds.), *The Sage Handbook of Media Studies* (pp. 41–64). Thousand Oaks, CA: Sage.

Hafez, K. (2008). *The myth of media globalization.* Malden, MA: Polity Press.

Hafkin, N., & Taggart, N. (2001). Gender, information technology and developing countries: An analytic study. US AID Office of Women's Development. Retrieved from http://www.usaid.gov/our_work/cross-cutting_programs/wid/pubs/hafnoph.pdf

International Telcommunications Union. (2010). *Measuring the Information Society 2010.* Geneva, Switzerland: Author. Retrieved from http://www.itu.int/ITU-D/ict/publications/ idi/2010/Material/MIS_2010_without_annex_4-e.pdf

Isbister, J. (2001). *Promises not kept: The betrayal of social change in the third world.* West Hartford, CT: Kumerian Press.

Melkote, S., & H. L. Steeves (2001). *Communication for development in the Third World: Theory and practice for empowerment* (2nd ed.). Thousand Oaks, CA: Sage.

Mody, B. (2000). The contexts of power and the power of the media. In K. G. Wilkins (Ed.), *Redeveloping communication for social change: Practice theory and power* (pp. 185–195). New York, NY: Rowman and Littlefield.

O'Brien, K. J. (2010, November 1). Nokia taking a rural road to growth. *The New York Times.* Retrieved from http://www.nytimes.com/2010/11/02/technology/02nokia.html

OECD GID-DB. (2009). *Gender, Institutions and Development Database 2009 (GID-DB): Composite Indicators.* Retrieved from http://stats.oecd.org/Index.aspx?DataSetCode=GID2

Patterson, P. J. (2000). Twenty-fourth meeting of the Miami Conference on the Caribbean and Latin America. In D. Franklyn (Ed.), *A Jamaican voice in Caribbean and world politics P. J. Patterson selected speeches 1992–2000.* Kingston, Jamaica: Ian Randle.

Patton, M. Q. (1990). *Qualitative evaluation methods* (2nd ed.). Newbury Park, CA: Sage.

Planning Institute of Jamaica/Statistical Institute of Jamaica. (2010). *Jamaica Survey of Living Conditions 2008/2009.* Report presented to the United Nations Development Programme Jamaica. Retrieved from http://www.jm.undp.org/files/JSLC%2008-09%20Launch.pdf

Remedi, G. A. (2004). The production of local public spheres: Community radio stations. In A. del Sarto, A. Ríos, & A. Trigo (Eds.), *The Latin American cultural studies reader* (pp. 513–534). Durham, NC: Duke University Press.

Rice, X. (2007, March 20). Kenya sets world first with money transfers by mobile. *The Guardian*. Retrieved from http://www.guardian.co.uk/money/2007/mar/20/kenya.mobilephones

Rodriguez, C. (2000). Civil society and citizen's media: Peace architects for the new millennium. In K. G. Wilkins (Ed.), *Redeveloping communication for social change: Practice theory and power* (pp. 147–160). New York, NY: Rowman and Littlefield.

Rogers, E. (2003). *Diffusion of innovations* (5th ed.). New York, NY: Free Press.

Sarantakos, S. (1993). *Social research*. Melbourne: Macmillan Education Australia.

Sardar, Z. (2007, February 19). Not such a brave new world. *New Statesman*. Retrieved from http://www.newstatesman.com/religion/2007/02/technology-mobile-world

Tanner, E. (2001). Chilean conversations: Internet forum participants debate Augusto Pinochet's detention. *Journal of Communication, 51*(2), 383–403.

Tanner Hawkins, E. (2005). Creating a national strategy for Internet development in Chile. *Telecommunications Policy, 29*, 351–365.

UNESCO Institute for Statistics. (2010). *Education in Jamaica*. Retrieved from http://stats.uis.unesco.org/unesco/TableViewer/document.aspx?ReportId=121&IF_Language=eng&BR_Country=3880&BR_Region=40520

United Nations Development Program. (2010). *The human development concept*. Retrieved from Human Development Reports Web site: http://hdr.undp.org/en/humandev/

Wilkins, K. (2000). *Redeveloping communication for social change: Theory, practice and power*. Lanham, MD: Rowman and Littlefield.

Zukang, S. (2009). *Statement by Mr. Sha Zukang, Under-Secretary-General for Economic and Social Affairs to the World Telecommunication Policy Forum*. Statement presented at the World Telecommunication Policy Forum, Lisbon, Portugal. Retrieved from http://www.un.org/esa/desa/ousg/statements/2009/20090422_policy_forum.html

18. Cultural Peculiarities of Russian Audience Participation in Political Discourse in the Era of New Technologies

IRINA PRIVALOVA, PHD

Introduction

Mediated communication in Russia can be described both in terms of a neo-authoritarian model of conventional mass media (Becker, 2004) and in terms of an evolving model of new mass media. Counteracting and interacting with one other, these two models have their own peculiarities and characteristic features. This chapter examines the ways in which new media technologies are changing political participation in Russia. Specifically, it investigates the ways in which Russia's Web 2.0 differs from its Western and American analogues, as well as the role it plays in the civic engagement of the audience and its impact on development of political awareness in Russian citizens. First, I describe the importance of taking a sociocultural approach in investigating the function of information and communication technologies (ICTs) and give a brief historical overview of the progress of different Internet communication platforms in Russia. I then consider the very important question of the status of user-generated content (UGC) in media environments in relation to the official media format in this country. Last, I explore two case studies in which UGC triggered significant and new forms of political discourse in Russia.

Sociocultural shaping of mediated communication

Without diminishing the importance of the reciprocal impact of ICTs on a society's development, I can state that sociocultural factors are pivotal in shaping the style and character of mediated communication in any given society.

Mediated communication has cultural peculiarities and national colorings. My approach to understanding mediated communication in Russia and my efforts to develop models to describe it therefore employ as a theoretical underpinning a perspective on the technology–media relationship that focuses on "the socio-cultural shaping of technology" (Preston, 2009, p. 114), building on the theories of Carey (2005, 2007) and Williams (1974).

McLuhan (1962, 1964) refers to two products of the media: *"literary man"* and "*electronic man*," describing them as two types of mentality that are formed under the influence of printed technologies and electronic gadgets, respectively. With the enrichment of the human senses via media of various kinds, argues McLuhan, cognitive organization is becoming more complex, and, consequently, societal organization is undergoing profound change. "McLuhan's perspective on "technology" reveals its "sociocentric" and "anthropocentric" core, since it highlights changes happening on the cognitive level of a human being, and the further ramifications of these at the societal level.

Although McLuhan was writing long before the Internet existed (and indeed predicted the advent of the World Wide Web 30 years before its invention), some scholars have extended his analysis to encompass the possibilities and realities of the Internet era (Levinson, 1999). Web 1.0 pushed users to the screen of a computer, providing "read-only" access to all types of information sources; it also introduced new realities such as hypertexts with electronic content. To use McLuhan's terminology, Web 1.0 permitted an enormous extension of the senses but in a social (lack of) context that promoted individualism and isolation. A much richer field of investigation is the influence of Web 2.0 on the cognitive organization of individuals and societies. Web 2.0 offers a profound paradox, in which we discover that the medium that—in McLuhan-esque terms—is argued to extend the senses is, at the same time, being generated by Internet users themselves.

The overwhelming pace of development of new technologies in recent decades has encouraged some theorists to adopt a technologically deterministic perspective when considering the discourse practices that are mediated by ICTs. Granted, a small set of basic principles can be identified that characterize different kinds of digitalized discourse, but in spite of the "universal features" of Web-based text structures, their content is culturally and socially specific (Danet & Herring, 2007). Nonetheless, new technologies do transmit and disseminate "globalized" ideas, notions, and concepts. Preston (2009) writes about "technology-tempered times" (p. 115) in relation to the astounding array of technologies and in an earlier text (2001) discusses the changes in modes of thinking and influential discourses brought about by technological change.

Proponents of techno-centric and technological determinist approaches to understanding new media and its social effects tend to emphasize examples that illustrate the globalization and homogenization of selected communicative practices. A wider analysis, however, demonstrates that no "unified mediated communication model" is in operation across all countries of the world. In reality, successful mediated communication processes always evolve through interaction with local or national values and practices; for instance, experimental investigations demonstrate that individuals respond differently to in-group and out-group communicators depending on their uses of culturally specific symbols (Lwin, Stanaland, & Williams, 2010), and research demonstrates the absence of homogeneity in news reporting in European countries (Preston, 2009).

I therefore suggest using the term *Internet communicative behavior* in order to describe behavioral patterns of electronic communication that are typical of a particular culture. This Internet communicative behavior is a part of the everyday behavior reflecting the culture, which consists of concepts, values, and assumptions about life that guide behavior and are widely shared by people (Brislin & Yoshida, 1994).

Culture shapes communicative behavior, including Internet communicative behavior. From my observations, mediated communication depends not only on culture but also on sociopolitical context. For instance, in China, where the Internet is strongly controlled by the government, government-sponsored producers of Web-based information play a disproportionate role in shaping the content and style of the mediated communications that will influence large numbers of citizens. European users (if such a generalization is possible) have embraced a variety of Web-based platforms that provide avenues for expression and debate of individual opinions and points of view in addition to embracing the opportunities provided by mainstream media. Russians, on the other hand, currently consider the Internet to be the only possible venue for self-expression, given the strong censorship of conventional mass media.

History of mass media development in Russia

The modern Russian media play an important role in conveying information to the general public. Yet, the question is still urgent: Is this information complete and does everyone have access to it? To answer this question, one has to take a closer look at the history of media development in Russia in recent decades.

Mainstream Russian media fit the "neo-authoritarian model"

In the past, media information in socialist societies was strongly ideologized and censored. In the USSR, the mass media were the "spokesmen" of official ideology. The State had monopoly control of television, which propagated

Figure 1. Number of free TV channels and viewers (% population) from 1990 to 2009

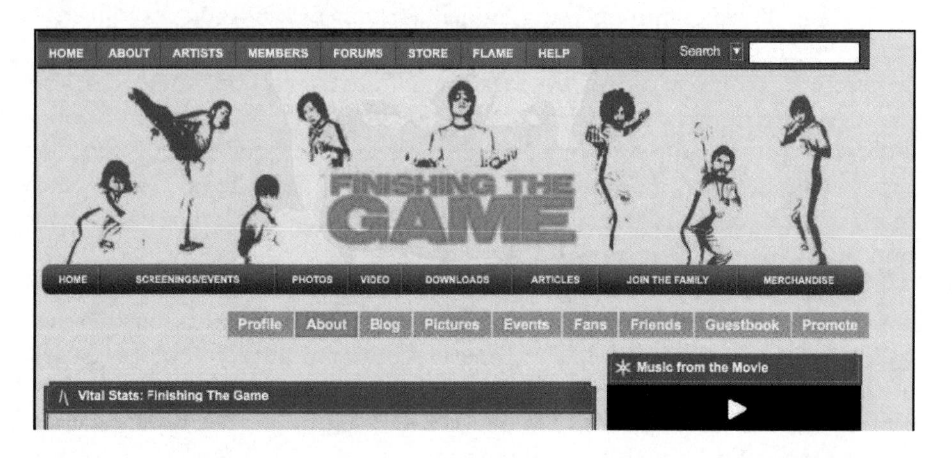

the philosophy of the ruling Communist Party (Mickiewicz, 1997; Sparks & Reading, 1994). In 1989, almost 98% of the adult Russian population watched television; however, since 1990 this figure has been declining steadily despite the substantial growth of the number of federal TV channels (Figure 1).

By 2010, only 70% of the adult Russian population watched television; of this population, only one third agree that they trust TV news. Poluekhtova (2010) explains the decrease in television audience as a reaction to the proliferation of poor quality programs and to the increase in imported programs containing inappropriate cultural messages for a Russian audience. I would like to argue that two further factors are reducing viewership: the significant spread of Internet communication and the continuing centralized control of television broadcasting by Russian government and business. By the middle of the first decade of the new millennium, there were 14 main free-access TV channels in Russia, of which five were controlled by the government; the remainder were in the hands of Russian oligarchs (Dashevskaya, 2006). All international and domestic events covered by these channels are reported from the perspective of the ruling party United Russia, which performs ideological framing (Goffman, 1974; Entman, 1993) by choosing to emphasize some issues over the others, thereby affecting the audience's awareness. Leaders of the Russian opposition, either Communists or Democrats, had and have few broadcasting opportunities, if any. In addition, socioeconomic changes in Russia and the countries of the former Soviet Union have transformed the media landscape (Corcoran & Preston, 1995) and the whole philosophy of news reporting. Gorbachev's introduction of the concepts of *glasnost* (openness) and *perestroika* (restructuring)

Figure 2. Numbers of periodicals (in thousands) and subscribers (% population) from 1990 to 2009

ushered in new principles of realistic coverage of political, social, and economic affairs, including information about past wars and current military operations. Under Gorbachev (1985–1990) and Yeltsin (1991–1999), the media tried to present a variety of ideas, opinions, and criticisms. Under Putin (2000–2008), printed media had relative freedom, whereas TV representations of the events most crucial to the state, including elections and war campaigns in Chechnya, were strictly controlled. The current media situation in Russia can therefore be described as neo-authoritarian (Becker, 2004). All of the Russian newspapers, just like the TV channels, are private enterprises and are owned by oligarchs who are loyal to the government. There is no chance that they will publish information inconsistent with the official point of view.

In recent decades, print media and television have shown similar development trends: An upward trend in production is contrasted with a downward trend in its consumption. In the year 1990, the country had the greatest number of newspaper subscribers, almost 65% of the adult population, with the lowest number of periodical items—8,489. Since 1990, the number of subscribers has declined steadily, and this downward trend is forecast to continue (Figure 2; Zhebit & Zykov, 2010; Krasnoboka, 2010; Pietiläinen, 2010).

Under such circumstances, it is unsurprising that new media are regarded as the only venue for free expression of ideas that are not aligned with official perspectives.

Russian Internet users have enthusiastically embraced the new virtual opportunities for self-expression and freedom of speech and see new media as a venue for presenting new facts and reassessing “known” information.

A new media model for new Russian media

In Russia, Internet technologies are currently most significantly involved in the political sphere, rather than in business—exactly the opposite of the situation in the United States and in Europe. The Russian language Internet, RuNet, emerged in the mid-1990s, when the country was in economic and political turmoil. At that time, RuNet offered policy makers one more tool for shaping public opinion. Very few citizens had Internet access, but years of Soviet ideology had taught Russians that the most accurate information was information that was difficult to access. For this reason, early online information portals such as Lenta.ru, Газета.ru, Страна.ru, Вести.ru, Грани.Ru, СМИ.ru, NTVRu.com, and АПН.ru were extremely influential, and their content much sought after. Russian news makers appreciated immediately that Internet communication was easy, fast, cheap, and (increasingly) accessible. They considered another important characteristic to be its "pantophagy," or its ability to disseminate any and all kinds of content, including material not appropriate for conventional mass media. In the early stages of Web 1.0 development, political discourse existed via the presentation of text-based content, and content management systems were employed for the mere transmission of information.

The advent of Web 2.0 in 2004, however, was the trigger for a massive increase in Russian citizen engagement in political discussions. The number of Internet users in Russia doubled that year from 5.9 million to 18.5 million (Figure 3; Public Opinion Foundation, 2010).

Having emerged about six years ago in Russia, Web 2.0 is currently going through a "conditioning stage." Ethical norms and regulations for information flow are actively under negotiation by users, and UGC is having

Figure 3. Internet use in Russia since 1996

a dramatic societal impact. To some extent, Russian Web 2.0 is developing similarly to its Western analogue, although it appears to be taking two paths. On the one hand, modern Russian social networking is integrating to some extent with some of the well-known international blogospheres (i.e., Facebook, Myspace, Twitter); Russian sections of Wikipedia and YouTube are active, and Russian Internet users are making thorough use of American Web 2.0 resources. In 2010, almost seven million unique users from Russia visited the US Internet each month—close to 5% of the Russian population (Skobkin, 2010).

On the other hand, new Russian social networks with clear signs of Russian cultural identity have appeared (i.e., My Former Classmates, Professionals, etc.), and some now occupy leading positions on the World Wide Web in terms of numbers of annual visitors. Current data compiled by Google (2011) indicate that four Russian sites are among the top 100 most popular sites worldwide in terms of the number of unique monthly visitors. Two of these are pure social networks: Vkontakte.ru (ranked number 70) and Odnoklassniki.ru (ranked number 82). One site, Mail.ru, ranked at number 35, is a Web mail, messaging, and search engine/resource site with the elements and functions of Web 2.0. And Yandex.ru (ranked at number 48) is a search engine/resource site.

The Russian social networks Vkontakte.ru and Odnoklassniki.ru have not become carbon copies of their American analogues; they are different conceptually and are marked by national colorings. One possible explanation of their exploding membership numbers is that they exploit such cultural Russian values as "communalism." As Richmond (1996), a great connoisseur of Russian culture, explains, group affinity

> has deep roots in Russian culture and can still be seen today in everyday life . . . *Sobornost* (communal spirit, togetherness) distinguishes Russians from Westerners, for whom individualism and competitiveness are more common characteristics. Russian communalism was not an invention of communists, although its traditions were exploited under the Soviets. (p. 14)

In the era of new media technologies, social networking sites are, to some extent, replacing some styles of communication such as "soul-to-soul talks"—interactions that most Soviet citizens were fond of. In those days, the only place where people could openly express their views was in a modestly furnished kitchen. Conversation topics were chosen by a group of good friends, and their discourse was not censored. (Doesn't this somehow resemble the discourse of modern social networks, with the face-to-face element missing?) Soul-to-soul talks might have stemmed from gatherings of the members of a commune—a collective form of living deeply rooted in Russian society. Richmond (1996) explains, "This communal way of life persisted well into the twentieth century, lasting longer in Russia than elsewhere in Europe" (p. 16). Conceiving the project, the "founding fathers" of Russian social networks were well aware of communalism as a core Russian cultural value, and this understanding helped their project succeed.

In contrast with American and European networking, the discussion topics on Russian social networks tend to be more politically oriented, and Internet users employ them frequently as arenas for political debate. These networks are highly responsive to the most important political events: new law amendments, reforms, or coverage of man-made disasters. In May 2009, when 90 workers were killed as a result of the gas blast in the mine Raspadskaya, Russian social networks were filled with comments concerning the behavior of governmental officials and emergency agencies. Detailed information about the rescue operation could be found there, as well as eyewitness evidence of the protest demonstrations organized by the miners' relatives. The official media meanwhile failed to present any of these events or perspectives.

Engagement of citizens in political discourses through new media platforms is exceptionally high in countries like Russia that have limited press freedom as a result of the interactive and participatory affordances of Web 2.0. In situations where authoritarian government has full control over the media, UGC in Web 2.0 offers the only avenue for alternative political debate and self-expression. To some extent, Web 2.0 sites meet citizens' expectations for freedom of speech, because they allow them to interact with each other in a political dialogue. They create virtual communities and generate digital content, in place of being members of actual political parties and political speeches.

It is interesting that Russia's leadership appears to have accepted the challenge of Web 2.0, with the last two years seeing some critical events. The year 2009 witnessed the opening of President Medvedev's blog on the LiveJournal Web site and the President's Channel on the YouTube Web site. On June 24, 2010, the president, who is known to be a great fan of different new media and technologies, opened an account on Twitter. It is apparent, then, that even though governing authorities continue to regulate the mainstream media in Russia, they have also recognized the capacity and reach of Web 2.0 and UGC for shaping public opinion.

An axiological analysis of two Russian case studies

As the previous discussion illustrates, the political environment may predetermine the content of user-generated products, and social networking may be based different on different concepts in different countries. Another factor that conceptually shapes the character of Web 2.0 products is culture. Because culture penetrates all spheres of life and acts as the lens through which all actions are viewed or performed, it is possible to offer the following axiological approach—a philosophical study of any possible correspondence between the Internet communicative behavior of participants involved in a political discourse and their (national or cultural) values.

Such an approach is premised on the assumption that Internet communicative behavior reflects one of the most significant cultural concepts and correlates with national values. Compiling a list of national values is, however, a very unrewarding task. Kohls (1988) notes that most Americans have never undertaken such a task in relation to their own culture and believe that each individual is so unique that the same list of values could never be applied to all. The same could be said about Russians, although I have elsewhere addressed some aspects of Russian culture that reveal national values (Privalova, 2006, pp. 229–230). I present some details of Russian culture drawn from my earlier work in Table 1 and will refer to it in the following analyses of two case studies of modern political discourse in Russia. Both cases offer examples of contemporary political discourse in that they illustrate moments of interaction between citizens and either local authorities or governmental officials.

Case study 1. An example of multidirectional political discourse

This regional example deals with a case in which one message posted on the LiveJournal Web site created serious panic in the city of Saratov in December 2009.

Table 1. Some aspects of Russian culture

1. Life follows a preordained course. The individual can influence the future by making a choice within the frames of his or her own destiny or fate.
2. People are intended to adjust to the physical environment rather than to alter it.
3. Ideals (the meaning of life, the concept of truth) are to be pursued regardless of what is "reasonable."
4. Hard work is not the only prerequisite for success; wisdom, luck, and time are also required.
5. A commitment might be superseded by a conflicting request, or an agreement may only signify intention and have little or no relationship to the capacity of performance.
6. Group considerations, friendship, personal characteristics, and other considerations may determine employment practices.
7. The removal of a person from a professional position involves a great loss of prestige and will rarely be done.
8. Stabilization is viewed as an improvement and a guarantee of safety in a constantly changing world.
9. Ideals (concepts of the meaning of life) and the process of pursuing them may be most valued.

Historical context. At the beginning of December 2009, an epidemic of swine flu was in full swing in Saratov. There had been more than 40 deaths, and vaccine was in short supply. Local drugstores were acutely short of flu medications and disposable dressings. All educational establishments and recreational centers were closed, and the atmosphere was nervous and tense. As a result, when someone posted a comment on LiveJournal about possible spraying of plague vaccine over the city from the sky in order to prevent other cases of the dangerous disease, many people believed it. Even though the news sounded absurd, the majority of the residents accepted it as true. The news spread with lightning speed by SMS message, e-mail, phone calls, and word of mouth. Friends, relatives, and acquaintances were trying to warn each other.

At the time that the operation was supposed to start, the city was empty, all shops and markets were closed, and all bottled water had been sold out. Ominous silence and sinister emptiness turned the streets of the city into the backdrop of a horror movie. Local authorities tried to set things straight—through television and printed press, they denied any idea that the city was going to be "decontaminated" by this supposed method. Despite the assurances of local authorities in the conventional media that they were keeping the situation in check, however, city dwellers believed the "news" of a possible lung plague epidemic. All of the governor's attempts to persuade people to go back to work were in vain; it was his political Waterloo. The next day, the alleged author of that ill-fated comment (a second-year student at the local medical university) was reported to have been arrested. He was accused of disseminating rumors that had destabilized the situation in Saratov.

Axiological Commentary. Why did the dialogue between the municipal administration and local residents fail? Why did the political discourse follow a divergent perspective? Residents of Saratov made no effort to discuss the situation with local agencies, although in principle they could have sent a petition demanding some explanations or could have launched social and political debates through mass media channels in relation to the case.

Both political and cultural explanations for these questions should be considered. Two important political outcomes are noteworthy: Saratov residents demonstrated the lowest possible level of confidence in local authorities, while these same authorities displayed a total inability to ensure control over the situation in the region. The audience opted to avoid political discourse rather than to participate in it. People chose to believe the UGC disseminated via new media because they did not trust the information from government authorities. Some federal media demanded the governor resign, because they viewed residents' reactions as a vote of no confidence. Political polarization reached its peak when the most influential journalists, representatives of

Figure 4. T-shirt satirizing the alleged aerial decontamination plan

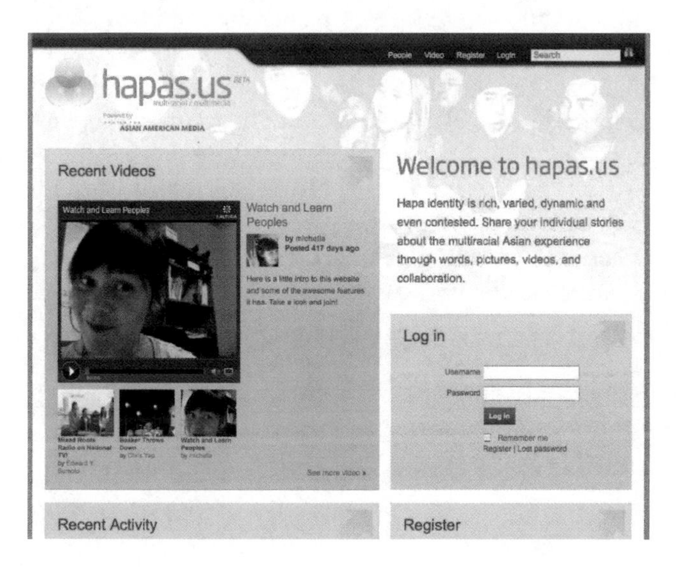

Source: Image reproduced with permission from Vzglyad-Info (2009).

NGOs and local businesses, protested violently against the arrest of the blogger—the designated scapegoat. A local clothing manufacturer even started the production of T-shirts with the inscription "I've survived disinfection" (Figure 4); (Vzglyad-Info, 2009).

These T-shirts soon became a top seller and wore down government officials' denials of this event. In the end, consolidated protests by the community forced local police to admit that there was no case against the blogger, and he was released. New technologies contained the most complete information concerning this political crisis, and in this way they played the crucial role in consolidating community protest.

Cultural considerations are also helpful for analyzing this case. I suggest that the unexpected character of this political nondialogue can be partially explained by referring to some aspects of Russian culture presented in Table 1, in specific, Items 1, 2, and 8. First, residents of Saratov accepted the situation as presented on the Web and adopted a fatalistic approach. They stocked up on bottled water and consumer goods as a means of adjusting to the crisis. Nevertheless, most residents elected to stay in Saratov rather than to leave, reflecting a belief that bad stability is better than good change.

This case reveals two more features of Russian culture that Richmond (1996) has also explored: caution and pessimism. He notes: "Americans expect things to go well and become annoyed when they do not. Russians expect things to go poorly and have learned to live with misfortune" (p. 40). With regard to the "decontamination panic," residents of Saratov may well have decided not to interfere with the natural course of events, believing that any efforts made to change things for the better would yield poor results. Moreover, reciprocal political dialogue can only occur in a conceptually transparent situation. On the other hand, Richmond argues that Russians typically hide or distort facts to present a "brave face": "Russians can fudge the facts, a national characteristic called *vranyo* (fibs) . . . an inability to face the facts, particularly when the facts do not reflect favorably on Russia" (p. 123). And he suggests that "related to *vranyo* is *pokazukha,* the tradition of staging an event 'for show,' especially for high officials and visiting foreign dignitaries" (p. 126).

Having concealed some crucial facts about the swine flu epidemic, the municipal authorities of Saratov guaranteed that no reciprocal political dialogue was possible.

Case Study 2. An example of interdependent political discourse

The following example illustrates the way in which a message distributed via new media *united* the Russian audience on a federal level. In this case, three YouTube videos of an ordinary *militsiya* (police) major provoked debate at all levels of society and prompted serious reforms of the national police agency.

Historical Context. On November 24, 2009, a 33-year-old major of the Russian police, Alexey Dymovsky, posted his first video on YouTube, addressed to the president. In it, he outlines common practices of the Russian police force: endemic corruption, framing and accusing of innocent people, torturing of suspects, and making deals with criminals. In addition, he shows the seamy side of work in the police force: fraudulent reporting of crime clearance figures, low salaries of workers, poor medical care, and lack of possibility of sick leave. The presentation was a bombshell. Almost 725,000 viewers watched the first part of Dymovsky's video appeal on YouTube, and his messages triggered a wave of videos from other Russian police officers highlighting corruption and cases of innocent people who were framed. These YouTube appeals had a massive societal impact. It was hard for the Ministry of Internal Affairs to admit that all of the revealed facts were true and were common practice. On December 24, 2009, however, the president signed a degree for the reform of the Federal Agency of Internal Affairs. He suggested that "militsiya" should be renamed "police" being that the old name had become a synonym for corruption, fraud, injustice, and the low morale of employees.

Axiological Commentary. Russian Internet users have adapted to the possibilities of Web 2.0 for creating a new, culturally specific form of communication with government officials. Dymovsky's video appeal has created a precedent, and the "Dear Mr. President!" style of presentation remains the most popular means of personal appeal to the president. Because all types of demonstrations and rallies are illegal, Russian people have now found the way to present their grievances via YouTube.

The Dear Mr. President! style of presentation is based on the fictional assumption that a kind tsar/ruler/boss must be misinformed by his incompetent counselors or wicked advisers. The apparent belief is that were he correctly informed, things would go well at once. Belief in a "kind Tsar" is still strong among Russians, and this faith underpins actions that may seem to be politically naïve.

One aspect of Russian culture relevant to this case is the belief that a rational and pragmatic approach is not useful when truth is being pursued (Table 1, Item 3). Major Dymovsky was fired from the police soon after posting his first YouTube appeal; he was later subjected to a psychiatric examination and stood trials on the accusations of fraud, embezzlement, and disclosure of state secrets. It is remarkable that at the trial, he pleaded not guilty and was acquitted. In spite of a new awareness of the consequences of such video appeals, however, Russian users continue to post them on YouTube! For Russians, the search for truth is closely linked to the search for the meaning of life (see Item 9, Table 1). Apparently illogical actions can make some sense when considered against a backdrop of a further unique element of Russian culture—*dusha:* "A romantic ethos, *dusha* appeals to feeling rather than fact, sentiment over certainty, suffering instead of satisfaction, and nostalgia for the past as opposed to the reality of the present" (Richmond, 1996, p. 46).

It is interesting that a textual analysis of political video appeals posted to YouTube offers little support for the assertions that Russian communicative behavior is *high context* (Hall, 1977). As in *low-context* cultures, the presenters of these video appeals use a direct style of verbal expression in which important information is usually conveyed through explicit verbal messages; they tend to directly express their opinions and intend to persuade others to accept their viewpoints (Chen & Starosta, 1998, p. 50). An occasional lack of verbal fluency and eloquent speech is evident, and presenters also frequently attempt to address too many issues at a time, making the connections between ideas not immediately clear. For instance, the second and third sections of Dymovsky's presentation raise such issues as his professional relations with his boss and immediate subordinates; he also demands second hearings of such politically significant cases for Russian society as the terrorist attacks at the Beslan School and at the Nord-Ost Recreational Center in Moscow.

Conclusion

"Not much has changed the world more than the blogs," declares Hiemstra (2009). Without question, with the advent of Web 2.0, mediated communication has undergone serious metamorphoses. Nevertheless, these transformations have not set humanity on a path toward either monoculturality or McLuhan's prosperous "global village." Web 2.0 technologies are universal, but UGC has *context*, because it is being produced by individuals who imbue it with their cultural and social understandings. For this reason, the contemporary World Wide Web is filled with country-specific and culture-specific content, which become especially interesting in states like Russia where freedom of speech is a challenge. New technologies are politically challenging, and at the same time, they have an enlivening and stimulating effect on civil society.

This study has demonstrated that two factors predetermine the influence of UGC on the social participation of the audience: (a) the level of access to Internet-based communication and (b) the level of liberty around discourse in society. Russia falls behind its Western neighbors in the number of computers per capita, as well as in the technological literacy of its citizens. The current Internet penetration in Russia stands at 43% of the population, compared with 80% in Australia, 77% in the USA, 83% in the UK, 78% in Japan, 79% in Germany, and 52% in Italy (Internet World Stats, 2010). Internet information is varied and uncensored, in contrast to "official" media. Examination of the cases discussed here and other instances of UGC suggest that an inverse relationship can explain the impact of new media on an audience. It appears that the lower the levels of Internet access and freedom of discussion, the greater the potential impact of new mass media on a society. Web 2.0 has not only revolutionized communication processes but has also generated many theoretical questions that yet have to be answered. Further exploration of the practical experiences of the development of new mass media in different cultures and societies will continue to reveal their complex effects.

References

Becker, J. (2004). Lessons from Russia: A neo-authoritarian media system. *European Journal of Communication, 19*, 139–163.

Brislin, R. W., & Yoshida, T. (Eds.). (1994). *Intercultural communication training: An Introduction*. Thousand Oaks, CA: Sage Publications.

Carey, J. (2005). Historical pragmatism and the Internet. *New Media & Society, 7*(4), 443–455.

Carey, J. (2007). A short history of journalism for journalists. *The Harvard International Journal of Press/Politics, 12*(3), 3–16.

Chen, G. M., & Starosta, W. (1998). *Foundations of intercultural communication*. Boston, MA: Allyn & Bacon.

Corcoran, F., & Preston, P. (Eds.). (1995). *Democracy and communication in the new Europe: Change and continuity in East and West*. Cresskill, NJ: Hampton Press.

Danet, B., & Herring, S., (Eds.). (2007). *The multilingual Internet: Language, culture, and communication online*. New York, NY: Oxford University Press.

Dashevskaya, I. (2006). *Market Mass Media in Russia: Television (Part 6)*. Retrieved from http://www.advertology.ru/article57914.htm

Entman, R. M. (1993). Framing: Toward clarification of a fractured paradigm. *Journal of Communication*, *43*(4), 51–58.

Goffman, E. (1974). *Frame analysis: An essay on the organization of experience*. Cambridge, MA: Harvard University Press.

Google. (2011). *The 1000 most-visited sites on the web*. Retrieved from http://www.google.com/ adplanner/static/top1000/

Hall, E. T. (1977). *Beyond cultures*. Garden City, NY: Anchor Press.

Hiemstra, G. *Future talks: User generated content part 1*. Retrieved from http://www.youtube.com/watch?v=eef4Fgrb5c4

Index Mundi. (2008). *Internet users*: Russia. Retrieved from http://www.indexmundi.com/ russia/internet-users.html

Internet World Stats. (2010). *Internet World Stats*. Retrieved from http://www.internetworldstats.com

Kohls, L. R. (1988). *The values Americans live by*. Washington: International Centre of the Peace Corps.

Krasnoboka, N. (2010). *Media landscape: Russia*. The European Journalism Centre. Retrieved from http://www.ejc.net/media_landscape/article/russia/

Levinson, P. (1999). *Digital McLuhan: A guide to the information millennium*. London, United Kingdom: Routledge.

Lwin, O. M., Stanaland, A., & Williams, J. D. (2010). American symbolism in intercultural communication: An animosity/ethnocentrism perspective on intergroup relations and consumer attitudes. *Journal of Communication*, *60*(3), 491–515.

McLuhan, M. (1962). *The Gutenberg galaxy: The making of typographic man*. Canada: University of Toronto Press.

McLuhan, M. (1964). *Understanding media: The extensions of man*. New York, NY: McGraw-Hill.

Mickiewicz, E. (1997). *Changing channels. Television and the struggle for power in Russia*. New York, NY, & Oxford, United Kingdom: Oxford University Press.

Pietiläinen, J. (2010). Social class and media use in Russia. In J. Nikula & M. Chernysh (Eds.), *Social class in the Russian society: Studies in the social classes and social change of contemporary Russia*. Saarbrücken, Germany: Lambert Academic.

Poluekhtova, I. (2010). Dynamics of the Russian TV audience. *Sociological Research*, *1*, 66–77.

Preston, P. (2001). *Reshaping communication: Technology, information and social change*. Thousand Oaks, CA: Sage.

Preston, P. (2009). An elusive trans-national public sphere?: Connectivity versus journalism and news cultures in the EU Setting. *Journalism Studies, 10*(1), 114–130.

Privalova, I. (2006). *Interculture and a verbal sign: Linguocognitive aspects of intercultural communication.* Moscow, Russia: Gnozis.

Public Opinion Foundation. (2010). *The Internet in Russia.* Retrieved from http://corp.fom.ru/projects/23.html&usg=ALkJrhhuwbtLxVASoPueuQjs_0qXse42bQ#2

Richmond, Y. (1996). *From nyet to da: Understanding the Russians.* Yarmouth, ME: Intercultural Press.

Skobkin, A. (2010, January 27). U.S. networks attract Russians. Retrieved September 2010 from http://libymax.ru/?p=3833

Sparks, C., & Reading, A. (1994). Understanding media change in East and Central Europe. *Media, Culture and Society, 16,* 243–270.

Vzglyad-Info. (2009). The release of t-shirts with the inscription "I survived the disinfection" is announced (trans. Editors). Retrieved from http://www.vzsar.ru/news/2009/12/03/soobschaetsya_o_vypuske_futbolok_s_nadpisyu_ya_perezhil_dezinfekciyu.html

Williams, R. (1974). *Television, technology and cultural form.* London, United Kingdom: Collins.

Zhebit, M., & Zykov, T. (2010, September 6). Newspaper statistics remain stable. Print media readers in no hurry to abandon the habit of reading the newspaper. *Rossiyskaya Gazeta.* Retrieved from http://www.rg.ru/2010/06/09/gazety.html

19. The Vienna Unibrennt Platform: Hidden Pitfalls of the Social Web

HERBERT HRACHOVEC, PHD

Introduction

University education has been perceived to be fairly similar across the Atlantic and Pacific beginning with the 19th-century emphasis on the pursuit of scholarship promoted by people like Abraham Flexner, Cardinal Newman, and Wilhelm von Humboldt (Clark, 1983). There appears to be no profound intercultural trenches here, yet it is easy to overlook some important differences. Von Humboldt was, among other things, a Prussian bureaucrat who established a thoroughly centralized system, which shapes Continental European universities even today. Against this backdrop, university reforms in some European countries have forced dramatic rearrangements. The following account will bring out the considerable cross-Atlantic differences in what might at first glance be considered a homogeneous field.

Reforms of the higher education sector in Austria will be recounted in some detail, but this is not primarily a local story. I will show that the political confrontation that took place between the administration of Vienna University and a group of student protesters who were opposing new legislation was profoundly shaped by the availability of Web 2.0 technologies. And I will argue that the unprecedented efficiency of those technologies is generally a mixed blessing with regard to the political agenda of this kind of protest.

Background

In October 2009, Austrian universities were hit by a surprising development. In 2002, a new national law had mandated that the higher education system be reorganized according to the guidelines of the so-called "Bologna Follow-Up Group" (Benelux Bologna Secretariat, 2010), a voluntary association of European (consisting of EU and non-EU) governments designed to implement a "European Higher Education Area" (EHEA, 2011). Threatening to severely cut funding, the Federal Ministry for Science and Research had pushed Austrian universities to dramatically rearrange their traditional curricula. Despite initial dismay, most universities complied fairly quickly, and their courses of study were to a considerable extent switched to a new set of curricula, when a demonstration organized by the students of one of the last noncompliant institutions, the Academy of Fine Arts in Vienna, triggered a dramatic nationwide wave of belated protest. Student demonstrators occupied the biggest lecture hall (the Auditorium Maximum, or Audimax) of the University of Vienna, Austria's largest institution of higher learning with more than 80,000 students, on the afternoon of an anti-Bologna protest march, and it remained in the hands of students until the end of the year. Anti-Bologna activities subsequently erupted on many other Austrian campuses and spread to Germany, Switzerland, and other countries operating under the Bologna rules.

In previous decades, occupations of the Audimax had developed into something of a ritual, as students repeatedly mobilized against cuts in the education budget. But the 2009 protest turned out to be quite different from earlier occupations, as Web 2.0 technologies were, for the first time, available and used extensively. Live streaming, Twitter channels, wiki pages, and blogging were set up almost immediately, significantly enhancing the visibility and impact of the protest movement. Students had a big advantage by using these comparatively new means of communication with which university administrators were not especially familiar. This fact was widely noted in the traditional mass media, which drew parallels between these students' use of technology and the roles blogs and Twitter had played in recent Chinese and Iranian political confrontations. And indeed, there had never before been a prolonged occupation of the symbolically important Audimax that was accompanied by a daily program with continual updates regarding the political and cultural activities offered to the general public by the occupants and their sympathizers. Given the interactive and multimedia capacities of Web 2.0 applications, the occupation turned into something like political reality TV on the Internet.

Recounting these developments, a report by members of the IT team (Heissenberger et al., 2010) of the Audimax protesters stresses the similarity between the students' *modus operandi* and the Internet procedures:

> Considering a model of networked protest consisting of the Internet and participants themselves, an interesting observation can be made: the most important medium of networking (as well as of the publicity aroused by the movement) by means of which networking could only be established in such a quick, inexpensive

> way, free of barriers, can—in a certain respect—be regarded as similar to the movement itself. Its organization lacks a center and is openly accessible to everyone, causing considerable redundancy of data and information in the process. (p. 221, trans. H. Hrachovec)

The narrative of the Internet as a liberating force is manifest in this assertion. Rather than address the claim directly, this chapter will examine some basic presuppositions for the success achieved by the *unibrennt* (which translates into "the university burns") platform and relate them to the political aims asserted by the students. In order to assess the impact of their protest, an account of the politics of Austrian educational reform is given. The chapter contends that compared to more traditional procedures, the administration of Vienna University has significantly changed its governing structure. The Internet activities mentioned can, in a surprising way, be shown to follow a pattern quite similar to this shift. *Both* contestants were operating within the more general framework of a global sea change of business practices. This essay will conclude by raising the question how well the narrative of Internet-induced liberation stands up in view of the case study presented here.

Austrian universities and the European context

The entire structure of national higher education was changed by the *Universitätsgesetz* (UG02) introduced by the center-right Austrian government in 2002. Historically, and according to Central European tradition, the state had had a dominant influence in university affairs - not just paying their bills, but also (in many cases) managing their internal business. Professors had been, in fact, civil servants (like policemen or diplomats) with the appropriate government department being their legal employer." In 2002, this time-honored setup was discarded in favor of an "autonomy" of universities constructed along the standard business model of a presiding board (*Universitätsrat*) and a chief executive officer (*rektor*; for context, see European Universities Institute, 2011). As a concession to the venerable practice of academic self-government in nonfiscal affairs, a senate was introduced as a third element. Its main responsibility consisted in supervising and approving the curricula governing teaching under its jurisdiction. (Curricula are still considered legally binding documents subject to certain national standards.) Details of this major shake up do not concern us here, but one law must be stressed: Before 2002, it was required that curricula templates were issued by the Austrian national parliament (!), and every single curriculum had to be approved by the governmental education authorities. One consequence of the new autonomy mandated in 2002 is that each university was now entitled to define its own courses of study, participating in the education market and thereby designing its own profile.

The state has been forced, to a significant extent, to withdraw from meddling in university affairs. But one crucial proviso imposed upon curricular development has turned this newly granted freedom into a special new form of dependency. (It will eventually be argued that a common pattern linking this transition in education to students' recourse to Internet resources can be detected.) Austrian universities were now required to follow the Bologna guidelines developed by the Bologna group to which European states have voluntarily acceded. In particular, they were obliged to reorganize their existing (usually four-year) diploma courses into a pattern of 3-year bachelor's and 2-year master's degrees. It was required that all teaching units be measured in European Credit Transfer System (ECTS) points in place of the customary hours taken by the various teaching units offered. Individual courses (in North America, this would be translated as "programs") were to be abandoned as the basic curricular element, giving way to "modules" (or North American "courses") that group a number of courses into a substructure of the entire degree. These requirements have resulted in a number of predictable and some unpredictable problems that triggered the student movement being examined. Numerous lively discussions have been provoked by these developments (HRK, n.d.; Hrachovec, 2009). Yet, probably the most remarkable tectonic shift has gone largely unnoticed.

In introducing the reorganization of higher education, the policy announcements of the responsible government agencies all emphasized the withdrawal of state influence and an increase in available opportunities for local and private actors. But they largely failed to mention that the state had simply switched its role. In ridding itself of the increasingly cumbersome burden of responsibility for the higher education sector, the state had not just "set universities free"—it had delivered them to another normative regime. Instead of shouldering the responsibility of managing the national planning of academic development, the state claimed to be empowering universities by enforcing the rules of the more general European recommendations. Once parliament had been relieved of its authority to determine the set of acceptable disciplines and their teaching modalities on a national scale, the state started relying on European meta-statelike committees to ensure a modicum of consistency across the board of Austrian curricula. (No mechanism facilitating national consultations between universities had been established.)

It has often been observed that big multinational companies have developed into economic agents more powerful than many state governments. The European Union has a special way of dealing with this problem. Introducing monetary unions and enhancing pan-European control over many areas (including educational policy), which were the former prerogatives of nation-states, it aims at gaining political clout on a global scale. Because we are dealing with a local Austrian political disturbance, what is the point of recounting these developments? The abandonment of parts of Austrian educational sovereignty is a remarkable fact; states do not easily hand over such prerogatives. In order to understand the legal rearrangement previously outlined, it might be helpful to change the frame of reference from the legal to the economic realm. A useful distinction has been made between merchant- and two-sided markets (Rochet & Tirole, 2006; Evans, 2006; Hagiu, 2007; Gawer, 2009).Traditional department stores, for instance, have supplied their customers with a variety of goods under one management. Their operations are unidirectional, but in recent decades, big stores have switched to a two-sided platform: Instead of presenting themselves as one big supplier of goods, they have reinvented themselves as mediating platforms, enabling exchange between shops (often brand names) and their customers. The former pattern of comprehensive management has been replaced by an approach of matching the needs of selected sellers to their clientele.

The restructuring of the Austrian higher education sector has followed a similar pattern. Just as the previously state-owned railways, mail services, and energy providers had been transformed into companies sui generis, universities were given the means for self-determination, albeit within an environment of unexplored complexity. The Austrian state switched from the mono-directional model of having ultimate responsibility over its academic sector to a mediating role as a platform provider, correlating national institutions of research and education with the overarching aims of the so-called Lisbon Declaration (European University Association, 2007), which envisaged a leading role for Europe in bringing about a future "knowledge society."

The unibrennt platform

There are numerous examples of two-sided platforms in Austria; for example, travel agents and credit card companies. The present concern is with one of the most pervasive cases. The Internet can, to a significant extent, be thought of as such a mediating realm. The World Wide Web, in its early years, did not yet fit the bill. More often than not, it was used to simply extend existing public relations policies or to provide hitherto unavailable opportunities for publicity, as seen, for example, in the case of the Web presence of the University of Vienna (Universität Wien), which hosted a complex, multilayered site whose main purpose was to inform university staff, students, and visitors of the enterprise. In an area of the Web site that is graphically distinct but technically integrated, the university's student body has had its own bundle of service pages under the univie.ac.at domain. The arrangement is reminiscent of (and in fact an expression of) the conventional understanding of the role of the student body in European institutions of higher education, who are conceived of as "colleagues," not just to be instructed but also to be regarded as the future generation of scholars, sharing some of the privileges of academia. Student representatives, elected every second year, are mandated by national law and are supposed to allow the voice of the up-and-coming academic elite to be heard within the framework of a well-entrenched social partnership.

However, when the long-term equilibrium between the state and its universities was shattered, new modes of political activism became prominent, and these protests have been considerably enhanced by the availability of a new suite of Web 2.0 technologies. The software designed to enable networked cooperation are multisided platforms, which serve as a mediating interface for activists and their audience. Rather than simply announcing external events, social sites draw users into the spell of those events themselves, fashioning them as participants in a large interactive community. In past days, students could sign petitions circulated in school, in the office or in the neighborhood, which were then collected by hand and subsequently submitted to the authorities. Today, a Facebook group is established and "friends" are notified in a more convenient and far-reaching manner, facilitating the meteoric growth of protest communities such as unibrennt.

When the Audimax was occupied on October 22, 2009, detailed information immediately spread on the Internet and was swiftly picked up by online news sites:

> The fight centering on the Audimax of Vienna University is not only fought face to face. Rather, it is shifting to the net. Twitter and Facebook are instruments of mobilization, results are uploaded to YouTube. Numerous clips from the occupied Audimax are already available from the portal. (OE24, 2009, para. 1, trans. H. Hrachovec)

It took only four days for "a perfect homepage" to be established for the demonstrations. The online version of the Viennese daily *Wiener Zeitung* titled its report of the occurrences "http://unibrennt.at" and applauded their professional Web presence, adding "Organizing a strike by means of such a diversity of new tools of communication is unprecedented in Austria" (Weber, 2009). By October 26, the unibrennt Facebook group numbered 8,400 fans, live streaming was in place, wikis and forums had been set up, and "Unibrennt occupied 5th rank as search term on German language Twitter" (Weber, 2009). Almost as an afterthought, the report added another item: that a press office had been established. Before the week had ended, the students' virtual activities had been noted and extensively discussed on a national and international level (Oppelt, 2009; Leonhard, 2009).

A vivid account of how various tools combined to trigger this forceful media effect is given by Luca Hammer (Schwenk, 2009):

> Friday, when I was on site, I almost habitually streamed some of the speeches using Qik. When I noticed that there was a big demand on Twitter, but sound quality was unsatisfactory, on Saturday I took my laptop and a camera to offer a better stream. . . . This stream was widely distributed by Twitter in a very short time so that after a few minutes several hundred viewers were online. . . . On Tuesday we had almost 3000 parallel views which was the top result up to now. Altogether in those four days there were 140,000 views of the stream and 2.5 million minutes viewed. It was important to me to make events in the Audimax available outside to show that this is not just a party, as was claimed by some big TV stations. (para. 6, trans. H. Hrachovec)

The echo local activists produced in cyberspace is easily measured by an examination of log files and their graphical rendering. Tweets, for example, can readily be collected and processed in order to visualize channel activity (Heissenberger et al., 2010, p. 247). It is beyond doubt that the occupations of the auditorium maximum that happened in 1987, 2000 and 2002 (see http://unibrennt.at/wiki/index.php/Geschichte) lacked the glamor and international reach that newly emerging Web 2.0 technologies could confer. One should not, of course, mistake instant excitement about unprecedented opportunities to spread one's cause, with sustainable political impact.

Although the establishment of the unibrennt portal provides a first look at technical novelties, we have to dig deeper to comprehend the more ominous aspects of those developments. A closer look at the composition of the portal's entry page is instructive. The popular open-source WordPress blogging software was used, with extensive links to partner initiatives and to Facebook and Twitter. Live audio and video was streamed via ustream.tv. Flickr and Feedburner provided photos and RSS feeds. A Google analytics script tracked traffic on the site, and unibrennt@gmail.com was established with Google as the e-mail address that would be used by the activists. It should be noted that, with the exception of the open source WordPress installation, all providers mentioned here are commercial entities. In contrast with the Web site of the official student association, whose IP address is administered by the university's IT division, unibrennt is registered by Netmonic, a branch of Timewarp IT Consulting, GmbH, which, on its home page, explicitly mentions its expertise in taking on assignments outsourced by middle-sized enterprises. Considering this setup, it is obvious how the site was up and running in a very short time. Rather than depend on building a net presence from scratch, students plugged into the existing Web site framework an extensive suite of globally available Web 2.0 tools and services offering top-level functionality at little cost. Unibrennt is, in fact, a good example of a two-sided platform, joining a wide audience interested in the Viennese events with an array of big-league Internet companies thriving on the provision of social Web applications. I have pointed out that the Austrian government has adopted a two-sided platform policy in forcing its higher education reform upon the nation. It is no small irony that the Web platform that was central to the protest movement should employ a very similar strategy in meshing its visitors with global business corporations.

To quote the advertising pitch of just two of the firms mentioned: "Promote your business to millions of viewers for only $1 a month" (Ustream, 2011). And:

> Google Analytics is the enterprise-class web analytics solution that gives you rich insights into your website traffic and marketing effectiveness. Powerful, flexible and easy-to-use features now let you see and analyze your traffic data in an entirely new way. With Google Analytics, you're more prepared to write better-targeted ads, strengthen your marketing initiatives and create higher converting websites. (Google Analytics, 2011, para. 1)

To savor the irony of these developments, one has to keep in mind that one of the most prominent complaints the protestors directed against government and university administration was that the future of higher education was being abandoned to the interests of big capital. Heissenberger et al. (2010) point out that the significance of the protests was in drawing attention to "the fundamental failure of neoliberal capitalism and its appropriation of all ranges of life" (p. 189). According to these authors:

> . . . criticism and demands for transcending the immediate context of education and universities were dominant within the protest movement. They included an awareness of the pervasive infiltration and commodification of all realms of life by the neoliberal capitalistic logic of the markets. An awareness of how this logic isolates people by racist and sexist politics of exclusion, and destroys all collective efforts, has been tentatively established by the protests. (p. 191, trans. H. Hrachovec)

The student movement was much admired for its deft use of newly available communication technologies. Its message was clear and appealing: that market Darwinism should be stopped from interfering with academia. Yet, the very foundation upon which the protest built was a well-established set of global companies. Their contribution added a crucial surplus to plain information circulating in mailing lists and SMS messages. Multimedia savvy users wanted peer discussion, pictures, videos, and live streaming. The print media and TV were eager to report not so much the events taking place in a particular lecture hall but rather the purportedly innovative digital distribution of the events. The entertainment industry had entered the university by the backdoor.

I have focused on the networking aspect of the occupation. These activities were, at the time, not seen as a main concern but just as convenient support strengthening the political agenda. Members of the Internet team were cautious about the hype elicited by their work. As Heissenberger et al. (2010) point out: "Even given all the opportunities created and used on the web: the uni-protests are neither exclusively nor causally a 'Facebook-revolution' just as the protests following the 2009 Iranian elections were no 'Twitter-revolution'" (p. 242). Counting hits on a Web site was felt to be a derivative delight compared to getting publicity for the far-reaching demands put forward by the occupants: pervasive antidiscrimination, redemocratization of university governance, free access to higher education, and abolition of the whole curricular apparatus imposed by the Bologna reform (Heissenberger et al., 2010, p. 177). It could be objected that the present emphasis on auxiliary techniques distracted from the main thrust of the students' political engagement, as observed by Schwenk (2009):

> For many students it is probably the first time they are discovering how various tools from the social web can be employed and how they work. The "real-time web" is not just a playground for leisure and hanging out with friends, it is also a forceful weapon in the fight for improved conditions at the universities. There is, however, one thing the Internet cannot compensate for: the task of coming up with an agenda and entering into negotiations with political representatives. (Para. 11, trans. H. Hrachovec)

In the next section, I examine this concern and argue that the students' successful two-sided platform communication strategy ultimately presented a serious drawback with regard to achieving the actual aims the movement had set itself.

Some problems with Internet-assisted politics

The recent disclosure of 250,000 documents from the US diplomatic service (for overview and related references see Wikipedia (2011)) highlights a basic feature of Internet communication: Although information can be distributed rapidly and globally, this opportunity simultaneously rests on a foundation of hardware and IT rules that are under the control of a few experts, being that it requires just a single decision of a server's administrator to gain access to an official address. (There are, it is well known, counter experts at work to balance this asymmetry.) Given the dominant excitement of free access to information online, it is often forgotten that these achievements would be impossible without some fairly strict security provisions. The world must not know the password of the WordPress blog running on the machine with the IP address 131.130.199.36, and it had better be a more recent version not susceptible to the earlier exploits that such Web sites were subjected to. The backbone of the frequently touted mutual cooperative Web 2.0 opportunities is a strict technocratic regime of IT controls. There is nothing wrong with this per se; freedom to travel extensively depends, in a similar way, on (increasingly strict) airport surveillance. Yet, it is worth considering the impact that the system administration had in the case of the Audimax occupants. In prior settings, these experts were at the service of a (bureaucratic) university hierarchy that determined the requirements their software had to meet. In this case, the official student associations were led by representatives that had been elected on the basis of a certain political agenda, which in turn governed their public relations policy. Once in a while, protests erupt and the representational legitimacy of those students is called into question. What happens next?

Disruptive political actions in the 20th century have followed a certain pattern in which after local unrest has triggered resistance like marches and occupations, spokespersons step forward (i.e., Martin Luther King, Jr., Václav Havel, Aung San Suu Kyi), voicing the crowd's grievances and securing support from the movement's participants. These representatives are consequently featured in the news, interviewed by journalists, and persecuted (and sometimes accepted) by the authorities. Media remains a supporting character in these social dramas, as TV reporting lacks the global interactive capabilities available to Internet users today. In contrast, the highly visible unibrennt protest Web site was in operation within days of the initial spark of the protests, erected by a few students knowledgeable about the machinery of Web 2.0. There was, in particular, no need for a leader to emerge or for a common political denominator to be determined. Numerous activities were started, and various lists of demands were quickly circulated and were voted for in the Plenum—the assembly of students present at certain predetermined instants as the ultimate authority of the movement. There was no need for frontline speakers. In fact, the *refusal* to designate any representatives was crucial to the self-image of the activists. Success, in the sense of public recognition and active indications of support, had come without a standard hierarchical or nonstandard "revolutionary" structure. To a large degree, Internet support had been spontaneous, feeding on its own success. Political consciousness followed suit. In principle, the rejection of "any claim to leadership on part of political groups or single persons" was proclaimed, "something the ÖH [that is, the Österreichische Hochschülerschaft, the official student association] as well as various left-wing cadres had to feel the consequences of" (Heissenberger et al., 2010, p. 203, trans. H. Hrachovec). Political struggle, however, does not work like a clever logistical assignment of moments of attention across given arrays of interests.

As such, it is important here to note the inherent, perennial tension between grassroots and representational democracy. The occupants had, with a single stroke, drawn public attention to their cause and to the idea that the slow-moving, bureaucratically entrenched ÖH was no match for their dynamic presence. But who was to present the Plenum's demands—and to whom? Political action had, to a certain degree, gone into a self-reinforcing loop, with the mere event of live streaming already constituting the public appeal that usually has to be established separately. Given this setup, the catalog of demands drawn up by the Plenum resembled more of a litany (listing 37 items; see http://philo.at/wiki/index.php/Herbert_Hrachovec_(MuD_09)/Audimax-Besetzung) than a basis for negotiations. It lacked priorities, reaching from the most general calls for free education to the minutiae of curricular deadlines. Even if there had been someone appointed to put this document on the table and even if there had been someone receiving it, this list of demands would have been woefully vague. Locked into its instant success, the Audimax occupation proved to be incapable of entering a process of give and take. It is true that previous student protests had also come up with highly rhetorical pronouncements, achieving very little. The difference is that on prior occasions, the outcome was measured by the number of participants actually present, the press reports filed, and the comments being made in the national media. This time, the movement had somehow supplied itself with its own response.

Given these circumstances, it was not an entirely hostile decision on the part of the university administration to refuse to enter into direct negotiations with the Audimax occupants. The haughty attitude exhibited by the university's *rektor* did not help, but even if he had been more forthcoming, there could hardly have been agreement reached between the two sides. In any case, the administration followed another strategy. It turned to the legitimate student representatives, namely the ÖH, and offered negotiations on some improvements of the status quo. The ÖH was the very group that had been thoroughly de-legitimized by events, yet who else was there to address, as there were no negotiating teams with which to attempt to reach a middle ground? The Audimax occupants had created a highly attractive point of reference for the Viennese intelligentsia that was featured in regular evening events: lectures, podium discussions, and poetry readings. The occupation had drawn international attention to the grievances of the Austrian system of higher education. The business model of two-sided platforms served it well. The logic of mutual enforcement of the interests of student activists and global service providers had replaced the antagonistic struggle of more traditional politics.

Moreover, the university administration, in a belated fashion, came to realize that a new game was being played and initiated a "forum for sustainable strategies of university development." The main purpose of this move was to counter the impact of the Audimax occupation by offering a live stream of a different kind. The entire university *nomenclatura* was assembled around a table, facing two group of students: the elected ÖH student representatives and the student participants (emphatically not the "delegates") from the Audimax Plenum. The underlying purpose of this enterprise was to offer some concessions in exchange for an agreement from the occupants to leave the Audimax. It came as no surprise that this could not be accomplished, being that the Audimax group had received no negotiation mandate. But this event included a suggestive vignette exemplifying the curious political stand off that had developed in the course of just over a month.

A handout was distributed by one of the student members of the Plenum (2009, trans. H. Hrachovec) group, which reads:

> In this context, please regard the occupants, but also the ÖH colleagues, as a kind of consultancy team, helping you to understand the concerns of the occupants and to fulfill their demands. The more you are prepared to cede authority to the students the easier implementation of the demands conforming to the students' ideas will be. . . . The views expressed in this leaflet have to be regarded as persuasions of individuals. A number of participants from the side of the occupants have contributed in writing this handout. It is meant to inform you and to assist in understanding the structure of the occupation. (Para. 6)

The style of this leaflet is painfully precocious, but that is not the point here. Some students, trying to take advantage of the situation, had instinctively come up with an astonishing verbal reshuffle of the political process. Their proposal, notwithstanding its almost surreal touch, can be regarded as a vivid example of a postpolitical attitude.

This is my reading of the pamphlet's business model: As negotiations between the conflicting interests were, by design, impossible, several members of the opposing groups offered their services to the other side. They would be willing to explain the "strange beast" of the Audimax and facilitate not a compromise but an adequate understanding of the opposite side. This offer was not covered by any decision or vote by the Plenum, which had excluded anything but acceptance of its demands in toto. Still, the establishment of the "forum for a sustainable strategy" was apparently too tempting an opportunity to gain publicity for some clever strategists, who came up with a strange procedural move: Just like the Internet experts who had prepared a state-of-the-art public relations environment, these students appeared to slip into a supporting role as a kind of service provider, offering political advice and strategic counsel to the management of what is, after all, a mid-sized business, namely Vienna University. The logic of the two-sided platforms had definitely extended from IT support to the actors' self-description. To those accustomed to thinking in conventional political categories, this looked like a crushing defeat and a glaring irony, folded into one.

Conclusion

Numerous types of political activity have developed over the course of history, from tribal leadership to Western-style democracies, which are presently driven, to a large extent, by broadcasting mass media. Internet activists, and in particular, proponents of Web 2.0, have argued that this trend can be reversed if the participatory infrastructure provided by recent social software products is fully utilized. These technologies can indeed have a staggering impact on the public, as the preceding account shows. Yet, it is by no means clear whether such an impact is a countervailing force against mass-media consumerism. It is even doubtful whether Internet technology should be regarded as politically helpful if it is understood as conducive to improved representational participation. Instead, the events reported here suggest that the social Web may have a corrosive influence on the tradition of civil commitment and community activism within the framework of established interest groups.

References

Benelux Bologna Secretariat. (2010). *About the Bologna process.* Retrieved from http://www.ond.vlaanderen.be/hogeronderwijs/bologna/about/

Bologna Process. (2011). *Welcome to the EHEA official website!* Retrieved from http://www.ehea.info/

Clark, B. R. (1983). *The higher education system: Academic organization in cross-national perspective.* Berkeley & Los Angeles: University of California Press.

European University Association. (2007). Lisbon Declaration. Europe's Universities beyond 2010: Diversity with a common purpose. Retrieved from http://www.eua.be/fileadmin/user_upload/files/Lisbon_Convention/Lisbon_Declaration.pdf

European University Institute. (2011). *Austria, academic career structure.* Retrieved from http://www.eui.eu/ProgrammesAndFellowships/AcademicCareersObservatory/AcademicCareersbyCountry/Austria.aspx

Evans, D. S. (2006). Two-sided platforms and analysis of single firm conduct. Retrieved from The United States Department of Justice Web site: http://www.justice.gov/atr/public/hearings/single_firm/comments/219673_a.htm

Gawer, A. (Ed.). (2009). *Platforms, markets and innovation.* Cheltenham, United Kingdom: Edward Elgar.

Google Analytics. (2011). *Enterprise-class web analytics made smarter, friendlier and free.* Retrieved from http://www.google.com/intl/en/analytics/

Hagiu, A. (2007). Merchant or two-sided platform? *Review of Network Economics, 6*(2), article 3.

Heissenberger, S., Mark, V., Schramm, S., Sniesko, P., & Süss, R. S. (2010). *Uni brennt. Grundsätzliches—kritisches—atmosphärisches* (2nd ed.). Vienna, Austria: Turia & Kant.

Hrachovec, H. (2009). Checkpoint Bologna, Universität Wien 2008. In I. Schrittesser (Ed.). *University goes Bologna: Trends in der Hochschullehre: Entwicklungen, Herausforderungen, Erfahrungen* (pp. 7–185). Vienna: University Press.

HRK. (n.d.). *Welcome to the Bologna Centre of the German Rectors' Conference (HRK).* Retrieved from http://www.hrk.de/bologna/de/home/4110_4276.php

Leonhard, R. (2009, October 28). *Österreich im ausnahmezustand. Uni brennt weiter.* Retrieved from http://www.taz.de/1/zukunft/wissen/artikel/1/uni-brennt-weiter/

OE24. (2009, October 23). *"Die uni brennt": Eindrücke aus dem besetzten Audimax der Uni Wien.* Retrieved from http://www.oe24.at/oesterreich/politik/uni/Die-Uni-brennt/663611

Oppelt, N. (2009, October 26). Österreichs Studenten organisieren sich virtuell—"Die Uni brennt." *Readers Edition.* Retrieved from http://www.readers-edition.de/2009/10/27/oesterreichs-studenten-organisieren-sich-virtuell-die-uni-brennt/

Plenum. (2009). *Benutzer:Anna/MuDO9* [Collaborative statement developed by Plenum participants and presented at a Forum with university administrators]. Retrieved from http://philo.at/wiki/index.php/Benutzer:Anna/MuD09

Rochet, J.-C., & Tirole, J. (2006). Two-sided markets: A progress report. *The RAND Journal of Economics, 35*(3), 645–667.

Schwenk, M. (2009, October 28). Studentenproteste in Österreich: Mit allen Registern des Netzes. *Carta.* Retrieved from http://carta.info/17167/studentenproteste-in-oesterreich-mit-allen-registern-des-netzes/

Ustream. (2011). Retrieved from http://www.ustream.tv/

Weber, I. (2009, October 27). Uni-Streik kam durch neue Kommunikationsmitteln ins Rollen. *Wiener Zeitung.* Retrieved from http://www.wienerzeitung.at/themen_channel/wzwissen/forschung/231901_http-Unibrennt.at.html

Wikipedia. (2011). United States diplomatic cables leak. *Wikipedia, The Free Encyclopedia.* Retrieved from http://en.wikipedia.org/w/index.php?title=United_States_diplomatic_cables_leak&oldid=442257374

Contributing Authors

Gado Alzouma (PhD, Southern Illinois University, USA) is Associate Professor of Anthropology at the School of Arts and Science in the American University of Nigeria (AUN), Yola. His research and publications focus on information and communication technologies (ICTs) for development; science, technology, and society; and globalization and identities. Before joining AUN, he taught sociology and anthropology courses for 12 years at the Abdou Moumouni University (formerly the University of Niamey) in Niamey, Niger. He also worked as coordinator, evaluating the learning systems in the Africa and the Information Society Program of the International Development Research Center (IDRC, Dakar, Senegal) and as a research fellow in the Global Media Research Center of Southern Illinois University, Carbondale.

Guo-Ming Chen (PhD, Kent State University, USA) is Professor of Communication Studies at the University of Rhode Island and was the founding president of the Association for Chinese Communication Studies. He has also served as chair of the ECA Intercultural Communication Interest Group and coeditor of the *International and Intercultural Communication Annual.* In addition to serving as an editorial board member on several professional journals, Dr. Chen is the Executive Director of the International Association for Intercultural Communication Studies and the Coeditor of China Media Research. His primary research interests are in intercultural, organizational, and global communication. His published books include *Foundations of Intercultural Communication* (coauthored with William J. Starosta), *Introduction to Human Communication* (coauthored with Hsueh-Hua Chen), *Communication and Global Society* (coedited with William J. Starosta), *A Study of Intercultural Communication Competence, Chinese Conflict Management and Resolution* (coedited with Ringo Ma), and *Theories and Principles of Chinese Communication.*

Pauline Hope Cheong (PhD, University of Southern California, USA) is Associate Professor of Communication at the Hugh Downs School of Human Communication at Arizona State University, USA. She is also a graduate faculty member of the School of Justice and Social Inquiry and School of Women and Gender Studies, as well as an affiliate faculty with the Department of Film and Media Studies and the Center for the Study of Religion and Conflict. She has co-led multidisciplinary and comparative research projects on religious and extremist groups of multiple faiths and backgrounds and is Lead Editor of *Digital Religion, Social Media and Culture* (forthcoming, 2012) and Coauthor of *Narrative Landmines* (forthcoming, 2012); a book about mediated rumors in global strategic communication. She has presented more than 60 papers at international conferences and has published in multiple key journals on communication technologies. For further details, see http://www.paulinehopecheong.com

Andrew Jared Critchfield (PhD, Howard University, USA) is a Consultant with Communication and Culture Consulting and has worked in China, Japan, Mongolia, and Taiwan. During his doctoral studies at Howard University, he was a Ryoichi Sasakawa Young Leaders Fellow (SYLFF). His research interests include intercultural and organizational communication theories and practice(s). He is one of the founders of the Women's Leadership Institute at The George Washington University.

Xiaodong Dai (PhD, Fudan University, China) is Associate Professor in the Foreign Languages College of Shanghai Normal University. He presently serves as the Executive Chief of the Intercultural Research Center. His major research interests are cultural identity, identity negotiation, and intercultural communication theory. As a 2007 Fulbright scholar, he conducted research at the Department of Communication Studies of University of Rhode Island. He has published numerous articles, which have appeared in the *Chinese Journal of European Studies, American Quarterly, Journal of World Economy and Politics, Contemporary International Relations, International Survey*, and *China Media Research*. His recent book publications include *Canada: Cultural Security in the Setting of Globalization* and *Identity and Intercultural Communication: Theoretical and Contextual Construction* (coedited with Steve J. Kulich).

Charles Ess (PhD, Pennsylvania State University, USA) is a Distinguished Research Professor of Interdisciplinary Studies and a Professor of Philosophy and Religion at Drury University in Springfield, Missouri. He is immediate past-president of the Association of Internet Researchers and currently serves as Guest Professor of Information and Media Studies at Aarhus University, Denmark (2009–2012). Recent publications include: *Information Technology Ethics: Cultural Perspectives* (2007), *Digital Media Ethics* (2009; with Soraj Hongladarom, coeditor), and *The Blackwell Handbook of Internet Studies* (2010; with Mia Consalvo, coeditor). Dr. Ess serves on the editorial boards of *New Media & Society* and the *Journal of Computer-Mediated Communication*. Since 1998, Dr. Ess has co-chaired the biannual conference series Cultural Attitudes Towards Technology and Communication (CATaC) with Fay Sudweeks.

Cátia Ferreira is a PhD candidate in Communication Sciences at the Catholic University of Portugal and a Junior Researcher at the Research Center for Communication and Culture. Her area of research is new media, particularly digital social platforms, with a focus on social change in and through the virtual platform Second Life. Her master's degree dissertation focused on the relationship established between Portuguese emigrants in the USA and the homeland using new technologies.

Carla Ganito (PhD, Catholic University of Portugal) is a Lecturer in New Media and Marketing at the Catholic University of Portugal and a researcher at the Research Center for Communication and Culture on issues concerning new media, gender and technology, new media literacy, and entertainment. Dr. Ganito also holds an MBA and a MA on information management focused on mobile entertainment; her recent doctoral research focused on the gendering of the mobile phone.

Nickesia S. Gordon (PhD, Howard University, USA) is Assistant Professor of Communication at Barry University in Miami Shores, Florida, where she teaches media programming, media management, communication research, mass communication theory, and television production courses. She also teaches introductory communication courses. Her current research interests examine the intersections among gender, mass media, and popular culture. She also has an active research agenda in new media technologies, media globalization, media consolidation and convergence, and the implications that these subjects have for programming and regulation in the Caribbean. In addition to research articles, Dr. Gordon's recent book is titled *Media and the Politics of Culture: The Case of Television Privatization & Media Globalization in Jamaica (1990–2007)*.

Lelia Green (PhD, Murdoch University, Australia) is Professor of Communications at the School of Communications and Arts at Edith Cowan University and a Chief Investigator with the Australian Research Council Centre of Excellence for Creative Industries and Innovation. The author of *The Internet: An Introduction to New Media* (2010), Dr. Green's research interests focus on audiences, users, and the social and cultural aspects of communication technologies. She is pleased to acknowledge her foundation partners in this research project, The National Heart Foundation Australia (Western Australian Division) and the Australian Research Council.

Beth Bonniwell Haslett (PhD, University of Minnesota, USA) is Professor in the Department of Communication at the University of Delaware. Her primary research and teaching interests are in discourse analysis, cross-cultural communication, and organizational communication, especially interaction in multinational organizations and virtual groups. Her forthcoming book, *Communicating and organizing in context: The theory of structurational interaction* (2011), integrates the work of Giddens and Goffman into a new theoretical approach to analyzing organizations. Dr. Haslett is former chair of the Language and Social Interaction Division of the National Communication Association (USA) and serves on the editorial board of *Communication Studies.*

Herbert Hrachovec (PhD, University of Vienna, Austria) is Lecturer in the Department of Philosophy, at the University of Vienna, Austria. He has held scholarships and visiting appointments at the University of Oxford (United Kingdom), the University of Münster (Germany), Harvard University, Cambridge, MA (USA), Freie Universität Berlin, Universität Duisburg-Essen (Germany), Bauhaus-Universität Weimar (Germany), University of Bergen (Norway), and The University of Klagenfurt (Austria). His areas of specialization are analytic philosophy, aesthetics, and media theory. He was a member of the Academic Senate and Chair of the Central Committee for Curriculum Planning at the University of Vienna from 2005 to 2010. For further details, see http://hrachovec.philo.at

Joonsoeng Lee (PhD, Ohio University, USA) is Assistant Professor in the Department of Communication at California State University, San Marcos, where he teaches courses on media and society. In addition to studying media, religion, and culture in Korea, his research centers on new media and the technologies of governmental dominance, feminist media studies, and empowerment in cyber-cultures. He is currently working on a book project about the construction of digital spirituality in Korean shamanism.

Azi Lev-On (PhD New York University, USA) is Lecturer and Head of the New Media Track in the Ariel University Center of Samaria in Israel. Dr. Lev-On's studies explore behaviors and collective action in computer-mediated environments, employing a variety of methods, such as link analysis, surveys, and laboratory experiments. His recent research analyzes how and why computer-mediated communication impacts monetary transfers and trusts, how people rank news stories online, and Internet usage by candidates in the Israeli municipal elections 2008 and by evacuees from Gush Katif. For further detail see http://www.azilevon.com

Leah P. Macfadyen (PhD, The University of British Columbia, Canada) is a Researcher and Instructor in the Faculty of Arts at The University of British Columbia in Vancouver, Canada. She is interested in the many and varied intersections (and collisions) between the social and the technological, and her research agenda includes educational technology studies, global and transformative education, and Internet research. She is particularly interested in the ways that the Internet and new media are expanding the horizons of local communities and the role of new media in contemporary political and social activism and transformation. Her most recent writings on aspects of culture, identity, and online education have appeared in edited collections such as *Digital Differences: Perspectives on Online Education* (2010) and *Learning*

Cultures in Online Education (2010). For further details see http://www.changingeye.com/

Judith N. Martin (PhD, Pennsylvania State University, USA) is Professor of Intercultural Communication in the Hugh Downs School of Human Communication at Arizona State University, USA. Her principle research interests focus on the role of culture in online communication, interethnic and interracial communication, and sojourner adaptation and reentry. She has published numerous research articles in communication journals as well as other disciplinary journals and has coauthored three textbooks in intercultural communication with Thomas K. Nakayama: *Intercultural Communication in Contexts*, *Experiencing Intercultural Communication: An Introduction*, and *Readings in Intercultural Communication*. In 2001–2004, she was selected as the Jeanne Herberger Professor of Human Communication. She has developed and taught various communication courses (including intercultural communication) online for the past 10 years.

Rivka Neriya-Ben Shahar (PhD, Hebrew University of Jerusalem, Israel) is Lecturer at Sapir Academic College in Sderot, Israel. Her doctoral research is entitled "Ultra-Orthodox Women and Mass Media in Israel: Exposure Patterns and Reading Strategies." Dr. Neriya-Ben Shahar studies mass media and Internet usage from the perspectives of religion and gender. Her current research project addresses the tensions between values and new technologies in various religious populations.

Konrad Ng (PhD, University of Hawai'i, USA) is Assistant Professor of Creative Media at the University of Hawai'i at Mānoa. He has a long history of working in Asian and Asian American cultural institutions, programs, and organizations and has published articles and curated film programs on Asian and Asian American film, politics, and culture. Dr. Ng is a member of the Center for Asian American Media's board of directors and in 2010, he was named Acting Director of the Smithsonian Asian Pacific American Program.

Bolanle A. Olaniran (PhD, University of Oklahoma, USA) is Professor and interim Chair in the Department of Communication Studies at Texas Tech University in Lubbock. His research foci include organizational communication, cross-cultural communication, crisis communication, and communication technologies. Dr. Olaniran is the author of a number of peer-reviewed articles in discipline, focus, and interdisciplinary journals as well as chapters in edited collections in each of these areas. He also serves as a consultant to organizations and universities at the local, national, international, and government level. He is a recent recipient of an award from the American Communication Association for "Outstanding Scholarship in Communication."

Irina Privalova (PhD, Saratov State Technical University, Russia) is Professor and Chair of Business Communication at Saratov State Technical University. She completed her doctoral research in psycholinguistics at

Saratov State University and subsequently defended a postdoctoral thesis entitled "Ethno-Cultural Peculiarities of Language Consciousness" at the Russian Academy of Sciences, Moscow, in the Department of Psycholinguistics and the Theory of Communication. Dr. Privalova was extensively involved in the cross-cultural training for the United States Peace Corps during the period that it operated in Russia (1992–2000). She is a member of the Russian Communication Association, the European Communication Research and Education Association, and the International Communication Association and has authored more than 70 articles published in Russian and international journals.

Debbie Rodan (PhD, Murdoch University, Australia) is a Senior Lecturer in Media & Cultural Studies at Edith Cowan University in Perth, Australia. She is the author of *Identity and Justice: Conflicts, Contradictions and Contingencies* (2004), and her most recent writings are published in *Australian Journalism Review, Continuum: Journal of Media & Cultural Studies*, and the *Journal of Intercultural Studies)*. Dr. Rodan is currently one of the chief investigators on an Australia Research Council linkage grant, studying the construction of the subject as heart patient in the context of the gift economy. Her ongoing research centers on online communicative spaces, such as Australian political blog sites and GetUp!, and aims to advance citizen participation.

Natalia Rybas (PhD, Bowling Green State University, USA) is Assistant Professor of Communication Studies at Indiana University East in Richmond. Her research examines the interactions happening at the intersections of online and offline communication. Her most recent project in critical cyber-culture and feminist studies explores the performativity of technology and gender in the context of a Facebook group. Her work has been published in *Qualitative Inquiry, Feminist Media Studies*, and edited collections.

Kirk St. Amant (PhD, University of Minnesota, USA) is Associate Professor of Technical and Professional Communication and of International Studies, at East Carolina University in Greenville, North Carolina. He has a background in cultural anthropology, international government, and technical communication and has taught online and on-site courses in technical and professional communication and in intercultural communication for a number of different US universities. Dr. St. Amant has also taught courses in e-commerce, distance education, and business communication in Ukraine with the USAID-sponsored Consortium for the Enhancement of Ukrainian Management Education (CEUME) and has engaged in collaborative online and on-site teaching with the Aarhus School of Business in Denmark. He coedited the recent books *Culture, Communication, and Cyberspace: Rethinking Technical Communication for International Online Environments* and

Computer-Mediated Communication Across Cultures: International Interactions in Online Environments (forthcoming, 2012).

Robert Shuter (PhD, Northwestern University, USA) is Professor in the Department of Communication Studies at Marquette University in Milwaukee, Wisconsin, and Chair of the US National Communication Association's International and Intercultural Communication Division. A pioneer in intercultural communication, Dr. Shuter has written widely about communication and culture, with more than 60 articles and books published in leading journals, including *Communication Monographs, Journal of Communication, Journal of Applied Communication Research, Management Communication Quarterly,* and *Journal of Social Psychology.* He has served on the editorial boards of many journals and recently edited a forum on new media across cultures for the *Journal of International and Intercultural Communication.*

Kristin Sorensen (PhD, Indiana University, USA) is Assistant Professor of Global Studies at Bentley University in Waltham, Massachusetts. Her research investigates the manner in which cultural identities, historical memories, and their associated traumas circulate through the media and public culture in contemporary Latin America, and her book *Media, Memory, and Human Rights in Chile* was published by Palgrave Macmillan in 2009. Dr. Sorensen has published articles in the *Journal of Human Rights, Cuadernos de Información, Peace & Change,* and *SECOLAS Annals.* She has also contributed chapters to the following edited books: *Democracy in Chile: The Legacy of September 11, 1973, Global Memoryscapes: Contesting Remembrance in a Transnational Age,* and *The 1980s: A Critical and Transitional Decade.*

Fay Sudweeks (PhD, Murdoch University, Australia) is Emeritus Associate Professor of the School of Information Technology at Murdoch University. Widely published herself, she sits on the editorial boards of the *Journal of Computer-Mediated Communication, New Media & Society, WebNet Journal, Journal of Electronic Commerce Research,* and the *Journal of Electronic Commerce in Organizations.*

Wei Sun (PhD, Howard University, USA) is Assistant Professor in the Department of Communications at Bowie State University in Prince George's County, Maryland. Her research on intercultural communication, crisis communication, media and culture, conflict management in intercultural contexts, and minority invisibility have appeared as book chapters and in academic journals, such as *Human Communications, Howard Journal of Communications, Journal of World Communication,* and *Intercultural Communication Studies.* Her book *Minority Invisibility: An Asian American Experience* was published in 2007.

Lynsey Uridge is a PhD candidate at Edith Cowan University in Perth, Australia, and a postgraduate researcher on the Australia Research Council linkage grant entitled *Construction of the Subject as Heart Patient in the*

Context of the Gift Economy. She is currently the moderator of HeartNET (http://www.heartnet.com.au), an online community based at Edith Cowan University and supported by the National Heart Foundation (Western Australian Division) for people affected by heart disease. Dr. Uridge has a background in nursing and information services and has worked with people in remote and rural communities and people with disabilities and currently works with those affected by cardiovascular disease. With an interest in health promotion, online identity formation, and netnography, her aim is to make sure that ALL the participants in her current PhD research and the members of HeartNET are given a voice.

Maja van der Velden (PhD, University of Bergen, Norway) is a Researcher and Lecturer at the Department of Informatics at the University of Oslo, Norway. She is currently investigating the relationship between human autonomy and automation, including the way patient privacy is negotiated via the Internet and continues her ongoing exploration of the relations between information technology and knowledge. For further details, see http://www.globalagenda.org

Ping Yang (PhD, Arizona State University, USA) is Assistant Professor in the Department of Communication at Denison University in Granville, Ohio. Her scholarship focuses on the intersections of culture, communication, and technology. Her research interests include identity expression, cultural adaptation, media representation, heritage language education, and intercultural online communication. She has published book chapters and journal articles in most of these and related areas.

Acknowledgments

We would like to thank Ms Elizabeth Cantu, a doctoral candidate at the Hugh Downs School for Human Communication, for her kind administrative assistance in the editorial process.

Pauline wishes to thank her husband Simon, family (especially Peter, Margaret & Paul) and friends for their love and encouragement. She also gratefully acknowledges her colleagues at the Hugh Downs School of Human Communication, Arizona State University, for their dedicated scholarship and support. Special thanks to colleagues of the International Communication Association, National Communication Association, Association of Internet Researchers, as well as Charles Ess and Fay Sudweeks for their kind leadership of CATaC, which has been a wonderful context for reflecting on the themes related to this book. To Judith & Leah, thanks for being part of this editorial team; it was a rich and joyous collaboration.

Judith wishes to thank Ron once again for his continued unwavering support and to Pauline and Leah for their inspiration, intellectual insights and great energy. Thanks also to the CATaC scholars for suggesting (and pursuing) innovative and productive areas of research.

Leah wishes to thank partner Lynx for her endless interest in and support for this project. She is especially grateful to Charles Ess for his mentorship in this field, and for the opportunity to participate in this project, and to Pauline and Judith for their insight and patience.

Index

A

accountability 82–3
acculturation 140–1, 149
activism 64, 151, 168, 253, 256–7, 263, 286, 312–13, 316–19, 324
Africa 3, 7, 14, 25, 68–9, 193–6, 279
African Americans 52, 259, 270
agriculture 23, 71
anonymity 50, 109–11, 113–15, 117, 120, 150
appropriation, cultural 92, 125, 127–9, 136
Asia 3, 8, 13, 68, 78, 210, 281
Asian Americans 50, 255–63, 265–70
assemblage 29, 31, 143, 151
asymmetry 123, 125, 127–9, 131, 133, 135, 137
Australia xvi, 25, 27, 33, 139–40, 304
Austria 253, 307–9, 311–12
authenticity 48, 98–9, 113, 115, 117, 120, 151, 204, 269
avatars 39, 45–6, 48–50, 80, 143–4

B

blogs 26, 45–6, 49, 51, 92, 109, 156–9, 161–3, 165–7, 256, 260–1, 284, 301, 304, 308
Butler, Judith 102–3

C

CAAM (Center for Asian American Media) 252, 257, 263–5
camera, digital 240, 242, 245–6
cancer patients 92, 152, 155–69
capitalism 176–8, 189
care drain 203
CATaC (Cultural Attitudes towards Technology and Communication) xi–xii, xiv, 4
Center for Asian American Media *see* CAAM
chat rooms 117, 119–20, 142, 148
Chile 282–6
China 84, 92, 120, 129–30, 155–7, 164–8, 293
class, social 280
CMC (computer-mediated communication) xii–xviii, 1, 4, 40–1, 44–8, 50–3, 61, 109–10, 113, 118–19, 195
collective identities 125
collectivism xv-xvi, 51–2, 63–5, 70, 156–7
communication
  cross-cultural 75, 84
  high-context 64
  intercultural *see* intercultural communication
  low-context 64
  mediated 4–6, 9–10, 41, 43–5, 47, 291–3, 304
  online 7, 47, 53, 83, 87, 92, 109, 116, 118, 120, 150
communication style 51, 64, 66, 156
communicative behavior 293, 299, 303

communicative rationality 131–2
community, online or virtual 41, 47, 49, 92, 94, 102, 109–10, 124–7, 139–40, 143, 145–6, 149–51, 158, 257, 298
computer-mediated communication *see* CMC
conservative communities 232, 234
contact zone 29–30, 33–4
context
  e-learning 61–2, 65
  economic 13, 23, 43, 62–3, 68, 168, 176, 189, 193–4, 200, 277, 280, 285, 296
  importance of 40–1, 53, 75–87, 101–2, 104, 109–12, 125, 142–4, 146, 149–50, 218, 304
  off-line xv, 100
  online xv, 86, 102, 105, 113, 118, 143–4, 146, 237
  political 13, 171, 253, 276, 293
  social 2, 13, 40, 51, 63, 172, 219
  sociohistorical 18, 40, 50, 52, 252
convergence 62, 74, 124, 127, 238, 257, 323
credibility 49, 75–6, 85–6, 110, 165–6
cultural differences 47, 50, 52, 64–7, 116, 119
cultural diversity 124, 130
cultural expectations 75, 240
cultural groups 10, 82, 110, 114, 116, 140–1, 149
cultural hybridization 129
cultural spaces 115, 118, 120, 195, 268
culture
  definitions of 10–13, 31, 39–41, 44, 46, 52, 62–3, 77–8, 94, 129, 178, 293
  dominant 13, 129–30
  high-context 64, 66, 78, 82–3, 156, 303
  low-context 44, 78, 303
  low power distance 65, 70
  mainstream 92–3, 130, 156, 167
  national 11, 44, 121
  popular 268, 323
cyberspace 45, 80–1, 87, 96, 111, 118, 120, 126, 142, 269

D

database design 21, 30, 33
de-centering science 22
design xiii, 4, 6, 8, 21, 28–34, 72, 193
design process 28, 30–2, 34
deterritorialization 176–7
developing countries 120, 210, 219, 275–6, 278–80, 282, 286
development 26–7, 252, 276–9, 281
  human 9–10, 277, 286
  national 275–8, 281, 286
  socioeconomic 24, 62–3, 275
dialectical perspective 9, 112, 120
dialectics 2, 5, 112–13, 115, 118, 120, 131
  mediated intercultural 5–6, 13
digital divide 9, 13, 18, 43, 68–70, 111, 128–9, 211, 257
dis-embeddedness 40
discussion forums 26, 51, 112, 115, 117, 141–3, 223, 227, 232, 252, 283–4, 312
distanciation 40, 47, 54
divergence 62, 127
diversity 2, 23, 25, 28, 85, 132, 263

E

e-learning 18, 61–74, 77, 79, 81, 83, 85, 87, 89
  cultural challenges 67, 72, 74
e-mail 66, 80, 110, 114–16, 119, 256, 268, 279, 281–2, 300
economic context *see* context, economic
economically developed countries *see* EDCs
ecopoiesis 33–4
EDCs (economically developed countries) 5, 68, 70–1

empowerment 2, 65, 92, 111, 120, 122, 130, 137, 151, 156, 171, 210, 215, 256, 260, 268, 283
enclave culture 224, 230
equality, digital 113–14
ethics 31, 34, 263
ethnicity 18, 43, 50–1, 109, 111, 113, 116, 119, 264
Europe 3, 8, 195, 205, 238, 296, 298, 311

F

face xv, 42, 45–6, 51–2, 302
face-to-face (FtF) communication 41–2, 45–7, 64, 116
Facebook 82–4, 93–9, 101–5, 113–15, 117, 119–20, 218, 242, 312–13, 315
family
  extended 7, 172, 193–4, 197–203, 205, 214–16, 218
  fragmentation of 7, 172
  transnational 199, 203
family photos 7, 237–9, 244, 246
family ties 47, 197–200, 202–3, 205
feminist perspectives 21–4
financial institutions, international 176, 277
Foucault, Michel 144–5, 152, 173, 177–8, 237, 240
fragmentation 241, 243
FtFcommunication *see* face-to-face (FtF) communication
functionalism 10
funeral culture 177, 179–81, 186, 189

G

Giddens, Anthony 40–3, 45–6, 50, 52–3, 128, 131
global village 9, 251, 304
globalization xiii, 3, 40, 61–2, 123–8, 176, 195, 201, 211, 219
Goffman, Erving 40–2, 45–50, 53, 156, 294
governmentality 171, 177–9, 190

H

Habermas, Jürgen 132
Hall, Edward T. xv, 5, 77–8, 303
Hall, Stuart 11–13, 95, 98, 105
hapa identity 264–5
Haraway, Donna 22–3, 30–1, 33
Haredi (ultra-Orthodox Jewish) society 223–4
HeartNET 92, 139–40, 142, 144–50
higher education 253, 310, 312, 314–15, 317
Hofstede, Geert xv, 5, 41, 62–3, 65–6, 68

I

ICA (International Communication Association) 4
identity 6, 39–54, 75, 79–86, 94–8, 100, 104–5, 109–15, 117–21, 123–33, 139–51, 200, 203, 211, 215, 241, 255
  changing 46, 110
  construction of 42, 47, 51, 53, 91, 100, 110, 112, 130, 140, 142–4, 151
  cultural 44, 50, 109–10, 117, 119, 123–6, 130, 132, 239, 297
  development of 92, 111, 144
  distributed 45
  expression of 48, 112–13, 115, 120–1
  heart patient 92, 139–40, 142, 150
  multiple 42, 45–7, 50, 53
  national 44, 62
  negotiation of 47, 92, 114, 123, 125, 127–8, 130–3, 322
  presentation of 41, 47, 52–3, 105, 110, 118–20
  religious 117, 149

India 27, 29, 209–21
Indian women 209–20
indigenous communities 17, 26–7, 35
Australia 25, 33
indigenous knowledge 21–7, 29–30, 33
management 21, 26–7, 35–6
systems 23
individualism xv, 18, 51, 63, 65, 195, 200, 203, 292, 297
individualization 188, 246
industrialization 155, 179–80
information society 10
insecurity 43
interaction order 41, 56
interactivity 71, 124, 187
intercultural adaptation 131
intercultural awareness 132–3
intercultural communication 1–2, 4–6, 8, 10, 13, 52, 112–13, 127, 129–30, 132, 252, 268, 270
asymmetry 123, 127–8, 130
mediated 3, 6–8, 12, 92
scholars 4–5, 8, 109, 123
intercultural interactions 109–10, 116, 127
intercultural sensitivity 132–4
interculturality 133
Internet access 168, 210, 226, 232, 280–2, 285, 296, 304
Internet penetration 68, 196, 279
Internet use 7, 68, 226, 276, 279–81, 285–6, 296
by gender 45, 210, 223, 281
by level of education 50, 111, 194, 210, 244, 281
in Russia 296
intra-action 30, 34
invisibility 109, 111, 113, 116, 158, 167
Israel 223–6, 231

J

Jamaica 278–82, 286
Jewish identity 118
justice 8, 10, 282–3

K

kinship ties 204
knowledge 6, 21–30, 32–5, 39–40, 42, 65, 69–70, 96, 105, 109, 133, 215
diversity 23, 28, 34
non-Indigenous 21, 24
situated 22–3
knowledge management (KM) 21, 25–30, 34
software 21, 29–30, 34
knowledge production 22, 25, 33
knowledge systems 23, 29
Korea 51, 88, 116, 175–9, 182, 184, 186, 188–90, 248, 324

L

language and language barriers 67, 117–18, 120, 122, 128–9, 196
Latin America 3, 78, 210, 277, 281
Latour, Bruno 95
learning 65, 68, 70–2, 79
LEDCs (less economically developed countries) 5, 68–71
less economically developed countries *see* LEDCs
Levinas, Emmanuel xvii, 31–2
literacy xvi, 193, 196–7, 202, 204–5, 210–11, 278
technological 61, 69, 304

M

marginalization 111, 113, 120, 130, 156, 263, 284
mass media 124, 225, 291, 293, 296
new 291, 304
massively multiplayer online game (MMOG) 47, 58
McLuhan, Marshall 9, 292, 304
metadesign 32–4
methodology 11, 111
Middle East xviii, 3, 13, 78, 210, 279
migrants 194–205
African 194

minorities 257, 270
minority invisibility 158, 167, 285
MMOG *see* massively multiplayer online game (MMOG)
mobile camera phone 238, 240, 243–6
mobile communication 47, 172, 195, 201
mobile devices xvi-xvii, 7, 193–7, 199–205, 210, 214–15, 219–20, 237–49, 256, 258, 266
  by gender 203, 210, 238–40, 243–4, 246–7
mobility xvi
  in cyberspace 126
  of workers 172, 194, 205
modernity 39–40, 43, 47, 128–30, 178, 189, 200, 233
MySpace 50, 101, 104, 115, 117, 260, 297

N

nation-states 41, 43–4, 176, 277, 311
networking 28, 84, 95, 100–1, 104, 157, 242, 308
Niger 172, 193, 195–7, 199–202, 205
Nigeria 172, 194–7, 201–2
non-Western 129–30
non-Westerners 130–2
nonverbal cues 49, 80, 110, 113, 119
North America xii, xiv, 3, 310
noso-politics 178

O

online forums *see* discussion forums
online interactions 51, 53, 62, 75, 80, 84–5, 87, 96–7, 115, 117, 142
online learning *see* e-learning
ontology 12, 33
orality xvi, 39, 71, 194, 205, 245
Other xvii, 31–2, 34

P

participation 10, 73, 187, 253, 257–9, 285
patriarchy 172, 219
performativity 95
  of technologies 95, 326
phenomenology 109, 111–12
photographs 144, 188, 239–40, 242, 245–6
photography 238–42
  digital 238, 242, 246
poetry 141, 147–8
political engagement 256, 278, 286, 315
politics viii, xvii, 31, 34, 129, 199, 253, 256, 260, 276, 314–15, 317
Portugal 238, 243
postcolonial theory 23–4
poverty 195–7, 205, 210, 286
power 12, 23–4, 63, 66, 105, 109, 111–12, 118–19, 129–30, 178, 189, 270, 277, 283, 285
power distance (PD) 18, 63, 65
power dynamics 101, 110, 277, 280
privacy 18, 48, 53, 120, 150, 163, 204, 228, 240, 328
production, cultural 257, 265, 268, 283
protests, student 253, 317

R

race 18, 51, 80, 96, 102, 106, 109, 111, 113, 116, 118, 139, 149, 257–8, 264, 268
radio 41, 199–200, 225, 239, 284
religion 18, 22, 41, 52, 117, 120, 224
remittances 194, 202–3, 205, 211
reterritorialization 177–8, 188
risk 40, 43, 46–7, 53
rituals 129, 242, 246, 308
Russia 291–9, 302, 304
Russian culture 297, 299, 301, 303

S

script 17, 28, 30
Second Life 39
self
  authentic 6, 96–8, 104
  presentation of 41, 44, 47–8, 52–3, 142

self-disclosure 47, 51, 100, 144, 147, 150
self-esteem 144
self-images 41, 46, 124, 127, 131, 316
self-transformation 127–8, 144–8
selves
  constellation of 143, 149
  cultural 109, 112, 114, 116–20
sexuality 96
SIP (social information processing) 41
situated knowledges 22–3
SNSs *see* social networking services/systems/sites
social capital 102
social change 111, 256, 258, 277
social media 1, 91, 218, 239, 260, 322
social networking services/systems/sites 45, 48, 50, 82–4, 93–8, 102–5, 109–10, 114–15, 117, 120, 218, 228, 244, 251–2, 256, 265–6, 281, 297–8
social positioning 43, 46, 50, 52
social presence 18, 41–2, 44–5, 49, 51, 53–4
social psychology 4–5, 121, 327
social relations 43, 95, 101, 104, 194
societies
  patriarchal 211
  traditional collective 215
stereotypes 7, 49, 114, 118–20, 132, 255–7, 266, 268
storytelling26, 28, 33, 95, 103, 165, 187, 237, 239–40, 245–6, 264
  family 237, 239–40, 246
structurational interaction 40–3, 45, 47, 49–55
  theory of 40, 42, 50, 53
students 47–8, 65–6, 68–70, 76–7, 79, 101, 111–17, 119–20, 308–13, 315–18
  international 116, 118
Sub-Saharan Africa 195
Suchman, Lucy 30–1
Switzerland 308

T

TAMI 32–3
teams *see* virtual teams
techno-culture 94, 104
technological hothouses 225, 227, 229, 232
technologies of domination 177–8, 189
technologies of the self 152, 178, 189, 237
television 200, 226, 230, 240, 270, 293, 295, 300
text messaging 116, 172, 209–20, 244
  and gender 210, 213–16, 218–20, 225, 227, 229, 232
  Indian women 172, 209–19
  interpersonal norms 209–10, 218–21
  power 172, 210
textiquettes 172, 209, 218–19
third-person effect *see* TPE
third space 33, 52
Third World 277
TPE (third-person effect) 223, 227–8, 230–4
trust 18, 39, 43–4, 46–7, 49, 51, 53, 83, 86, 113, 118–19, 205, 300
Turkle, Sherry 45–6, 53, 110, 142
Twitter 242, 297–8, 308, 312–13, 315

U

UGC (user-generated content) 253, 291, 296, 298, 300, 304
UGM (user-generated media) 47
ultra-Orthodox
  community 223–6, 228–30, 232–3
  lifestyle 173, 230, 232
  women 223–5, 227–9, 232–3
user-generated content *see* UGC

V

values, cultural xiii–xiv, 46, 52–3, 65, 67, 128, 130, 209, 257, 298

videos 33, 45, 99–100, 102, 256, 264, 267, 302, 313–14
virtual space 117, 124, 126, 146
virtual teams 61–2, 67, 75, 85
  international 75
visibility 109, 111, 156–7, 308
vlogs 256, 260, 268

W

Web 2.0 1, 27, 46, 65, 82, 85, 93, 251–3, 291–2, 296–8, 303–4, 307–8, 312–13, 315–16, 319
Western 7, 17, 22–5, 29, 33, 68, 71, 125, 127–31, 140, 167, 197, 224, 291, 297, 304, 318
Westerners 128, 130–2, 166, 297
Wikipedia 13, 297

Y

YouTube 7, 47, 117, 256–7, 266–8, 285, 297, 302–3, 312

General Editor, Thoı

Critical approaches to the study of intercultural communication have arisen twentieth century and are poised to flourish in the new millenium. As cultures coı driven by migration, refugees, the internet, wars, media, transnational caı imperialism, and more—critical interrogations of the ways that cultures interact are needed to understand culture and communication. This series will interrogate- perspective—the role of communication in intercultural contact, in both domestic a contexts. This series is open to studies in key areas such as postcolonialism, transnati race theory, queer diaspora studies, and critical feminist approaches as they relate communication, tuning into the complexities of power relations in intercultural c Proposals might focus on various contexts of intercultural communication such a advertising, popular culture, language policies, hate crimes, ethnic cleansing and conflicts, as well as engaging theoretical issues such as hybridity, displacement, multip orientalism, and materialism. By creating a space for these critical approaches, this se the forefront of this new wave in intercultural communication scholarship. Maı proposals are welcome that advance this new approach.

For additional information about this series or for the submission of manuscripts, please

Dr. Thomas K. Nakayama
Northeastern University
Department of Communication
360 Huntington Ave
Boston, Massachusetts 02115